Quantum Mechanics

Edited by Paul Bracken

Published in London, United Kingdom

IntechOpen

Supporting open minds since 2005

Quantum Mechanics
http://dx.doi.org/10.5772/intechopen.87908
Edited by Paul Bracken

Contributors
Betül Çalişkan, Ali Cengiz Çalişkan, Kenjiro Yanagi, Vahram Mekhitarian, Shiang-Yi Han, Ciann-Dong Yang, Islam Aynul, Anika Tasnim Aynul, Maria Esther Burgos, Paul Bracken, Aghaddin Mamedov, Eugen Sheregii, Bruno Chandrasekar L, K Gnanasekar, Marimuthu Karunakaran, Francis Yu

First published in London, United Kingdom, 2020 by IntechOpen
IntechOpen is the global imprint of INTECHOPEN LIMITED, registered in England and Wales, registration number: 11086078, 5 Princes Gate Court, London, SW7 2QJ, United Kingdom
Printed in Croatia

British Library Cataloguing-in-Publication Data
A catalogue record for this book is available from the British Library

Additional hard and PDF copies can be obtained from orders@intechopen.com

Quantum Mechanics
Edited by Paul Bracken
p. cm.
Print ISBN 978-1-83968-044-1
Online ISBN 978-1-83968-045-8
eBook (PDF) ISBN 978-1-83968-046-5

For EU product safety concerns:
IN TECH d.o.o., Prolaz Marije Krucifikse Kozulić 3, 51000 Rijeka, Croatia, info@intechopen.com or visit our website at intechopen.com.

Meet the editor

Paul Bracken, PhD, is a professor in the Department of Mathematics at the University of Texas, Edinburg. He received a BSc from the University of Toronto and a PhD from the University of Waterloo in Canada. His research interests include applications of differential geometry and partial differential equations to problems in mathematical physics, such as gravity, as well as problems in quantum mechanics and quantum field theory. He has published more than 160 papers in journals and books and presented talks at numerous meetings and conferences. This is the sixth volume he has contributed with IntechOpen.

Contents

Preface

Quantum mechanics has been remarkably successful at accounting for physical phenomena at very small scales. The area of quantum mechanics has become a fundamental part of modern physics and hence a very intensive subject for research. It is applicable to physical situations where matter can be studied on very small scales from atoms and molecules to physics of the solid state and elementary particles. It has also generated new areas of physics in its wake that require quantum mechanics for their understanding, such as Bose Einstein condensation and the quantum Hall effect.

This book presents 12 solid contributions that illustrate the range and diversity of topics in quantum mechanics. Of these 12 chapters, three reference solvable models such as hyperfine structure interactions. Other chapters cover some foundational issues such as complex space forms of quantum mechanics, entropy in quantum mechanics, and equations of relativistic quantum mechanics. The final chapters examine applications of quantum mechanics to more complicated situations. As illustrated by this book, the scope, influence, and domain of applicability of quantum mechanics have become very widespread.

This book is a result of the hard work of an international group of invited authors. It is a pleasure to thank them for their efforts and scientific contributions. I am grateful to acknowledge with many thanks the assistance of Mr Mateo Pulko who was the publishing coordinator and manager throughout the production process of this book. It is also a pleasure to thank IntechOpen for the opportunity to work on this volume, which serves to further study the fascinating subject of quantum mechanics.

Paul Bracken
Professor,
Department of Mathematics,
University of Texas,
USA

Section 1

Solvable Systems

Chapter 1

Dipolar Interactions: Hyperfine Structure Interaction and Fine Structure Interactions

Betül Çalişkan and Ali Cengiz Çalişkan

Abstract

The interaction between the nuclear spin and the electron spin creates a hyperfine structure. Hyperfine structure interaction occurs in paramagnetic structures with unpaired electrons. Therefore, hyperfine structure interaction is the most important of the fundamental parameters investigated by electron paramagnetic resonance (EPR) spectroscopy. For EPR spectroscopy the two effective Hamiltonian terms are the hyperfine structure interaction and the electronic Zeeman interaction. The hyperfine structure interaction has two types as isotropic and anisotropic hyperfine structure interactions. The zero-field splitting term (electronic quadrupole fine structure), the nuclear Zeeman term, and the nuclear quadrupole interaction term are among the Hamiltonian terms used in EPR. However, their effects are not as much as the term of the hyperfine structure interaction. The zero-field splitting term and the nuclear quadrupole interaction term are the fine structure terms. The interaction of two electron spins create a zero-field splitting, the interaction between the two nucleus spins form the nuclear quadrupole interaction. Hyperfine structure interaction, zero-field interaction, and nuclear quadrupole interaction are subclasses of dipolar interaction. Interaction tensors are available for all three interactions.

Keywords: dipolar interaction hyperfine structure, isotropic hyperfine structure, anisotropic hyperfine structure, the zero-field splitting, the nuclear quadrupole interaction, the electronic Zeeman interaction, the nuclear Zeeman term, EPR

1. Dipolar interactions

Dipolar interaction occurs due to the interaction between the two spins. If one spin becomes an electron spin and the other spin becomes a nucleus spin, this interaction is called a hyperfine structure interaction. If two of the spins are electron spin or both are nucleus spin, this interaction is called fine structure interaction. The dipolar interaction Hamiltonian is expressed as

$$\varkappa = \left[\frac{\overrightarrow{\mu_1}.\overrightarrow{\mu_2}}{r^3} - \frac{3\left(\overrightarrow{\mu_1}.\vec{r}\right)\left(\overrightarrow{\mu_2}.\vec{r}\right)}{r^5}\right] \quad (1)$$

where $\overrightarrow{\mu_1}$ and $\overrightarrow{\mu_2}$ are the magnetic dipole moments for each spin (electron spin or nucleus spin).

1.1 Hyperfine structure interaction

The interaction between the magnetic dipole moment of the nucleus and the magnetic dipole moment of the electron gives the hyperfine structure interaction. There are two types of hyperfine structure interaction. These are isotropic hyperfine interaction and anisotropic hyperfine interaction.

1.1.1 Isotropic hyperfine structure

Isotropic superfine interaction is also known as Fermi contact interaction. The Hamiltonian term of isotropic hyperfine structure interaction is expressed as

$$\varkappa = g_e g_N \beta_e \beta_N \left[\frac{8\pi}{3} \vec{S}.\vec{I}.\delta(r) \right] \tag{2}$$

where g_e = g-value of the electron, g_N = g-value of the nucleus, β_e = Bohr magneton, β_N = nuclear magneton, $\vec{S}$ = electron spin operator, $\vec{I}$ = nuclear spin operator, and $\delta(r)$ = Dirac delta function for the distance between the electron and the nucleus.

In a shorter way, it is expressed as

$$\varkappa = a\vec{S}.\vec{I} \tag{3}$$

The isotropic hyperfine constant is written as

$$a = \frac{8\pi}{3} g_e g_N \beta_e \beta_N \delta(r) \tag{4}$$

Here a is called the isotropic hyperfine constant, $\vec{S}$ is the spin angular momentum of the electron, and $\vec{I}$ is the spin angular momentum of the nucleus.

1.1.2 Anisotropic hyperfine structure

Anisotropic hyperfine interaction is also called dipolar interaction or dipole–dipole interaction. The Hamiltonian term of anisotropic hyperfine structure interaction is expressed as

$$\varkappa = g_e g_N \beta_e \beta_N \left[\frac{3\left(\vec{S}.\vec{r}\right)\left(\vec{I}.\vec{r}\right)}{r^5} - \frac{\vec{S}.\vec{I}}{r^3} \right] \tag{5}$$

More specifically, the expression of the anisotropic hyperfine interaction in the Cartesian coordinate is written as

$$\varkappa = g_e g_N \beta_e \beta_N \left[\frac{(3x^2 - r^2)}{r^5} I_x S_x + \frac{(3y^2 - r^2)}{r^5} I_y S_y + \frac{(3z^2 - r^2)}{r^5} I_z S_z + \frac{3xy}{r^5}(I_x S_y + I_y S_x) + \frac{3yz}{r^5}(I_y S_z + I_z S_y) + \frac{3xz}{r^5}\left(I_x S_z + I_z S_x\right) \right] \tag{6}$$

In a shorter way, it is expressed as

$$\varkappa = \vec{S}.\overrightarrow{\overrightarrow{A^0}}.\vec{I} \tag{7}$$

where $\overrightarrow{\overrightarrow{A^0}}$ is called the anisotropic hyperfine coupling tensor. The tensor is expressed in two ways as diagonal elements and non-diagonal elements.
The diagonal elements of the tensor is expressed as

$$A^0_{ii} = g_e g_N \beta_e \beta_N \left\langle \frac{3i^2 - r^2}{r^5} \right\rangle, i = x, y, z \tag{8}$$

The non-diagonal elements of the tensor is expressed as

$$A^0_{ij} = g_e g_N \beta_e \beta_N \left\langle \frac{3ij}{r^5} \right\rangle, i, j = x, y, z \tag{9}$$

The sum of the isotropic and anisotropic terms fully expresses the hyperfine structure interaction Hamiltonian and is expressed as

$$\varkappa = a\vec{S}.\vec{I} + \vec{S}.\overrightarrow{\overrightarrow{A^0}}.\vec{I} = \vec{S}.\overrightarrow{\overrightarrow{A}}.\vec{I} \tag{10}$$

where $\overrightarrow{\overrightarrow{A}}$ is the general hyperfine structure tensor.

Figure 1 shows the formation of the hyperfine structure splittings. **Figure 2** shows the formation of an EPR spectrum due to the hyperfine structure splittings.

1.2 Fine structure interaction

The fine structure is seen in two ways. The first is the fine structure interaction between two electron spins. The second is the fine structure interaction between the two nucleus spins. The fine structure interaction between two electron spin is also

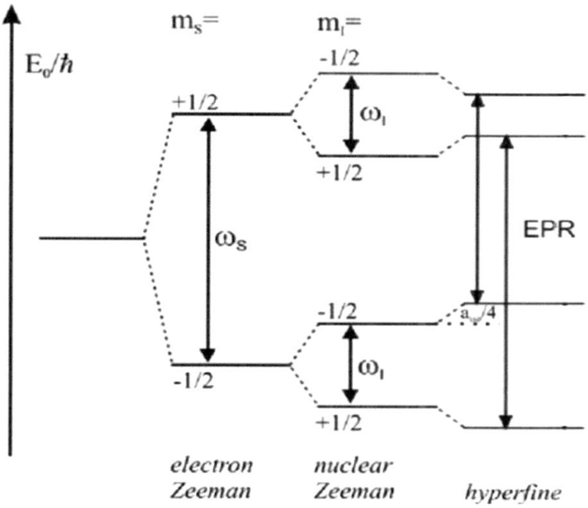

Figure 1.
The formation of the hyperfine structure splittings.

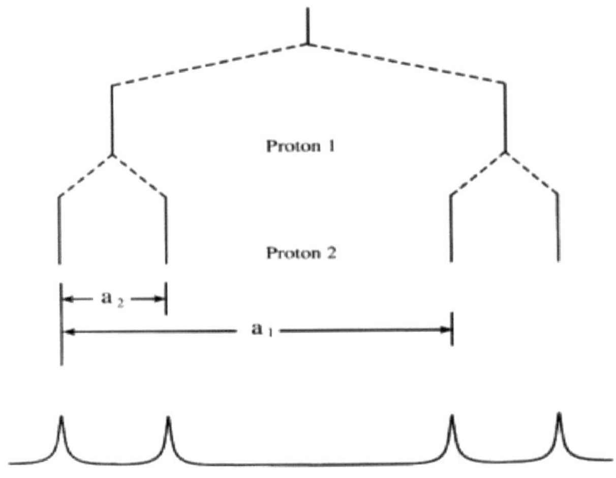

Figure 2.
The formation of an EPR spectrum due to the hyperfine structure splittings.

referred to as zero-field interaction or zero-field splitting. The interaction between two nuclear spin is called nuclear quadrupole interaction.

1.2.1 Zero-field splitting (interaction)

This interaction between two electron spins is the dipolar interaction. When writing Hamiltonian for zero-field interaction, the magnetic dipole moments in Eq. (1) are arranged for two electron spins. In this case, the Hamiltonian of the zero-field splitting is written as

$$\varkappa = g_e^{\,2}\beta_e^{\,2}\left[\frac{\overrightarrow{S_1}.\,\overrightarrow{S_2}}{r^3}-\frac{3\left(\overrightarrow{S_1}.\vec{r}\right)\left(\overrightarrow{S_2}.\vec{r}\right)}{r^5}\right] \tag{11}$$

More specifically, the expression of the anisotropic hyperfine interaction in the Cartesian coordinate is written as

$$\begin{aligned}\varkappa = g_e^{\,2}\beta_e^{\,2}\Bigg[&\frac{(r^2-3x^2)}{r^5}S_{1x}S_{2x}+\frac{(r^2-3y^2)}{r^5}S_{1y}S_{2y}+\frac{(r^2-3z^2)}{r^5}S_{1z}S_{2z}\\ &-\frac{3xy}{r^5}\left(S_{1x}S_{2y}+S_{1y}S_{2x}\right)-\frac{3yz}{r^5}\left(S_{1y}S_{2z}+S_{1z}S_{2y}\right)-\frac{3xz}{r^5}\left(S_{1z}S_{2x}+S_{1x}S_{2z}\right)\Bigg]\end{aligned} \tag{12}$$

In a shorter way, it is expressed as

$$\varkappa = \overrightarrow{S_1}.\overrightarrow{\overrightarrow{D}}.\overrightarrow{S_2} \tag{13}$$

In general, the Hamiltonian of the zero-field splitting is written as

$$\varkappa = \vec{S}.\overrightarrow{\overrightarrow{D}}.\vec{S} \tag{14}$$

where $\overrightarrow{\overrightarrow{D}}$ is called the zero-field splitting tensor or the spin–spin coupling tensor. The tensor is expressed in two ways as diagonal elements and non-diagonal elements. The diagonal elements of the tensor is expressed as

DOI: http://dx.doi.org/10.5772/intechopen.91791

$$D_{ii} = g_e^{\,2}\beta_e^{\,2}\left\langle \frac{r^2 - 3i^2}{r^5} \right\rangle, i = x, y, z \tag{15}$$

The non-diagonal elements of the tensor is expressed as.

$$D_{ij} = g_e^{\,2}\beta_e^{\,2}\left\langle \frac{3ij}{r^5} \right\rangle, i, j = x, y, z \tag{16}$$

The zero-field splittings for s = 1/2, s = 1, and s = 3/2 are shown in **Figure 3**.

1.2.2 Nuclear quadrupole interaction

The interaction between the nucleus spins is known as the nuclear quadrupole interaction. The effects of nuclear quadrupole interaction can be observed on the energy levels of the hyperfine structure for a nucleus with $I \geq 1$. The Hamiltonian of the nuclear quadrupole interaction is expressed as

$$\varkappa = \frac{eQ}{6I(2I-1)} \sum_{\alpha,\beta=x,y,z} V_{\alpha\beta}\left\{\frac{3}{2}\left(I_\alpha I_\beta + I_\beta I_\alpha\right) - \delta_{\alpha\beta} I^2\right\} \tag{17}$$

where $V_{\alpha\beta}$ is the component of the field gradient tensor and eQ is the nuclear quadrupole moment, and it is a measure of the deviation of charge distribution from spherical symmetry. The nuclear quadrupole moment is expressed as

$$eQ = \int \rho_N \left(3z^2 - r^2\right) dV \tag{18}$$

where e is the proton charge, ρ_N is the distribution function of the nuclear charge, z is the z-coordinate of the charge element a distance r from the origin. The integral was taken over the volume of the nucleus.

In general, the nuclear quadrupole interaction Hamiltonian is written as

$$\varkappa = \vec{I} . \vec{P} . \vec{I} \tag{19}$$

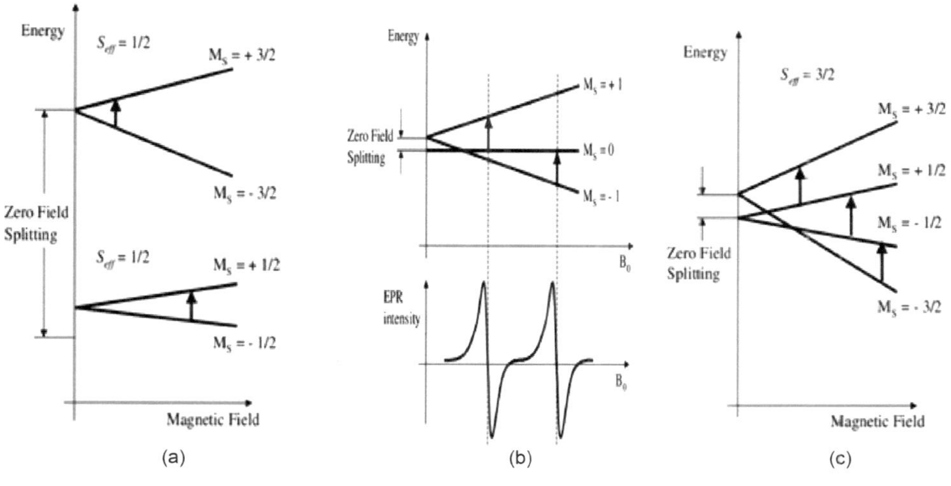

Figure 3.
The zero-field splittings for (a) s = 1/2, (b) s = 1, and (c) s = 3/2.

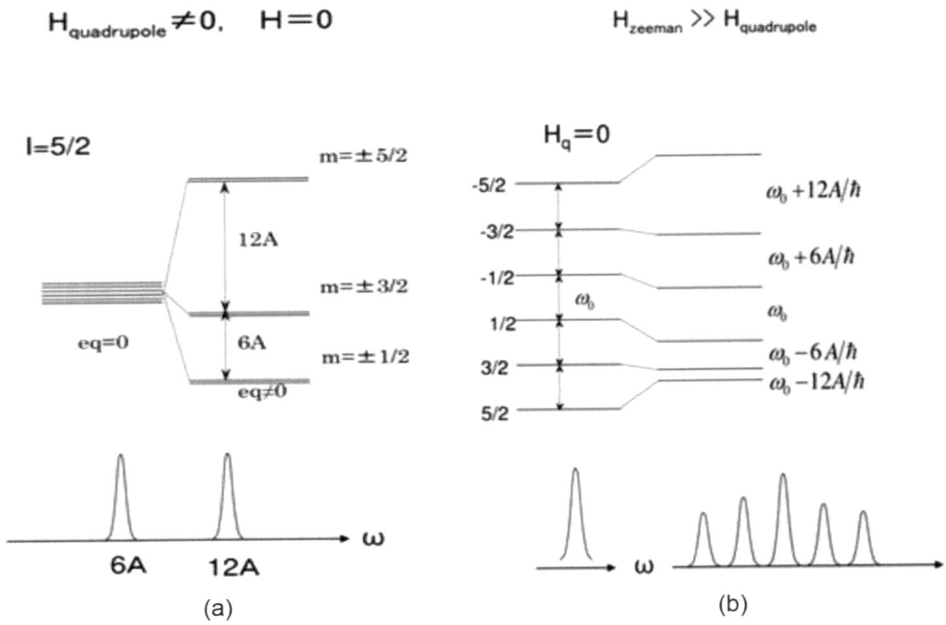

Figure 4.
The nuclear quadrupole splittings for (a) $H_{quadrupole} \neq 0$, $H = 0$ and (b) for $H_{Zeeman} \gg H_{quadrupole}$.

where $\overleftrightarrow{P}$ is called the nuclear quadrupole coupling tensor.
The nuclear quadrupole splittings are shown in **Figure 4**.

2. Effective Hamiltonian terms in electron paramagnetic resonance spectroscopy

The hyperfine structure Hamiltonian term, electron Zeeman Hamiltonian term, nuclear Zeeman Hamiltonian term, the term of the zero-field splitting, and the term of the nuclear quadrupole interaction are Hamiltonian terms in EPR Spectroscopy. However, in EPR spectroscopy, the electron Zeeman term and the hyperfine structure term are effective Hamiltonian terms. Therefore, the effect of the terms other than the electron Zeeman term and the hyperfine structure term is not taken into account, since the effect is minimal compared to these two terms. The electron Zeeman term and the nuclear Zeeman term have not been mentioned before. Therefore, it will be explained briefly below.

The electron Zeeman interaction occurs as a result of the interaction of the magnetic dipole moment caused by the spin of the electron with the applied magnetic field:

$$\varkappa = -\vec{\mu_s}.\vec{H} \tag{20}$$

$$\varkappa = -\left(\gamma_s \vec{S}\right).\vec{H} \tag{21}$$

where γ_s is the gyromagnetic ratio of electron spin and is written as.

$$\gamma_s = -\frac{g_s \beta_e}{\hbar} = -g_s \beta_e \ (\text{in the atomic unit system}, \hbar = 1) \tag{22}$$

where g_s is the spectroscopic splitting factor of the electron spin and is written as

$$g_s = 2 \tag{23}$$

$$\varkappa = -\left(-g_s\beta_e\vec{S}\right).\vec{H} \tag{24}$$

$$\varkappa = g_s\beta_e\vec{S}.\vec{H} \tag{25}$$

The nuclear Zeeman interaction occurs as a result of the interaction of the magnetic dipole moment caused by the spin of the nucleus with the applied magnetic field:

$$\varkappa = -\vec{\mu_I}.\vec{H} \tag{26}$$

$$\varkappa = -\left(\gamma_I\vec{I}\right).\vec{H} \tag{27}$$

where γ_I is the nuclear gyromagnetic ratio and is written as.

$$\gamma_I = \frac{g_I\beta_N}{\hbar} = g_I\beta_N \ (\text{in the atomic unit system}, \hbar = 1) \tag{28}$$

where g_I is the spectroscopic splitting factor of the nucleus spin and is written as

$$g_I = 1 \tag{29}$$

$$\varkappa = -\left(g_I\beta_N\vec{I}\right).\vec{H} \tag{30}$$

$$\varkappa = -g_I\beta_N\vec{I}.\vec{H} \tag{31}$$

The general spin Hamiltonian for EPR spectroscopy can be written as

$$\varkappa = g_s\beta_e\vec{S}.\vec{H} + \vec{S}.\vec{\vec{A}}.\vec{I} - g_I\beta_N\vec{I}.\overrightarrow{H} + \vec{I}.\vec{\vec{P}}.\vec{I} + \vec{S}.\vec{\vec{D}}.\vec{S} \tag{32}$$

The effective spin Hamiltonian for EPR spectroscopy can be written as [1–9].

$$\varkappa = g_s\beta_e\vec{S}.\vec{H} + \vec{S}.\vec{\vec{A}}.\vec{I} \tag{33}$$

3. Conclusion

Dipolar interaction can be seen in three ways. These are the hyperfine structure interaction, the zero-field splitting interaction, and the nuclear quadrupole interaction. Each interaction involves the interaction of two spins. The interaction between a nucleus spin and an electron spin is mentioned in the hyperfine structure interaction. The interaction of two electron spins is mentioned in the zero-field splitting interaction. The interaction of two nuclear spins is mentioned in the nuclear quadrupole interaction. The last two interactions are also known as fine structure interactions.

The hyperfine structure interaction is an important interaction for EPR spectroscopy. In EPR spectroscopy, the effect of the hyperfine structure interaction is taken into account together with the electron Zeeman interaction [10–24]. In addition, nuclear Zeeman interaction, the zero-field interaction, and the nuclear quadrupole interaction have an effect on EPR spectroscopy. However, their effects are negligible.

Author details

Betül Çalişkan[1*] and Ali Cengiz Çalişkan[2]

1 Faculty of Arts and Science, Department of Physics, Pamukkale University, Kinikli, Denizli, Turkey

2 Faculty of Science, Department of Chemistry, Gazi University, Ankara, Turkey

*Address all correspondence to: bcaliska@gmail.com

References

[1] Atherton NM. Electron Spin Resonance Theory and Applications. New York: John Wiley & Sons Inc.; 1993

[2] Weil JA, Bolton JR, Wertz JE. Electron Paramagnetic Resonance Elementary Theory and Practical Applications. New York: John Wiley & Sons Inc.; 1994

[3] Gordy W. Theory and Applications of Electron Spin Resonance. New York: John Wiley & Sons Inc.; 1980

[4] Symons M. Chemical and Biochemical Aspects of Electron-Spin Resonance Spectroscopy. New York: Van Nostrand Reinhold Company; 1978

[5] Bersohn M, Baird JC. An Introduction to Electron Paramagnetic Resonance. New York: W.A. Benjamin, Inc.; 1966

[6] Assenheim HM. Introduction to Electron Spin Resonance. New York: Plenum Press; 1967

[7] Ingram DJE. Free Radicals as Studied by Electron Spin Resonance. Butterworth's Scientific Publications; 1958

[8] Ranby B, Rabek JF. ESR Spectroscopy in Polymer Research. Berlin: Springer-Verlag; 1977

[9] Roylance DK. An EPR investigation of polymer fracture, PhD Dissertation, Department of Mechanical Engineering, University of Utah, Salt Lake City, Utah. August 1968

[10] Caliskan B, Caliskan AC, Er E. Electron paramagnetic resonance study of gamma-irradiated potassium hydroquinone monosulfonate single crystal. Radiation Effects and Defects in Solids. 2016;**171**(5–6):440-450. DOI: 10.1080/10420150.2016.1203924

[11] Caliskan B. EPR study of gamma irradiated cholestanone single crystal. Acta Physica Polonica A. 2014;**125**(1): 135-138. DOI: 10.12693/APhysPolA. 125.135

[12] Caliskan B, Caliskan AC, Yerli R. Electron paramagnetic resonance study of radiation damage in isonipecotic acid single crystal. Journal of Molecular Structure. 2014;**1075**:12-16. DOI: 10.1016/j.mol.struc.2014.06.030

[13] Caliskan B, Caliskan AC. EPR study of radiation damage in gamma irradiated 3-nitroacetophenone single crystal. Radiation Effects and Defects in Solids. 2017;**172**(5–6):398-410. DOI: 10.1080/10420150.2017.1320800

[14] Caliskan B, Caliskan AC, Er E. Electron paramagnetic resonance study of radiation-induced paramagnetic centers in succinic anhydride single crystal. Journal of Molecular Structure. 2017;**1144**:421-431. DOI: 10.1016/j. molstruc.2017.05.039

[15] Caliskan B, Caliskan AC. EPR study of free radical in gamma-irradiated bis (cyclopentadienyl)zirconium dichloride single crystal. Radiation Effects and Defects in Solids. 2017;**172**(5–6): 507-516. DOI: 10.1080/10420150. 2017.1346652

[16] Caliskan B, Aras E, Asik B, Buyum M, Birey M. EPR of gamma irradiated single crystals of cholesteryl benzoate. Radiation Effects and Defects in Solids. 2004;**159**(1):1-5. DOI: 10.1080/10420150310001604101

[17] Caliskan B, Tokgoz H. Electron paramagnetic resonance study of gamma-irradiated phenidone single crystal. Radiation Effects and Defects in Solids. 2014;**169**(3):225-231 DOI: 10.1080/10420150.2013.834903

[18] Caliskan B, Civi M, Birey M. Electron paramagnetic resonance

analysis of gamma irradiated 4-nitropyridine N-oxide single crystal. Radiation Effects and Defects in Solids. 2006;**161**(5):313-317. DOI: 10.1080/10420150600576049

[19] Caliskan B, Caliskan AC. Electron paramagnetic resonance study of the paramagnetic center in gamma-irradiated sulfanilic acid single crystal. Acta Physica Polonica A. 2019;**135**(3):480-484. DOI: 10.12693/APhysPolA.135.480

[20] Caliskan B, Civi M, Birey M. Electron paramagnetic resonance characterization of gamma irradiation damage centers in S-butyrylthiocholine iodide single crystal. Radiation Effects and Defects in Solids. 2007;**162**(2):87-93. DOI: 10.1080/10420150600907632

[21] Aras E, Asik B, Caliskan B, Buyum M, Birey M. Electron paramagnetic resonance study of irradiated tetramethyl-4-piperidion. Radiation Effects and Defects in Solids. 2004;**159**(6):353-358. DOI: 10.1080/10420150410001731820

[22] Asik B, Aras E, Caliskan B, Eken M, Birey M. EPR study of irradiated 4-chloromethyl pyridinium chloride. Radiation Effects and Defects in Solids. 2004;**159**(1):55-60. DOI: 10.1080/10420150310001639770

[23] Caliskan B, Caliskan AC. Electron paramagnetic resonance study of the radiation damage in phosphoryethanolamine single crystal. Journal of Molecular Structure. 2018;**1173**:781-791. DOI: 10.1016/j.molstruc.2018.07.045

[24] Caliskan B, Caliskan AC. Electron paramagnetic resonance study of the radiation damage in trans-chalcone single crystal. Acta Physica Polonica A. 2019;**136**(1):92-100. DOI: 10.12693/APhysPolA.136.92

Chapter 2

Exactly Solvable Problems in Quantum Mechanics

Lourdhu Bruno Chandrasekar,
Kanagasabapathi Gnanasekar and Marimuthu Karunakaran

Abstract

Some of the problems in quantum mechanics can be exactly solved without any approximation. Some of the exactly solvable problems are discussed in this chapter. Broadly there are two main approaches to solve such problems. They are (i) based on the solution of the Schrödinger equation and (ii) based on operators. The normalized eigen function, eigen values, and the physical significance of some of the selected problems are discussed.

Keywords: exactly solvable, Schrödinger equation, eigen function, eigen values

1. Potential well

The potential well is the region where the particle is confined in a small region. In general, the potential of the confined region is lower than the surroundings (**Figure 1**) [1, 2].

The potential of the system is defined as

$$V = \begin{cases} 0, & -L < x < L \\ \infty, & \text{Otherwise} \end{cases}$$

The one dimensional Schrödinger equation in Cartesian coordinate is given as

$$\frac{-\hbar^2}{2m}\Psi'' + V\Psi = E\Psi \Longrightarrow \Psi'' + \frac{2m}{\hbar^2}(E - V)\Psi = 0 \tag{1}$$

In the infinite potential well, the confined particle is present in the well region (Region-II) for an infinitely long time. So the solution of the Schrödinger equation in the Region-II and Region-III can be omitted for our discussion right now. The Schrödinger equation in the Region-II is written as

$$\Psi'' + \frac{2m}{\hbar^2}(E)\Psi = 0$$

$$\Psi'' + \alpha^2\Psi = 0, \text{where } \alpha^2 = \frac{2mE}{\hbar^2} \tag{2}$$

The solution of the Eq. (2) is

$$\Psi = A_1 \sin \alpha x + A_2 \cos \alpha x \tag{3}$$

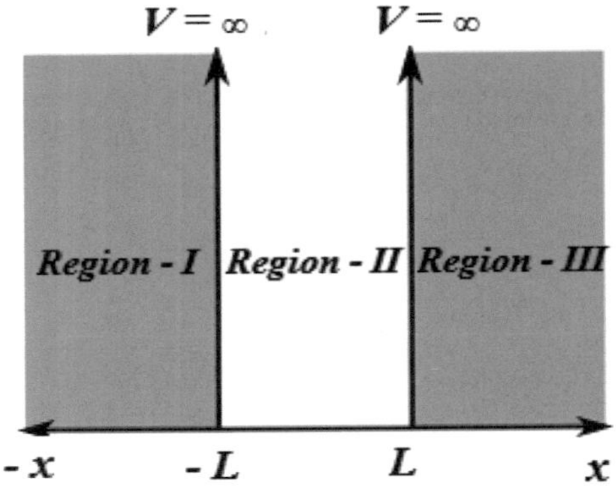

Figure 1.
Infinite potential well.

At $x = -L$, and at $x = L$, the wave function vanishes since the potential is infinite. Hence, At $x = -L$,

$$-A_1 \sin \alpha L + A_2 \cos \alpha L = 0 \tag{4}$$

Similarly, at $x = L$

$$A_1 \sin \alpha L + A_2 \cos \alpha L = 0 \tag{5}$$

The addition and subtraction of these equations give two different solutions.

i. $2A_2 \cos \alpha L = 0 \Longrightarrow \cos \alpha L = 0 \Longrightarrow \alpha L = n\pi/2 \Longrightarrow \alpha^2 = n^2\pi^2/4L^2; n = 1, 3, 5, \ldots \ldots$.
Since $\alpha^2 = \frac{2mE}{\hbar^2}, \frac{2mE}{\hbar^2} = n^2\pi^2/4L^2$, the energy eigen value is found as

$$E = n^2\pi^2\hbar^2/8mL^2 \tag{6}$$

The eigen function is $\Psi = A_1 \cos \alpha x$

$$= A_1 \cos\ (n\pi x/2L)$$

According to the normalization condition,

$$\int_{-L}^{L} \Psi^* \Psi \mathrm{d}x = 0 \Longrightarrow A_1 = L^{-1/2}$$

Hence the normalized eigen function for $n = 1, 3, 5, \ldots \ldots$ is

$$\Psi = L^{-1/2} \cos\ (n\pi x/2L) \tag{7}$$

ii. $2A_1 \sin \alpha L = 0 \Longrightarrow \sin \alpha L = 0 \Longrightarrow \alpha L = n\pi/2 \Longrightarrow \alpha^2 = n^2\pi^2/4L^2; n = 2, 4, 6, \ldots \ldots$.
For this case, $n = 2, 4, 6, \ldots \ldots$, the corresponding energy eigen value is

$$E = n^2\pi^2\hbar^2/8mL^2 \tag{8}$$

DOI: http://dx.doi.org/10.5772/intechopen.93317

The eigen function is $\Psi = A_2 \cos \alpha x$ and the normalized eigen function is

$$\Psi = L^{-1/2} \sin\ (n\pi x/2L) \tag{9}$$

In Summary, the eigen value is $E = n^2\pi^2\hbar^2/8mL^2$ for all positive integer values of "n." The normalized eigen functions are

$$\Psi = \begin{cases} L^{-1/2} \cos\ (n\pi x/2L), n = 1, 3, 5, \ldots \\ L^{-1/2} \sin\ (n\pi x/2L), n = 2, 4, 6, \ldots \end{cases} \tag{10}$$

The integer "n" is the quantum number and it denotes the discrete energy states in the quantum well. We can extract some physical information from the eigen solutions.

- The minimum energy state can be calculated by setting $n = 1$, which corresponds to the ground state. The ground state energy is

$$E_1 = \pi^2\hbar^2/8mL^2 \tag{11}$$

This is known as zero-point energy in the case of the potential well. The excited state energies are $E_2 = 4\pi^2\hbar^2/8mL^2$, $E_3 = 9\pi^2\hbar^2/8mL^2$, $E_4 = 16\pi^2\hbar^2/8mL^2$, and so on. In general, $E_n = n^2 \times E_1$.

- The energy difference between the successive states is simply the difference between the energy eigen value of the corresponding state. For example, $\Delta E_{12} = E_1 \sim E_2 = 3E_1$ and $\Delta E_{23} = E_2 \sim E_3 = 5E_1$. Hence the energy difference between any two successive states is not the same.

- Though the eigen functions for odd and even values of "n" are different, the energy eigen value remains the same.

- If the potential well is chosen in the limit $0 < x < 2L$ (width of the well is $2L$), the energy eigen value is the same as given in Eqs.(6) and (8). But if the limit is chosen as $0 < x < L$ (width of the well is L), the for all positive integers of "n," the eigen function is $\Psi = (2/L)^{1/2} \sin\ (n\pi x/L)$ and the energy eigen function is $E = n^2\pi^2\hbar^2/2mL^2$.

2. Step potential

Step potential is a problem that has two different finite potentials [3]. Classically, the tunneling probability is 1 when the energy of the particle is greater than the height of the barrier. But the result is not true based on wave mechanics (**Figure 2**).

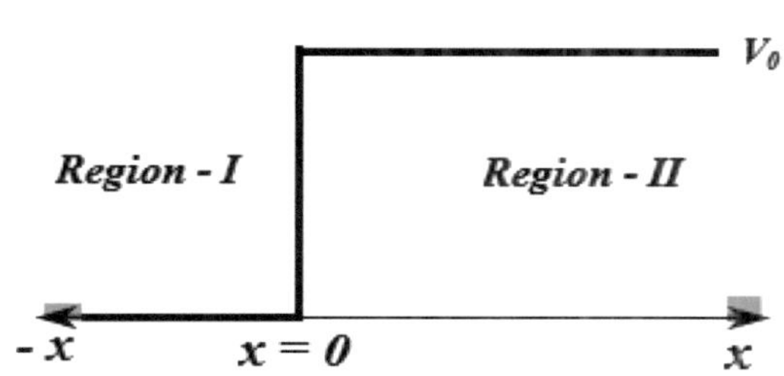

Figure 2.
Step potential.

The potential of the system

$$V = \begin{cases} 0, & -\infty < x < 0 \\ V_0, & 0 \leq x < \infty \end{cases}$$

The Schrödinger equation in the Region-I and Region-II is given, respectively as,

$$\Psi'' + \frac{2m}{\hbar^2}(E)\Psi = 0 \tag{12}$$

$$\Psi'' + \frac{2m}{\hbar^2}(E - V)\Psi = 0 \tag{13}$$

Case (i): when $E < V_0$, the solutions of the Schrödinger equations in the Region-I and Region-II, respectively, are given as

$$\Psi_1 = A_1 \exp(i\alpha x) + B_1 \exp(-i\alpha x) \tag{14}$$
$$\Psi_2 = A_2 \exp(-\beta x) + B_2 \exp(\beta x)$$

where $\alpha^2 = \frac{2mE}{\hbar^2}$ and $\beta^2 = \frac{2m(E-V_0)}{\hbar^2}$. Here, $B_2 \exp(\beta x)$ represents the exponentially increasing wave along the x-direction. The wave function Ψ_2 must be finite as $x \to \infty$. This is possible only by setting $B_2 = 0$. Hence the eigen function in the Region-II is

$$\Psi_2 = A_2 \exp(-\beta x) \tag{15}$$

According to admissibility conditions of wave functions, at $x = 0$, $\Psi_1 = \Psi_2$ and $\Psi_1' = \Psi_2'$. It gives us

$$A_1 + B_1 = A_2 \tag{16}$$

$$A_1 - B_1 = i\left(\frac{\beta}{\alpha}\right)A_2 \tag{17}$$

From these two equations,

$$A_2 = \left(\frac{2\alpha}{\alpha + i\beta}\right)A_1$$

$$B_1 = \left(\frac{\alpha - i\beta}{\alpha + i\beta}\right)A_1$$

The reflection coefficient R is given as

$$R = \frac{|B_1|^2}{|A_1|^2} = \left|\frac{\alpha - i\beta}{\alpha + i\beta}\right|^2 = 1 \tag{18}$$

It is interesting to note that all the particles that encounter the step potential are reflected back. This is due to the fact that the width of the step potential is infinite. The number of particles in the process is conserved, which leads that $T = 0$, since $T + R = 1$.

Case (ii): when $E > V_0$, the solutions are given as

$$\Psi_1 = A_1 \exp(i\alpha x) + B_1 \exp(-i\alpha x)$$
$$\Psi_2 = A_2 \exp(i\beta x) + B_2 \exp(-i\beta x)$$

where $\beta^2 = \frac{2m(E-V_0)}{\hbar^2}$. As $x \to \infty$, the wave function Ψ_2 must be finite. Hence $\Psi_2 = A_2 \exp(i\beta x)$ by setting $B_2 = 0$. According to the boundary conditions at $x = 0$,

$$A_1 + B_1 = A_2 \tag{19}$$

$$A_1 - B_1 = \left(\frac{\beta}{\alpha}\right) A_2 \tag{20}$$

From these equations,

$$A_2 = \left(\frac{2\alpha}{\alpha+\beta}\right) A_1$$

$$B_1 = \left(\frac{\alpha-\beta}{\alpha+\beta}\right) A_1$$

The reflection coefficient R and the transmission coefficient T, respectively, are given as

$$R = \frac{|B_1|^2}{|A_1|^2} = \left(\frac{\alpha-\beta}{\alpha+\beta}\right)^2 \tag{21}$$

$$T = \frac{|A_2|^2}{|A_1|^2} = \frac{4\alpha\beta}{(\alpha+\beta)^2} \tag{22}$$

From these easily one can show that

$$T + R = \frac{4\alpha\beta}{(\alpha+\beta)^2} + \left(\frac{\alpha-\beta}{\alpha+\beta}\right)^2 = 1 \tag{23}$$

The results again indicate that the total number of particles which encounters the step potential is conserved.

3. Potential barrier

This problem clearly explains the wave-mechanical tunneling [3, 4]. The potential of the system is given as (**Figure 3**)

$$V = \begin{cases} V_0, & 0 < x < L \\ 0, & \text{Otherwise} \end{cases}$$

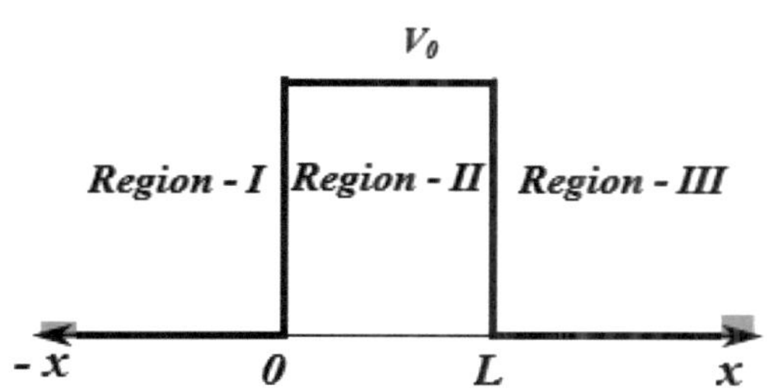

Figure 3.
Potential barrier.

In the Region-I, the Schrödinger equation is $\Psi'' + \alpha^2\Psi = 0$. The wave function in this region is given as

$$\Psi_1 = A_1 \exp(i\alpha x) + B_1 \exp(-i\alpha x) \text{ where } \alpha^2 = \frac{2mE}{\hbar^2} \tag{24}$$

In Region-II, if $E < V_0$, the Schrödinger equation is $\Psi'' - \beta^2\Psi = 0$. The solution of the equation is given as

$$\Psi_2 = A_2 \exp(\beta x) + B_2 \exp(-\beta x) \text{ where } \beta^2 = \frac{2m(\mathrm{E} - V_0)}{\hbar^2} \tag{25}$$

The Schrödinger equation in the Region-III is $\Psi'' + \alpha^2\Psi = 0$. The corresponding solution is $\Psi_3 = A_3 \exp(i\alpha x) + B_3 \exp(-i\alpha x)$. But in the Region-III, the waves can travel only along positive x-direction and there is no particle coming from the right, $B_3 = 0$. Hence

$$\Psi_3 = A_3 \exp(i\alpha x) \tag{26}$$

At $x = 0$, $\Psi_1 = \Psi_2$ and $\Psi_1' = \Psi_2'$. These give us two equations

$$A_1 + B_1 = A_2 + B_2 \tag{27}$$

$$A_1 - B_1 = \left(\frac{\beta}{i\alpha}\right)(A_2 - B_2) \tag{28}$$

At $x = L$, $\Psi_2 = \Psi_3$ and $\Psi_2' = \Psi_3'$. These conditions give us another two equations

$$A_2 \exp(\beta L) + B_2 \exp(-\beta L) = A_3 \exp(i\alpha L) \tag{29}$$

$$A_2 \exp(\beta L) - B_2 \exp(-\beta L) = A_3\left(\frac{i\alpha}{\beta}\right) \exp(i\alpha L) \tag{30}$$

Solving the equations from (27) to (30), one can find the coefficients in the equations. The reflection coefficient is R is found as

$$R = \frac{|B_1|^2}{|A_1|^2} = \left[\frac{V_0^2}{4E(V_0 - E)} \sinh^2(\beta L)\right]\left[1 + \frac{V_0^2}{4E(V_0 - E)} \sinh^2(\beta L)\right]^{-1} \tag{31}$$

The transmission coefficient T is found as

$$T = \frac{|A_2|^2}{|A_1|^2} = \left[1 + \frac{V_0^2}{4E(V_0 - E)} \sinh^2(\beta L)\right]^{-1} \tag{32}$$

From Eqs. (31) and (32), one can show that $T + R = 1$. The following are the conclusions obtained from the above mathematical analysis.

- When $E < V_0$, though the energy of the incident particles is less than the height of the barrier, the particle can tunnel into the barrier region. This is in contrast to the laws of classical physics. This is known as the tunnel effect.
- As $V_0 \to \infty$, the transmission coefficient is zero. Hence the tunneling is not possible only when $V_0 \to \infty$.

- When the length of the barrier is an integral multiple of π/β, there is no reflection from the barrier. This is termed as resonance scattering.
- The tunneling probability depends on the height and width of the barrier.
- Later, Kronig and Penney extended this idea to explain the motion of a charge carrier in a periodic potential which is nothing but the one-dimensional lattices.

4. Delta potential

The Dirac delta potential is infinitesimally narrow potential only at some point (generally at the origin, for convenience) [3]. The potential of the system

$$V = \begin{cases} -\lambda\delta(x), & x = 0 \\ 0, & \text{Otherwise} \end{cases}$$

Here λ is the positive constant, which is the strength of the delta potential. Here, we confine ourselves only to the bound states, hence $E < 0$ (**Figure 4**).

The Schrödinger equation is

$$\Psi'' + \frac{2m}{\hbar^2}(E - V)\Psi = 0 \Longrightarrow \Psi'' + \frac{2m}{\hbar^2}(E + \lambda\delta(x))\Psi = 0 \tag{33}$$

The solution of the Schrödinger equation is given as

$$\text{Region} - \text{I} : \Psi_1 = A_1 \exp(\beta x) \tag{34}$$

$$\text{Region} - \text{II} : \Psi_2 = A_2 \exp(-\beta x) \tag{35}$$

where $\beta^2 = \frac{-2mE}{\hbar^2}$. At $x = 0$, $\Psi_1 = \Psi_2$. So the coefficients A_1 and A_2 are equal. But $\Psi_1' \neq \Psi_2'$, since the first derivative causes the discontinuity. The first derivatives are related by the following equation

$$\Psi_2' - \Psi_1' = -\frac{2m\lambda}{\hbar^2} \tag{36}$$

This gives us

$$\beta = \frac{m\lambda}{\hbar^2} \tag{37}$$

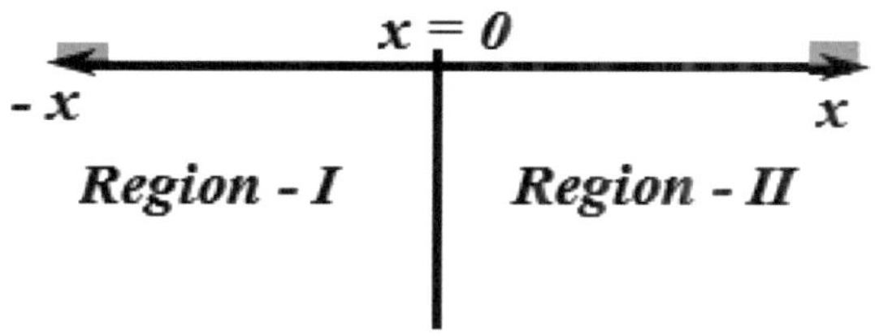

Figure 4.
Dirac delta potential.

Equating the value of β gives the energy eigen value as

$$E = -\frac{m\lambda^2}{2\hbar^2} \tag{38}$$

The energy eigen value expression does not have any integer like in the case of the potential well. Hence there is only one bound state which is available for a particular value of "m."

The eigen function can be evaluated as follows: The eigen function is always continuous. At $x = 0$ gives us $A_1 = A_2 = A$. Hence the eigen function is

$$\Psi = A \exp\left(\beta|x|\right)$$

To normalize Ψ,

$$\int_{-\infty}^{\infty} |\Psi|^2 dx = 1 \Rightarrow 2\int_{0}^{\infty} |\Psi|^2 dx = 1$$

This gives us $A = \sqrt{\beta} = \frac{\sqrt{m\lambda}}{\hbar}$.

5. Linear harmonic oscillator

Simple harmonic oscillator, damped harmonic oscillator, and force harmonic oscillator are the few famous problems in classical physics. But if one looks into the atomic world, the atoms are vibrating even at 0 K. Such atomic oscillations need the tool of quantum physics to understand its nature. In all the previous examples, the potential is constant in any particular region. But in this case, the potential is a function of the position coordinate "x."

5.1 Schrodinger method

The potential of the linear harmonic oscillator as a function of "x" is given as (**Figure** 5) [4–6]:

$$V = \frac{m\omega^2 x^2}{2} \tag{39}$$

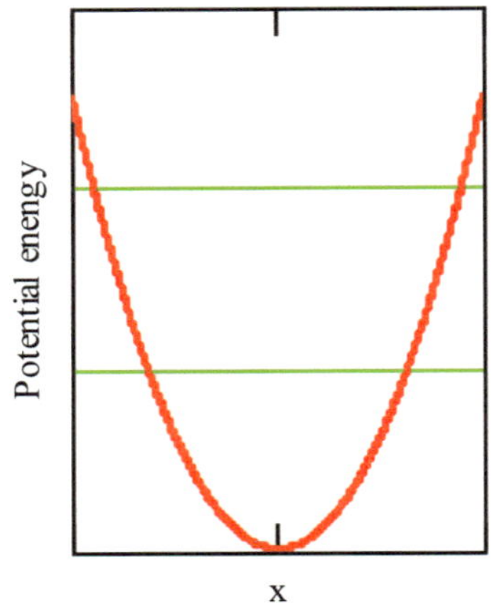

Figure 5.
Potential energy of the linear harmonic oscillator.

The time-independent Schrödinger equation is given as

$$\Psi'' + \frac{2m}{\hbar^2}\left(E - \frac{m\omega^2 x^2}{2}\right)\Psi = 0 \tag{40}$$

The potential is not constant since it is a function of "x"; Eq. (40) cannot solve directly as the previous problems. Let

$$\alpha = \left(\frac{m\omega}{\hbar}\right)^{1/2} x \text{ and } \beta = \frac{2E}{\hbar\omega.}$$

Using the new constant β and the variable α, the Schrödinger equation has the form

$$\frac{d^2\Psi}{d\alpha^2} + (\beta - \alpha^2)\Psi = 0 \tag{41}$$

The asymptotic Schrödinger equation $(\alpha \to \infty)$ is given as

$$\frac{d^2\Psi}{d\alpha^2} - \alpha^2\Psi = 0 \tag{42}$$

The general solution of the equation is $\exp(\pm a^2/2)$. As $\alpha \to \infty$, $\exp(+a^2/2)$ becomes infinite, hence it cannot be a solution. So the only possible solution is $\exp(-a^2/2)$. Based on the asymptotic solution, the general solution of Eq. (42) is given as

$$\Psi = \mathrm{H}_n(\alpha)\exp(-a^2/2)$$

The normalized eigen function is

$$\Psi = \left[\left(\frac{m\omega}{\hbar\pi}\right)^{1/2}\left(\frac{1}{2^n \times n!}\right)\right]^{1/2} \mathrm{H}_n(\alpha)\exp(-a^2/2) \tag{43}$$

The solution given in Eq. (43) is valid if the condition $2n + 1 - \frac{2E}{\hbar\omega} = C$ holds. This gives the energy eigen value as

$$E = \left(n + \frac{1}{2}\right)\hbar\omega \tag{44}$$

The important results are given as follows:

- The integer $n = 0$ represents the ground state, $n = 1$ represents the first excited state, and so on. The ground state energy of the linear harmonic oscillator is $E = \hbar\omega/2$. This minimum energy is known as ground state energy.

- The ground state normalized eigen function is

$$\Psi_0(x) = \left(\frac{m\omega}{\hbar\pi}\right)^{1/4} \exp\left(-\frac{m\omega x^2}{2\hbar}\right) \tag{45}$$

- The energy difference between any two successive levels is $\hbar\omega$. Hence the energy difference between any two successive levels is constant. But this is not true in the case of real oscillators.

5.2 Operator method

The operator method is also one of the convenient methods to solve the exactly solvable problem as well as approximation methods in quantum mechanics [5]. The Hamiltonian of the linear harmonic oscillator is given as,

$$H = \frac{p^2}{2m} + \frac{1}{2}m\omega^2 x^2 \tag{46}$$

Let us define the operator "a," lowering operator, in such a way that

$$a = (2m\omega\hbar)^{-1/2}(m\omega x + ip\ ) \tag{47}$$

and the corresponding Hermitian adjoint, raising operator, is

$$a^+ = (2m\omega\hbar)^{-1/2}(m\omega x - ip\ ) \tag{48}$$

$$\begin{aligned} a^+a &= (2m\omega\hbar)^{-1}(m\omega x - ip\ )(m\omega x + ip\ ) \\ &= (2m\omega\hbar)^{-1}\left(m^2\omega^2x^2 + p^2 + im\omega xp - im\omega px\right) \\ &= (2m\omega\hbar)^{-1}\left(m^2\omega^2x^2 + p^2 + im\omega[x,p]\right) \end{aligned} \tag{49}$$

Here, $[x,p]$ represents the commutation between the operators x and p. $[x,p] = i$ and Eq. (49) becomes

$$\begin{aligned} a^+a &= (2m\omega\hbar)^{-1}\left(m^2\omega^2x^2 + p^2 - m\omega\hbar\right) \\ &= \frac{1}{\omega\hbar}\left(\frac{1}{2}m\omega^2x^2 + \frac{p^2}{2m}\right) - \frac{1}{2} \\ &= \frac{H}{\hbar\omega} - \frac{1}{2} \end{aligned} \tag{50}$$

In the same way, one can find the aa^+ and it is given as

$$aa^+ = \frac{H}{\hbar\omega} + \frac{1}{2} \tag{51}$$

Adding Eqs. (50) and (51) gives us the Hamiltonian in terms of the operators.

$$H = \frac{\hbar\omega}{2}(aa^+ + a^+a) \tag{52}$$

Subtracting Eq. (50) from (51) gives, $aa^+ - a^+a = 1$. This can be simplified as

$$[a, a^+] = 1 \tag{53}$$

The Hamiltonian H acts on any state $|n\rangle$ that gives the eigen value E_n times the same state $|n\rangle$, that is, $H\ |n\rangle = E_n\ |n\rangle$.

The expectation value of a^+a is

$$\begin{aligned} \langle \mathrm{a^+a} \rangle \equiv \langle \mathrm{n}|\mathrm{a^+a}|\mathrm{n} \rangle &= \langle \mathrm{n}|\frac{\mathrm{H}}{\hbar\omega} - \frac{1}{2}|\mathrm{n}\rangle \\ &= \frac{1}{\hbar\omega}\langle \mathrm{n}|\mathrm{H}|\mathrm{n}\rangle - \langle \mathrm{n}|\frac{1}{2}|\mathrm{n}\rangle \\ &= \frac{1}{\hbar\omega}E_n\langle n|n\rangle - \frac{1}{2} = \frac{E_n}{\hbar\omega} - \frac{1}{2} \end{aligned} \tag{54}$$

Let us consider the ground state $|0\rangle$.

$$\langle 0|\mathrm{a^+a}|0\rangle = \frac{\mathrm{E_0}}{\hbar\omega} - \frac{1}{2}$$

Since $a\,|0\rangle = 0$, $\langle 0|a^+a|0\rangle = 0$. Thus,

$$\frac{E_0}{\hbar\omega} - \frac{1}{2} = 0 \Rightarrow E_0 = \frac{\hbar\omega}{2} \tag{55}$$

Similarly, the energy of the first excited state is found as follows:

$$\begin{aligned} \langle 1|\mathrm{a^+a}|1\rangle &= \frac{\mathrm{E_1}}{\hbar\omega} - \frac{1}{2} \\ \sqrt{1}\langle 1|\mathrm{a^+}|0\rangle &= \frac{\mathrm{E_1}}{\hbar\omega} - \frac{1}{2} \\ \sqrt{1}.\sqrt{1}\,\langle 1|1\rangle &= \frac{\mathrm{E_1}}{\hbar\omega} - \frac{1}{2} \\ 1 = \frac{E_1}{\hbar\omega} - \frac{1}{2} &\Rightarrow E_1 = \frac{3}{2}\hbar\omega \end{aligned} \tag{56}$$

In the same way, $E_2 = 5\hbar\omega/2$, $E_3 = 7\hbar\omega/2$, and so on. Hence, one can generalize the result as

$$E_n = \left(n + \frac{1}{2}\right)\hbar\omega \tag{57}$$

The uncertainties in position and momentum, respectively, are given as

$$\Delta x = \sqrt{\langle x^2\rangle - \langle x\rangle^2} \tag{58}$$

$$\Delta p = \sqrt{\langle p^2\rangle - \langle p\rangle^2} \tag{59}$$

In order to evaluate the uncertainties $\langle x^2\rangle$, $\langle x\rangle^2$, $\langle p^2\rangle$, and $\langle p\rangle^2$ have to be evaluated. From Eqs. (47) and (48) the position and momentum operators are found as

$$x = \left(\frac{\hbar}{2m\omega}\right)^{1/2}(a + a^+) \tag{60}$$

$$p = \left(\frac{m\omega\hbar}{2}\right)^{1/2}\left(\frac{a - a^+}{i}\right) \tag{61}$$

a. The expectation value of ‘x’ is given as,

$$\langle x\rangle \equiv \langle n|x|n\rangle = \left(\frac{\hbar}{2m\omega}\right)^{1/2} \langle n|(a+a^{+})|n\rangle$$

$$= \left(\frac{\hbar}{2m\omega}\right)^{1/2} (\langle n|(a)|n\rangle + \langle n|(a^{+})|n\rangle)$$

$$= \left(\frac{\hbar}{2m\omega}\right)^{1/2} (\sqrt{n}\,\langle n|n-1\rangle + \sqrt{n+1}\,\langle n|n+1\rangle)$$

Since the states $n-1, n, n+1$ are orthogonal to each other, $\langle n|n-1\rangle = 0$ and $\langle n|n+1\rangle = 0$. So $\langle x\rangle = 0$. The expectation value of the position in any state is zero.

b. The expectation value of momentum is

$$\langle p\rangle \equiv \langle n|p|n\rangle = \left(\frac{m\omega\hbar}{2}\right)^{1/2}\left(\frac{1}{i}\right)\langle n|a-a^{+}|n\rangle \Longrightarrow \langle p\rangle = 0.$$

Not only position, the expectation value of momentum in any state is also zero.

c.

$$\langle x^2\rangle \equiv \langle n|x^2|n\rangle = \frac{\hbar}{2m\omega}\langle n|(a+a^{+})(a+a^{+})|n\rangle$$

$$= \frac{\hbar}{2m\omega}\langle n|\left(a^2 + a^{+2} + aa^{+} + a^{+}a\right)|n\rangle$$

$$= \frac{\hbar}{2m\omega}\left(\langle n|a^2|n\rangle + \langle n|a^{+2}|n\rangle + \langle n|aa^{+}|n\rangle + \langle n|a^{+}a|n\rangle\right)$$

$$= \frac{\hbar}{2m\omega}\left(\sqrt{n}\sqrt{n-1}\,\langle n|n-2\rangle + \sqrt{n+1}\sqrt{n+2}\langle n|n+2\rangle + (n+1)\langle n|n\rangle + n\,\langle n|n\rangle\right)$$

$$= \frac{\hbar}{2m\omega}(2n+1)$$

d.

$$\langle p^2\rangle \equiv \langle n|p^2|n\rangle = -\left(\frac{m\omega\hbar}{2}\right)\langle n|(a-a^{+})(a-a^{+})|n\rangle$$

$$= -\left(\frac{m\omega\hbar}{2}\right)\left(\langle n|a^2|n\rangle + \langle n|a^{+2}|n\rangle - \langle n|aa^{+}|n\rangle - \langle n|a^{+}a|n\rangle\right)$$

$$= -\left(\frac{m\omega\hbar}{2}\right)\left(\sqrt{n}\sqrt{n-1}\,\langle n|n-2\rangle + \sqrt{n+1}\sqrt{n+2}\langle n|n+2\rangle - (n+1)\langle n|n\rangle - n\,\langle n|n\rangle\right)$$

$$= \left(\frac{m\omega\hbar}{2}\right)(2n+1)$$

From Eq. (58) and (59), the uncertainty in position and momentum, respectively are given as,

$$\Delta x = \left(\frac{\hbar}{2m\omega}(2n+1)\right)^{1/2} \tag{62}$$

$$\Delta p = \left(\left(\frac{m\omega\hbar}{2}\right)(2n+1)\right)^{1/2} \tag{63}$$

$$\Delta x.\Delta p = \frac{\hbar}{2}(2n+1) \tag{64}$$

6. Conclusions

- The minimum uncertainty state is the ground state. In this state, $\Delta x = \left(\frac{\hbar}{2m\omega}\right)^{1/2}$ and $\Delta p = \left(\frac{m\omega}{2}\right)^{1/2}$.

- Hence the minimum uncertainty product is $\Delta x.\Delta p = \frac{\hbar}{2}$. Since the other states have higher uncertainty than the ground state, the general uncertainty is $\Delta x.\Delta p \geq \frac{\hbar}{2}$. This is the mathematical representation of Heisenberg's uncertainty relation.

- Since $\Psi_0(x)$ corresponds to the low energy state, a $\Psi_0(x) = 0$. This gives us the ground state eigen function. This can be done as follows:

$$\text{a } \Psi_0(x) = 0$$

$$(2m\omega\hbar)^{-1/2}(m\omega x + ip\ )\ \Psi_0(x) = 0$$

$$\left(\left(\frac{m\omega}{2\hbar}\right)^{1/2} x + i\frac{(-i\hbar\partial/\partial x)}{(2m\omega\hbar)^{1/2}}\right)\ \Psi_0(x) = 0$$

$$\frac{\hbar}{m\omega}\frac{\partial\,\Psi_0(x)}{\partial x} = -x\ \Psi_0(x)$$

$$\frac{d\,\Psi_0(x)}{\Psi_0(x)} = -\frac{m\omega x}{\hbar}dx$$

Integrating the above equation gives,

$$\ln\ \Psi_0(x) = -\frac{m\omega}{\hbar}\left(\frac{x^2}{2}\right) + \ln A$$

$$\Psi_0(x) = A\ exp\left(-\frac{m\omega x^2}{2\hbar}\right)$$

The normalized eigen function is given as

$$\Psi_0(x) = \left(\frac{m\omega}{\hbar\pi}\right)^{1/4}\ exp\left(-\frac{m\omega x^2}{2\hbar}\right)$$

One can see that this result is identical to Eq. (45).

- The other eigen states can be evaluated using the equation, $\Psi_n(x) = \left((a^+)^n/\sqrt{n!}\right)\ \Psi_0(x)$.

7. Particle in a 3D box

The confinement of a particle in a three-dimensional potential is discussed in this section [4, 6]. The potential is defined as (**Figure 6**)

$$V = \begin{cases} 0, & 0 \leq x < a; 0 \leq y < b; 0 \leq z < c \\ \infty, & \text{Otherwise} \end{cases}$$

The three dimensional time-independent Schrödinger equation is given as

$$\nabla^2\Psi(x, y, z) - \frac{2m}{\hbar^2}V\Psi(x, y, z) = -E\Psi(x, y, z) \tag{65}$$

Let the eigen function $\Psi(x, y, z)$ is taken as the product of $\Psi_x(x)$, $\Psi_y(y)$ and $\Psi_z(z)$ according to the technique of separation of variables. i.e., $\Psi(x, y, z) = \Psi_x(x)\Psi_y(y)\Psi_z(z)$.

$$\Psi_y(y)\Psi_z(z)\frac{d^2\Psi_x(x)}{dx^2} + \Psi_x(x)\Psi_z(z)\frac{d^2\Psi_y(y)}{dy^2} + \Psi_x(x)\Psi_y(y)\frac{d^2\Psi_z(z)}{dz^2} - \frac{2m}{\hbar^2}V\Psi(x, y, z) = -\frac{2m}{\hbar^2}E\Psi(x, y, z)$$

Divide the above equation by $\Psi(x, y, z)$ gives us

$$\frac{1}{\Psi_x(x)}\frac{d^2\Psi_x(x)}{dx^2} + \frac{1}{\Psi_y(y)}\frac{d^2\Psi_y(y)}{dy^2} + \frac{1}{\Psi_z(z)}\frac{d^2\Psi_z(z)}{dz^2} = -\frac{2m}{\hbar^2}E \tag{66}$$

Now we can boldly write E as $E_x(x) + E_y(y) + E_z(z)$

$$\frac{1}{\Psi_x(x)}\frac{d^2\Psi_x(x)}{dx^2} + \frac{1}{\Psi_y(y)}\frac{d^2\Psi_y(y)}{dy^2} + \frac{1}{\Psi_z(z)}\frac{d^2\Psi_z(z)}{dz^2} = -\frac{2m}{\hbar^2}(E_x(x) + E_y(y) + E_z(z)) \tag{67}$$

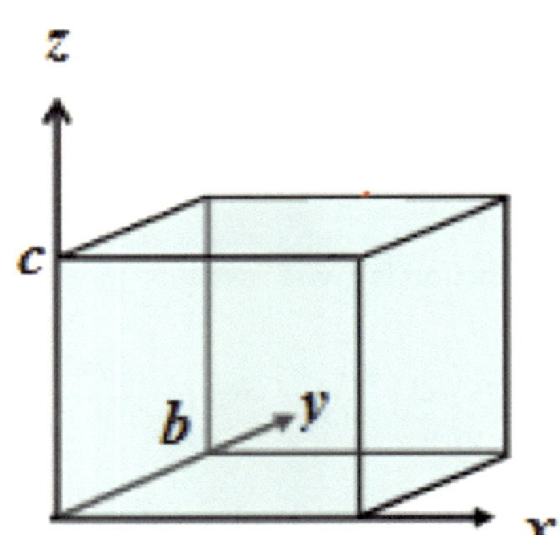

Figure 6.
Three-dimensional potential box.

Now the equation can be separated as follows:

$$\frac{d^2\Psi_x(x)}{dx^2} + \frac{2m}{\hbar^2}E_x(x)\Psi_x(x) = 0$$

$$\frac{d^2\Psi_y(y)}{dy^2} + \frac{2m}{\hbar^2}E_y(y)\Psi_y(y) = 0$$

$$\frac{d^2\Psi_z(z)}{dz^2} + \frac{2m}{\hbar^2}E_z(z)\Psi_z(z) = 0$$

The normalized eigen function $\Psi_x(x)$ is given as

$$\Psi_x(x) = \left(\frac{2}{a}\right)^{1/2} \sin\left(\frac{n_x \pi x}{a}\right)$$

In the same way, $\Psi_y(y)$ and $\Psi_z(z)$ are given as

$$\Psi_y(y) = \left(\frac{2}{b}\right)^{1/2} \sin\left(\frac{n_y \pi y}{b}\right)$$

$$\Psi_z(z) = \left(\frac{2}{c}\right)^{1/2} \sin\left(\frac{n_z \pi z}{c}\right)$$

Hence, the eigen function $\Psi(\mathrm{x, y, z})$ is given as

$$\Psi(\mathrm{x, y, z}) = \Psi_x(x)\Psi_y(y)\Psi_z(z) = \left(\frac{8}{abc}\right)^{1/2} \sin\left(\frac{n_x \pi x}{a}\right) \sin\left(\frac{n_y \pi y}{b}\right) \sin\left(\frac{n_z \pi z}{c}\right) \quad (68)$$

The energy given values are given as

$$\mathrm{E_x(x)} = \frac{\mathrm{n_x^2}\pi^2\hbar^2}{\mathrm{2ma^2}}$$

$$\mathrm{E_y(y)} = \frac{\mathrm{n_y^2}\pi^2\hbar^2}{\mathrm{2mb^2}}$$

$$\mathrm{E_z(z)} = \frac{\mathrm{n_z^2}\pi^2\hbar^2}{\mathrm{2mc^2}}$$

The total energy E is

$$E = E_x(x) + E_y(y) + E_z(z) = \frac{\pi^2\hbar^2}{2m}\left(\frac{n_x^2}{a^2} + \frac{n_y^2}{b^2} + \frac{n_z^2}{c^2}\right) \quad (69)$$

Some of the results are summarized here:

- In a cubical potential box, $a = b = c$, then the energy eigen value becomes,

$$E = \frac{\pi^2\hbar^2}{2ma^2}\left(n_x^2 + n_y^2 + n_z^2\right).$$

- The minimum energy that corresponds to the ground state is $E_1 = \frac{3\pi^2\hbar^2}{2ma^2}$. Here $n_x = n_y = n_z = 1$.
- Different states with different quantum numbers may have the same energy. This phenomenon is known as degeneracy. For example, the states (i) $n_x = 2; n_y = n_z = 1$, (ii) $n_y = 2; n_x = n_z = 1$; and (iii) $n_z = 2; n_x = n_y = 1$ have the same energy of $E = \frac{6\pi^2\hbar^2}{ma^2}$. So we can say that the energy $\frac{6\pi^2\hbar^2}{ma^2}$ has a 3-fold degenerate.
- The states (111), (222), (333), (444), ... has no degeneracy.
- In this problem, the state may have zero-fold degeneracy, 3-fold degeneracy or 6-fold degeneracy.

Author details

Lourdhu Bruno Chandrasekar[1*], Kanagasabapathi Gnanasekar[2] and Marimuthu Karunakaran[3]

1 Department of Physics, Periyar Maniammai Institute of Science and Technology, Vallam, India

2 Department of Physics, The American College, Madurai, India

3 Department of Physics, Alagappa Government Arts College, Karaikudi, India

*Address all correspondence to: brunochandrasekar@gmail.com

References

[1] Griffiths DJ. Introduction to Quantum Mechanics. 2nd ed. India: Pearson

[2] Singh K, Singh SP. Elements of Quantum Mechanics. 1st ed. India: S. Chand & Company Ltd

[3] Gasiorowicz S. Quantum Mechanics. 3rd ed. India: Wiley

[4] Schiff LI. Quantum Mechanics. 4th ed. India: McGraw Hill International Editions

[5] Peleg Y, Pnini R, Zaarur E, Hecht E. Quantum Mechanics. 2nd ed. India: McGraw Hill Editions

[6] Aruldhas G. Quantum Mechanics. 2nd ed. India: Prentice-Hall

Chapter 3

Transitions between Stationary States and the Measurement Problem

María Esther Burgos

Abstract

Accounting for projections during measurements is the traditional measurement problem. Transitions between stationary states require measurements, posing a different measurement problem. Both are compared. Several interpretations of quantum mechanics attempting to solve the traditional measurement problem are summarized. A highly desirable aim is to account for both problems. Not every interpretation of quantum mechanics achieves this goal.

Keywords: quantum measurement problem, transitions between stationary states, interpretations of quantum theory

1. Introduction and outlook

In 1930 Paul Dirac published *The Principles of Quantum Mechanics* [1]. Two years later John von Neumann published *Mathematische Grundlagen der Quantenmechanik* [2]. These initial versions of quantum theory share two characteristics, (i) the state vector $|\psi\rangle$ (wave function ψ) describes the state of an individual system, and (ii) they involve two laws of change of the state of the system: spontaneous processes, governed by the Schrödinger equation, and measurement processes, ruled by the projection postulate ([3], pp. 5–6).

Many other versions of quantum theory followed. Those where ψ describes the state of an individual system and where the projection postulate is included among its axioms are generally called standard, ordinary, or orthodox quantum mechanics (OQM), sometimes referred to as the Copenhagen interpretation, associated to Niels Bohr.

The most relevant differences between spontaneous processes (SP) and measurement processes (MP) are as follows [4]: in SP the observer plays no role, in MP the observer (or the measuring device) plays a paramount role; in SP the state vector $|\psi(t)\rangle$ is continuous, in MP $|\psi(t)\rangle$ collapses (jumps, is projected, is reduced); in SP the superposition principle applies, in MP the superposition principle breaks down; SP are ruled by a deterministic law, MP are ruled by probability laws; in SP every action is localized, in MP there is a kind of action-at-a-distance [5]; and in SP conservation laws are strictly valid, in MP conservation laws have only a statistical sense [6–8].

Since the projection postulate contradicts the fundamental Schrödinger equation of motion, some authors rushed to the conclusion that it was defective.

IntechOpen

Henry Margenau suggested in a manuscript sent to Albert Einstein on November 13, 1935, that this postulate should be abandoned. Einstein replied that the formalism of quantum mechanics inevitably requires the following postulate: "If a measurement performed upon a system yields a value m, then the same measurement performed immediately afterwards yields again the value m with certainty" ([3], p. 228). The projection postulate guarantees compliance with this principle.

The traditional measurement problem in quantum mechanics is how (or whether) wave function collapse occurs when a measurement is performed. Although *a similar measurement problem* is implied in transitions between stationary states (TBSS) induced by a time-dependent perturbation, *it is conspicuously absent* from the specialized literature on the subject.

The contents of this paper are as follows: time-dependent perturbation theory (TDPT) is summarized in Section 2. Section 3 shows that according to TDPT, measurements are required for TBSS to occur. Section 4 highlights the similarities and differences between the traditional measurement problem and that implied in TBSS. Section 5 includes several interpretations of quantum mechanics which attempt to solve the traditional measurement problem: Bohmian mechanics, decoherence, spontaneous localization, and spontaneous projection approach (SPA). Section 6 shows that SPA accounts for TBSS, and in cooperation with decoherence, it also accounts for the traditional measurement problem. Section 7 compiles conclusions.

2. The formulation of TDPT

TDPT was formulated by Dirac in 1930 ([1], Chapter VII). In his words: "In [TDPT] we do not consider any modification to be made in the states of the unperturbed system, but we *suppose* that the perturbed system, instead of remaining permanently in one of these states, is continually changing from one to another, or making transitions, under the influence of the perturbation" ([1], p. 167; emphasis added). The aim of TDPT is, then, to calculate the probability of TBSS which can be induced by the perturbation during a given time interval.

Dirac points out that "this method must ... be used for solving *all* problems involving a consideration of time, such as those about the transient phenomena that occur when the perturbation is suddenly applied, or more generally problems in which the perturbation varies with the time in any way (i.e. in which the perturbing energy involves the time explicitly). [It must also] be used in collision problems, even though the perturbing energy does not here involve the time explicitly, if one wishes to calculate absorption and emission probabilities, since these probabilities, unlike a scattering probability, cannot be defined without reference to a state of affairs that varies with the time" ([1], p. 168; emphasis added).

TDPT is a key ingredient of OQM. It has many applications and is at the basis of quantum electrodynamics, the extension of OQM accounting for the interactions between matter and radiation ([1], Chapter X; [9], Chapter 9). Without TDPT, OQM would hardly be such a powerful and successful theory.

To develop TDPT one starts by splitting in two the total Hamiltonian $\mathrm{H}(t)$ acting on the system:

$$\mathrm{H}(t) = \mathrm{E} + \mathrm{W}(t) \tag{1}$$

E is the Hamiltonian of an unperturbed system, which can be dealt with exactly. Every dependence on time is included in $\mathrm{W}(t)$. Dirac asserts that "the perturbing energy $\mathrm{W}(t)$ can be an arbitrary function of the time" ([1], p. 172).

The eigenvalue equations of E are

$$\mathrm{E}\,|\varphi_n\rangle = E_n\,|\varphi_n\rangle \tag{2}$$

where E_n (n = 1, 2, ..., N) are the eigenvalues of E and $|\varphi_n\rangle$ the corresponding eigenvectors. For simplicity we shall consider the spectrum of E to be entirely discrete and non-degenerate. All the E_n and $|\varphi_n\rangle$ are supposed to be known.

Let $|\psi(t)\rangle$ be the state of the system at time t. We assume that at the initial time t_0, the system is in the state $|\psi(t_0)\rangle = |\varphi_j\rangle$, the eigenvector of the non-perturbed Hamiltonian E corresponding to the eigenvalue E_j. If there is no perturbation, i.e., if the Hamiltonian were E, this state would be stationary. But the perturbation causes the state to change. At time t the state of the system will be

$$|\psi(t)\rangle = U_H(t,t_0)|\psi(t_0)\rangle = U_H(t,t_0)|\varphi_j\rangle \tag{3}$$

where $U_H(t,t_0)$ is the evolution operator, a linear operator independent on $|\psi\rangle$ and depending only on H, t, and t_0 ([1], p. 109).

The probability of a transition taking place from the initial stationary state $|\varphi_j\rangle$ to the final stationary state $|\varphi_k\rangle$ (respectively corresponding to the eigenvalues E_j and E_k of E) induced by the perturbation W(t) during the time interval (t_0, t) is then

$$P_{jk}(t_0,t) = |\langle\varphi_k|U_H(t,t_0)|\varphi_j\rangle|^2 \tag{4}$$

See, for instance, [1], Chapter VII; [9], Chapter 9; [10], Chapter XIII; [11], Chapter IV; [12], Chapter 19; and [13], Chapter XVII. Note: symbols used by these authors may have been changed for homogeneity.

3. TBSS require measurements

TDPT includes two clearly different stages. The first governed by the Schrödinger equation and the second ruled by probability laws [14]. Concerning this issue Dirac points out: "When one makes an observation on the dynamical system, the state of the system gets changed in an unpredictable way, but in between observations causality applies, in quantum mechanics as in classical mechanics, and the system is governed by equations of motion which make the state at one time determine the state at a later time. These equations of motion ... will apply so long as the dynamical system is left undisturbed by any observation or similar process ... Let us consider a particular state of motion through the time during which the system is left undisturbed. We shall have the state at any time t corresponding to a certain ket which depends on t and which may be written $|\psi(t)\rangle$ ... The requirement that the state at one time [t_0] determines the state at another time [t] means that $|\psi(t_0)\rangle$ determines $|\psi(t)\rangle$ ..." ([1], p. 108).

During the first stage of TDPT the process is ruled by the Schrödinger equation:

$$i\hbar\frac{\mathrm{d}}{\mathrm{dt}}|\psi(t)\rangle = \mathrm{H}(t)\,|\psi(t)\rangle \tag{5}$$

where H(t) is the total Hamiltonian of the system and $\hbar$ is Planck's constant divided by 2π. The solution of Eq. (5) corresponding to the initial condition $|\psi(t_0)\rangle = |\varphi_j\rangle$ is unique; $|\psi(t)\rangle$ is completely determined by the initial state $|\psi(t_0)\rangle$ and H(t), which includes the perturbation W(t). Since $|\psi(t)\rangle$ depends only on the initial state $|\varphi_j\rangle$ and on H(t), or if preferred on the perturbation W(t), then

$$|\psi(t)\rangle \equiv |\psi_{j,H}(t)\rangle = U_H(t,t_0)|\psi(t_0)\rangle = U_H(t,t_0)|\varphi_j\rangle \tag{6}$$

The evolution from $|\varphi_j\rangle$ to $|\psi_{j,H}(t)\rangle$ given by Eq. (6) is *automatic*. No transition from the initial state $|\varphi_j\rangle$ to a stationary state $|\varphi_k\rangle$ results until time t.

In the second stage of TDPT, *it is assumed* that at a time t_f, a measurement is performed. As a consequence, a projection from $|\psi_{j,H}(t_f)\rangle$ to $|\varphi_k\rangle$ takes place. In the words of Albert Messiah: "We suppose that at the initial time t_0 the system is in an eigenstate of E, the state $|\varphi_j\rangle$ say. We wish to calculate the probability that *if a measurement is made* at a later time t_f, the system *will be found* to be in a different eigenstate of E, the state $|\varphi_k\rangle$ say. This quantity, by definition the probability of transition from $|\varphi_j\rangle$ to $|\varphi_k\rangle$, will be denoted by $P_{jk}(t_0,t_f)$" ([13], p. 725; emphases added). Clearly

$$P_{jk}(t_0,t_f) = |\langle\varphi_k|U_H(t_f,t_0)|\varphi_j\rangle|^2 \tag{7}$$

Dirac does not explicitly mention measurements. He supposes that at the initial time t_0, the system is in a state for which E has the value E_j with certainty. The ket corresponding to this state is $|\varphi_j\rangle$. At time t_f the corresponding ket will be $U_H(t_f,t_0)\,|\varphi_j\rangle$ ([1], p. 172). The probability of E then *having* the value E_k is given by Eq. (7). For $E_k \neq E_j$, $P_{jk}(t_0,t_f)$ is the probability of a transition taking place from $|\varphi_j\rangle$ to $|\varphi_k\rangle$ during the time interval (t_0,t_f), while $P_{jj}(t_0,t_f)$ is the probability of no transition taking place at all. The sum of $P_{jk}(t_0,t_f)$ for all k is unity ([1], pp. 172–173).

Note that where Messiah says "the probability that *if a measurement* [of E] *is made* ... the system *will be found* to be in ... the state $|\varphi_k\rangle$ ... " Dirac says "the probability of E then *having* the value E_k ... " Dirac's assertion, however, has exactly the same meaning as Messiah's, as shown in the following quote from Dirac's book *The Principles of Quantum Mechanics*: "The expression that an observable 'has a particular value' for a particular state is permissible in quantum mechanics in the special case when a measurement of the observable is certain to lead to the particular value, so that the state is an eigenstate of the observable ... In the general case we cannot speak of an observable having a value for a particular state ... [but] we can go further and speak of the probability of its having any specified value for the state, meaning *the probability of this specified value being obtained when one makes a measurement of the observable*" ([1], pp. 46–47; emphases added). Hence Dirac's statement "the probability of E then *having* the value E_k is given by Eq. (7)" should be understood as "the probability of E_k being obtained when one makes a measurement of E is given by Eq. (7)." *Both Dirac (the author of TDPT) and Messiah place measurements at the very heart of TDPT.*

The following diagram illustrates the complete process leading the system from the initial state $|\varphi_j\rangle$ to the final state $|\varphi_k\rangle$:

$$|\psi(t_0)\rangle = |\varphi_j\rangle \longrightarrow |\psi_{j,H}(t_f)\rangle = U_H(t_f,t_0)\,|\varphi_j\rangle \longrightarrow |\varphi_k\rangle$$

First stage: during the interval (t_0,t_f) the evolution of the state is ruled by the Schrödinger equation	**Second stage**: $|\psi_{j,H}(t_f)\rangle$ jumps to $|\varphi_k\rangle$ with probability $P_{jk}(t_0,t_f)$

Let ε be the non-perturbed energy represented by the operator E. *Everything happens as if at time t_f a measurement of ε is performed* [14]. If no measurement of ε is performed, OQM states that the system continues to evolve in compliance with Schrödinger's equation.

4. Two kinds of measurement problems: similarities and differences

It is often overlooked that TDPT requires a measurement of ε in order to obtain the collapse $|\psi_{j,H}(t_f)\rangle \to |\varphi_k\rangle$, suggesting that TBSS are simply the result of perturbations [14]. A perturbation is something completely different from a measurement. When the perturbation $W(t)$ is applied, the Hamiltonian changes from E to E + $W(t)$, but the Schrödinger evolution is not suspended. By contrast, a measurement interrupts the Schrödinger evolution. According to TDPT the perturbation $W(t)$ applied during the interval (t_0, t_f) as well as the measurement of ε at t_f are necessary for the transition $|\varphi_j\rangle \to |\varphi_k\rangle$ to occur.

There are, then, two kinds of measurement problems: (i) the traditional measurement problem and (ii) the measurement problem related to TBSS. Both of them are measurement problems for in both the Schrödinger evolution is interrupted and the state of the system instantaneously collapses as established by the projection postulate.

i. In the traditional measurement problem, the experimenter chooses the physical quantity to be measured. This quantity can be, in principle, any physical quantity such as the position, a component of the angular momentum, the energy, etc. Measurements of these quantities have been performed many times, with different methods, by different people, and in different circumstances.

ii. In TBSS the system jumps to an eigenstate of E, the operator representing ε. The experimenter has no choice; the only physical quantity susceptible to be "measured" is the non-perturbed energy ε. We say "measured" because it seems difficult to admit that TBSS involve measurements of any physical quantity. It seems even more difficult to admit that ε is measured every time a photon is either emitted or absorbed by an atom, as TDPT requires TBSS could be considered "measurements" without observers or measuring devices.

"In most cases, the wave function evolves gently, in a perfectly predictable and continuous way, according to the Schrödinger equation; in some cases only (as soon as a measurement is performed), unpredictable changes take place, according to the postulate of wave packet reduction" [15]. TBSS, which are happening everywhere all the time, must also be included in *some of the cases* where unpredictable changes take place according to the projection postulate.

In previous papers we have pointed out the following contradiction: On the one hand, according to OQM *there is no room for the projection postulate* as long as we are dealing with spontaneous processes. On the other hand, to account for spontaneous processes involving a consideration of time OQM requires, through TDPT, *the application of the projection postulate*. This is a flagrant incoherence absent from the literature [14, 16].

Quantum weirdness has been associated with the traditional measurement problem. To solve it, several interpretations of quantum mechanics have been proposed. In the following section, we shall address a few of them. For a critical review of the most popular interpretations of quantum theory, see the interesting study of Franck Laloë *Do we really understand quantum mechanics?* [15].

5. Some alternative interpretations to OQM

5.1 Bohmian mechanics (BM)

It is also called the causal interpretation of quantum mechanics and the pilot-wave model. Its first version was proposed by Louis de Broglie in 1927, rapidly abandoned and forgotten, and reformulated by David Bohm in 1952 [17].

In BM it is assumed that particles are point-like. They have well-defined positions at each instant and thus describe trajectories. A system of N particles with masses m_k and *actual* positions $Q_k(t)$ ($k = 1, \ldots, N$) can be described by the couple $(Q(t), \psi(t))$, where $Q(t) = (Q_1(t), \ldots, Q_N(t))$ is the *actual* configuration of the system. The wave function of the system is $\psi = \psi(q, t) = \psi(q_1, \ldots, q_N; t)$, a function on the space of *possible* configurations q of the system. The wave function evolves according to the Schrödinger equation:

$$i\hbar \frac{\partial}{\partial t}\psi = H\psi \tag{8}$$

where H is the nonrelativistic Hamiltonian. The actual positions of the particles evolve according to the guiding equation:

$$\frac{d}{dt}Q_k(t) = \frac{\hbar}{m_k}\,\mathrm{Im}\left[\frac{\psi^* \,\partial_k\psi}{\psi^*\,\psi}\right] \tag{9}$$

where Im [] is the imaginary part of [] and $\partial_k = (\partial/\partial x_k, \partial/\partial y_k, \partial/\partial z_k)$ is the gradient with respect to the generic coordinates $q_k = (x_k, y_k, z_k)$ of the kth particle. For a system of N particles, Eqs. (8) and (9) completely define BM [18]. It is worth stressing that (i) BM is a nonlocal theory and (ii) BM is a deterministic theory: the initial couple $(Q(t_0), \psi(t_0))$ determines the couple at any time $t > t_0$.

BM accounts for all of the phenomena governed by nonrelativistic quantum mechanics, from spectral lines and scattering theory to superconductivity, the quantum Hall effect and quantum computing [18]. A proposed extension of BM describes creation and annihilation events: the world lines for the particles can begin and end [19]. For any experiment the deterministic Bohmian model yields the usual quantum predictions [18].

In BM the usual measurement postulates of quantum theory emerge from an analysis of the Eqs. (8) and (9). In the collapse of the wave function, the interaction of the quantum system with the environment (air molecules, cosmic rays, internal microscopic degrees of freedom, etc.) plays a significant role. Even if the Schrödinger evolution is not interrupted, replacing the original wave function for its "collapsed" derivative is justified as a pragmatic affair [18]. In this regard BM appeals for processes of decoherence.

5.2 Decoherence

Decoherence is an interesting physical phenomenon entirely contained in the linear Schrödinger equation and does not imply any particular conceptual problem [15]. It is a consequence of the unavoidable coupling of the quantum system with the surrounding medium which "looks and smells as a collapse" [20].

Decoherence is currently the subject of a great deal of research. To grasp how it works, let us consider the following case, taken from Daniel Bes' *Quantum Mechanics* ([9], pp. 247–248).

A quantum system in the state $|\Phi_i\rangle$ (i = 1, 2) interacts with the environment, initially in the state $|\eta_0\rangle$, resulting in

$$|\Phi_i\rangle\,|\eta_0\rangle \rightarrow |\Phi_i\rangle\,|\eta_i\rangle \tag{10}$$

If the initial state of the system is $|\Phi_\pm\rangle = \left(\frac{1}{\sqrt{2}}\right)(|\Phi_1\rangle \pm |\Phi_2\rangle)$, the linearity of the Schrödinger equation yields entangled states:

$$|\Phi_\pm\rangle\,|\eta_0\rangle \rightarrow \left(\frac{1}{\sqrt{2}}\right)(|\Phi_1\rangle\,|\eta_1\rangle \pm |\Phi_2\rangle\,|\eta_2\rangle) \tag{11}$$

The corresponding pure state density matrix is

$$\begin{aligned}\rho = &\frac{1}{2}\,|\Phi_1\rangle\,\langle\Phi_1|\,|\eta_1\rangle\,\langle\eta_1| \pm \frac{1}{2}\,|\Phi_1\rangle\,\langle\Phi_2|\,|\eta_1\rangle\,\langle\eta_2| \\ &\pm \frac{1}{2}\,|\Phi_2\rangle\,\langle\Phi_1|\,|\eta_2\rangle\,\langle\eta_1| + \frac{1}{2}\,|\Phi_2\rangle\,\langle\Phi_2|\,|\eta_2\rangle\,\langle\eta_2|\end{aligned} \tag{12}$$

Assuming that the environment states are almost orthogonal to each other, i.e., $\langle\eta_1|\eta_2\rangle \approx 0$ ([9], p. 248), the reduced density matrix becomes

$$\rho' \approx \frac{1}{2}\,|\Phi_1\rangle\,\langle\Phi_1| + \frac{1}{2}\,|\Phi_2\rangle\,\langle\Phi_2| \tag{13}$$

"Eq. (13) does not imply that the system is in a mixture of states $|\Phi_1\rangle$ and $|\Phi_2\rangle$. Since these two states are simultaneously present in Eqs. (11) and (12), the composite system + environment displays superposition and associated interferences. However, Eq. (13) says that such quantum manifestations will not appear as long as experiments are performed only on the system" ([9], p. 248).

It has been proven that for large classical objects, decoherence would be virtually instantaneous because of the high probability of interaction of such systems with some environmental quantum. Several models illustrate the gradual cancelation of the off-diagonal elements with decoherence over time. Experiments also show that, due to the interaction with the environment, superposition states become unobservable ([9], p. 251). "These experiments provide impressive direct evidence for how the interaction with the environment gradually delocalizes the quantum coherence required for the interference effects to be observed… We find our observations to be in excellent agreement with theoretical predictions" ([21], p. 265).

5.3 Spontaneous localization

The key assumption is that each elementary constituent of any physical system is subject, at random times, to spontaneous localization processes (called hittings) around random positions. The best known mathematical model stating which modifications of the wave function are induced by localizations, where and when they occur, is usually referred to as the Ghirardi-Rimini-Weber (GRW) theory [22, 23]. It holds as follows [24]:

Let $\psi(q_1, \ldots, q_N)$ be the wave function of a system of N particles. "If a hitting occurs for the ith particle at point x, the wave function is instantaneously multiplied by a Gaussian function (appropriately normalized)" [24]:

$$G(q_i, x) = \mathrm{K}\,\exp\left[-\left(\frac{1}{2d^2}\right)(q_i - x)^2\right] \tag{14}$$

where d and K are constants. Let

$$\Phi_i(q_1, \dots, q_N; x) = \psi(q_1, \dots, q_N)\, G(q_i, x) \tag{15}$$

be the unnormalized wave function immediately after the localization and $P(x)$ the density probability of the hitting taking place at x. Assuming that $P(x)$ equals the integral of $|\Phi_i|^2$ over the 3N-dimensional space implies that hittings occur with higher probability at those places where, in the standard quantum description, there is a higher probability of finding the particle. The constant K appearing in Eq. (14) is chosen in such a way that the integral of $P(x)$ over the whole space equals unity. Finally, it is assumed that the hittings occur at randomly distributed times, according to a Poisson distribution, with mean frequency f. The parameters chosen in the GRW-model are f = 10^{-16} s^{-1} and d = 10^{-5} cm [24].

GRW aims to a unification of all kinds of physical evolution, including wave function reduction. On the one hand, the theory succeeds in proposing a real physical mechanism for the emergence of a single result in a single experiment, which is attractive from a physical point of view, and solves the "preferred basis problem," since the basis is that of localized states. The occurrence of superposition of far-away states is destroyed by the additional process of localization [15]. On the other hand, it fails to account for TBSS referred to in TDPT. Similar theories to GRW, like the continuous spontaneous localization, confront the same problem. The reason is simple: localizations localize (see Eqs. (14) and (15)). They do not yield the system to a stationary state.

5.4 Spontaneous projection approach (SPA)

Two kinds of processes irreducible to one another occur in nature: those strictly continuous and causal, governed by a deterministic law, and those implying discontinuities, ruled by probability laws. This is the main hypothesis of SPA [25]. Continuous and causal processes are Schrödinger's evolutions. Processes implying discontinuities are jumps to the preferential states $|\varphi_j\rangle$ $(j = 1, \dots, N)$ belonging to the preferential set $\{N_\varphi\}$ (= $|\varphi_1\rangle, \dots, |\varphi_N\rangle$) of the system in a given state [26, 27].

In SPA conservation laws play a paramount role. The system has the tendency to jump to the eigenstates of every constant of the motion, while the jumps must respect the statistical sense of every conservation law [25].

The preferential set may or may not exist. If the system in the state $|\psi(\mathrm{t})\rangle$ has the preferential set $\{N_\varphi\}$, we can write

$$|\psi(t)\rangle = \sum_j \gamma_j(t)\, |\varphi_j\rangle \tag{16}$$

where $\gamma_j(t) = \langle\varphi_j|\psi(t)\rangle \neq 0$ for every $j = 1, \dots, N$ and $N \geq 2$.

Let us stress the following characteristics of the preferential set [26, 27]:

i. It depends on the state $|\psi(t)\rangle$.

ii. If it exists, the preferential set is unique. A system in the state $|\psi(t)\rangle$ cannot have more than one preferential set.

iii. Even if in the general case the Hamiltonian of the system can be written H(t) = E + W(t), the preferential set does not depend on W(t).

iv. At least $(N - 1)$ members of $\{N_\varphi\}$ are eigenstates of E. The exception, i.e., the case where a preferential state is not a stationary state, has been referred to elsewhere [28].

v. The relation

$$\langle\psi(t)|A|\psi(t)\rangle = \sum_j |\gamma_j(t)|^2 \langle\varphi_j|A|\varphi_j\rangle \tag{17}$$

must be fulfilled for every operator A representing a conserved quantity α when $W(t) = 0$. The validity of this relation ensures the statistical sense of the conservation of α [25].

If the system in the state $|\psi(t)\rangle$ does not have a preferential set, the Schrödinger evolution follows. By contrast, if it has the preferential set $\{N_\varphi\}$, in the small time interval $(t, t+dt)$, the system can either remain in the Schrödinger channel or jump to one of its preferential states. The probability that it jumps to the preferential state $|\varphi_k\rangle$ is

$$dP_k(t) = |\gamma_k(t)|^2 \frac{dt}{\tau(t)} = |\langle\varphi_k|\psi(t)\rangle|^2 \frac{dt}{\tau(t)} \tag{18}$$

where $\tau(t)\Delta E(t) = \hbar/2$ and $[\Delta E(t)]^2 = \langle\psi(t)|E^2|\psi(t)\rangle - [\langle\psi(t)|E|\psi(t)\rangle]^2$ [26, 27].

It is easily shown that in the interval $(t, t+dt)$, the probability that the system abandons the Schrödinger channel is $dt/\tau(t)$ and the probability that it remains in the Schrödinger channel is

$$dP_S(t) = 1 - \frac{dt}{\tau(t)} \tag{19}$$

So the dominant process in a small time interval $(t, t+dt)$ is always the Schrödinger evolution [25–27].

In cases where the system remains in the Schrödinger channel, the transformation of the state yielded by SPA exactly coincides with that yielded by OQM. It could be wrongly assumed that there is a complete correspondence (i) between OQM spontaneous processes and SPA processes where the preferential set is absent; and (ii) between OQM measurement processes and SPA processes where the system has its preferential set.

Certainly SPA processes where the preferential set is absent as well as OQM spontaneous processes are forcible Schrödinger evolutions. And unless the system is an eigenstate of the operator representing the quantity to be measured, CQM measurements entail projections. But if the system has its preferential set, according to SPA it can either be projected to a preferential state or remain in the Schrödinger channel [26, 27]. Differing from OQM, in SPA there is always room for Schrödinger evolutions.

In sum, SPA states that in general the wave function evolves gently, in a perfectly predictable and continuous way, in agreement with the Schrödinger equation; in some cases only, when the system jumps to one of its preferential states, unpredictable changes take place, according to the projection postulate. Assuming that projections are a law of nature, SPA succeeds in proposing a real physical mechanism for the emergence of a single result in a single experiment.

6. Facing both measurement problems

Measurement is a complicated and theory-laden business ([29], p. 208). When one talks about the measurement problem in quantum mechanics, one is not referring to a real and theory-laden process but just to the problem of *accounting in principle for projections resulting from measurements*, i.e., to the fact that the Schrödinger evolution is suspended when a measurement is performed.

SPA justifies Dirac's assertion: "in [TDPT] we do not consider any modification to be made in the states of the unperturbed system, but we suppose that the perturbed system, instead of remaining permanently in one of these states, is continually changing from one to another, or making transitions, under the influence of the perturbation" ([1], p. 167).

On the one hand, in general the preferential states of the system are the eigenstates of E, which do not depend on the perturbation W(t). Hence no modification of these states should be considered. On the other hand, if the initial state of the system is $|\psi(t_0)\rangle = |\varphi_j\rangle$, an eigenstate of E, the effect of the perturbation is to gently remove the state $|\psi(t_0)\rangle$ from $|\varphi_j\rangle$, and yield it to the linear superposition $|\psi(t)\rangle$ given by Eq. (16). Once the system is in this linear superposition, it can either suddenly jump to a stationary state or remain in the Schrödinger channel. If it jumps, it can either go to a state $|\varphi_k\rangle$ (where $k \neq j$) or come back to its initial state $|\varphi_j\rangle$. The result can be described as a system continually changing from one to another stationary state or making transitions, as Dirac asserts.

In principle SPA accounts for TBSS. By contrast, decoherence has little to contribute concerning this matter.

Assuming as valid the ideal measurement scheme, in previous papers we have addressed the traditional measurement problem as follows [4, 25].

Let A be the operator representing the physical quantity α referred to the system S. We shall denote by $|a_j\rangle$ the eigenvector of A corresponding to the eigenvalue a_j ($j = 1, 2, \ldots$); for simplicity we shall refer to the discrete non-degenerate case. If the initial state of S is $|a_j\rangle$ and the initial state of the measuring device M is $|m_0\rangle$, the initial state of the total system S + M (before the measurement takes place) will be denoted by $|a_j\rangle\, |m_0\rangle$. The final state of the total system (when the measurement is over) will be denoted by $|\Phi\rangle$.

According to the ideal measurement scheme the Schrödinger evolution results

$$|a_j\rangle\, |m_0\rangle \to |\Phi\rangle = |\Phi_j\rangle \tag{20}$$

This scheme is supposed to be valid in cases where the measured physical quantity is compatible with every conserved quantity referred to S + M [30].

If the initial state of S is $\sum_j \gamma_j |a_j\rangle$ (where $\gamma_j \neq 0$ for every $j = 1, \ldots, N$), the linearity of the Schrödinger equation yields entangled states:

$$\left(\sum_j \gamma_j\, |a_j\rangle\right) |m_0\rangle \to |\Phi\rangle = \sum_j \gamma_j\, |\Phi_j\rangle \tag{21}$$

The set $\{N_\Phi\} = \{|\Phi_1\rangle, \ldots, |\Phi_N\rangle\}$ can be considered the preferential set of S + M in the state $|\Phi\rangle$ (as a matter of fact, $\{N_\Phi\}$ clearly fulfills several of the requirements imposed to such a set). Hence, projections like $|\Phi\rangle \to |\Phi_1\rangle$, ... or $|\Phi\rangle \to |\Phi_N\rangle$ may result. This is SPA proposed solution to the traditional measurement problem.

Decoherence invokes an alternative solution to the traditional measurement problem. Once the expansion (21) is obtained, the density matrix corresponding to the state $|\Phi\rangle$ is replaced by the reduced density matrix as previously done in Section 5.2 (see Eqs. (12) and (13)). It is claim that "there has been a leakage of coherence from the system to the composite entity (system + environment). Since we are not able to control this entity, *the decoherence has been completed to all practical purposes*" ([9], p. 248; emphases added).

Laloë points out that "decoherence is not to be confused with the measurement process itself; it is just the process which takes place just before: during decoherence, the off-diagonal elements of the density matrix vanish ... " [15]. In his view "the crux of most of our difficulties with quantum mechanics is the question: what is exactly the process that forces Nature ... to make its choice among the various possibilities for the results of experiments?" [15]. SPA answers: spontaneous projections to the preferential states.

SPA and decoherence are not opposed theories competing for "an explanation" to the measurement problem but cooperating theories. Projections break down the Schrödinger evolution, but they are not frequent. If the system has its preferential set, projections can take place at the very beginning of the process or not (in SPA there is always room for Schrödinger evolutions). As long as projections do not take place, decoherence can make its work entangling the system with the environment. But nothing prevents the total, entangled system, to have its preferential set. This may be why a spontaneous projection finally breaks down the superposition of states of the total system. Nature makes its choice, and it is only then that decoherence is completed.

7. Conclusions

Carlton Caves declares: "Mention collapse of the wave function, and you are likely to encounter vague uneasiness or, in extreme cases, real discomfort. This uneasiness can usually be traced to a feeling that wave-function collapse lies 'outside' quantum mechanics: The real quantum mechanics is said to be the unitary Schrödinger evolution; wave-function collapse is regarded as an ugly duckling of questionable status, dragged in to interrupt the beautiful flow of Schrödinger evolution" [31].

If collapses implied in traditional measurement are regarded as an ugly duckling of questionable status, collapses implied in TBSS could result definitively unbearable. Neither observers nor measuring devices could be invoked to excuse their occurrence, but they are there, happening all the time, more or less everywhere, e.g., every time a photon is either emitted or absorbed by an atom.

The search for a solution to the traditional measurement problem is at the basis of most interpretations of quantum mechanics. In this paper we have summed up four of these interpretations which succeed in avoiding the quantum superposition of macroscopically distinct states, an important element of the traditional measurement problem. Every particular interpretation provides a particular point of view on the traditional measurement problem: (1) in Bohmian mechanics Schrödinger's evolution is not interrupted; replacing the original wave function for its "collapsed" derivative is just a pragmatic affair; (2) in decoherence the linear Schrödinger equation yields an unavoidable coupling of the quantum system with the surrounding medium, which is not a collapse but looks and smells as if it were; (3) in GRW collapses result from localizations; and (4) in SPA collapses result from jumps to preferential states.

By contrast, no different interpretations of quantum mechanics are invoked to account for TBSS, as if the corresponding measurement problem were immune to the different interpretations of the theory. We have shown, however, that at least one interpretation of quantum mechanics does not account for TBSS.

Every proposed solution to the measurement problem should apply to both measurement problems: the traditional and that implied in TBSS. A solution to just one of them is not good enough.

Acknowledgements

We are indebted to Professor J.C. Centeno for many fruitful discussions. We thank Carlos Valero for the transcription of formulas into Math Type.

Author details

María Esther Burgos
Independent Scientist, Ciudad Autónoma de Buenos Aires, Argentina

*Address all correspondence to: mburgos25@gmail.com

References

[1] Dirac PAM. The Principles of Quantum Mechanics. Oxford: Clarendon Press; 1930

[2] von Neumann J. Mathematische Grundlagen der Quantenmechanik. Berlin: Springer; 1932

[3] Jammer M. The Philosophy of Quantum Mechanics. New York: John Wiley & Sons; 1974

[4] Burgos ME. The measurement problem in quantum mechanics revisited. In: Pahlavani M, editor. Selected Topics in Applications of Quantum Mechanics. Croatia: IntechOpen; 2015. pp. 137-173. DOI: 10.5772/59209

[5] Burgos ME. Evidence of action-at-a-distance in experiments with individual particles. Journal of Modern Physics. 2015:6, 1663-1670. DOI: 10.4236/jmp.2015.6111

[6] Burgos ME, Criscuolo FG, Etter TL. Conservation laws, machines of the first type and superluminal communication. Speculations in Science and Technology. 1999;**21**(4):227-233

[7] Criscuolo FG, Burgos ME. Conservation laws in spontaneous and measurement-like individual processes. Physics Essays. 2000;**13**(1):80-84

[8] Burgos ME. Contradiction between conservation laws and orthodox quantum mechanics. Journal of Modern Physics. 2010;**1**:137-142. DOI: 10.4236/jmp.2010.12019

[9] Bes DR. Quantum Mechanics. 3rd ed. Berlin: Springer-Verlag; 2012. DOI: 10.1007/978-3-642-20556-9

[10] Cohen-Tannoudji C, Diu B, Laloë F. Quantum Mechanics. New York: John Wiley & Sons; 1977

[11] Heitler W. The Quantum Theory of Radiation. 3rd ed. New York: Dover Publications Inc.; 1984

[12] Merzbacher E. Quantum Mechanics. New York: John Wiley & Sons; 1961

[13] Messiah A. Quantum Mechanics. Amsterdam: North Holland Publishing Company; 1965

[14] Burgos ME. Success and incoherence of orthodox quantum mechanics. Journal of Modern Physics. 2016;**7**:1449-1454. DOI: 10.4236/jmp.2016.712132

[15] Laloë F. Do we really understand quantum mechanics? American Journal of Physics, American Association of Physics Teachers. 2001;**69**:655-701

[16] Burgos ME. Zeno of elea shines a new light on quantum weirdness. Journal of Modern Physics. 2017;**8**:1382-1397. DOI: 10.4236/jmp.2017.88087

[17] Bohm D. A suggested interpretation of the quantum theory in terms of hidden variables. Physical Review. 1952; **85**:166-179 180-193

[18] Bohmian GS. Mechanics. Stanford Encyclopedia of Philosophy; 2001. Revised 2017

[19] Dürr D, Goldstein S, Tumulka R, Zanghì N. Bohmian mechanics and quantum field theory. Physical Review Letters. 2004;**93**. Available from: https://arxiv.org/abs/quant-ph/0303156

[20] Tegmar M, Wheeler JA. 100 years of quantum mysteries. Scientific American. 2001;**284**(2):68-75

[21] Schlosshauer M. Decoherence and the Quantum-to-Classical Transition. Berlin: Springer-Verlag; 2007

[22] Ghirardi GC, Rimini A, Weber T. Unified dynamics for microscopic and

macroscopic systems. Physical Review D. 1986;**34**:470-491

[23] Ghirardi GC, Rimini A, Weber T. Disentanglement of quantum wave functions. Physical Review D. 1987;**36**: 3287-3289

[24] Ghirardi GC. Collapse theories. Stanford Encyclopedia of Philosophy. 2002. Revised 2016

[25] Burgos ME. Which Natural Processes Have the Special Status of Measurements? Foundations of Physics. 1998;**28**(8):1323-1346

[26] Burgos ME. Unravelling the quantum maze. Journal of Modern Physics. 2018;**9**:1697-1711. DOI: 10.4236/jmp.2018.98106

[27] Burgos ME. The contradiction between two versions of quantum theory could be decided by experiment. Journal of Modern Physics. 2019;**10**: 1190-1208. DOI: 10.4236/ jmp.2019.1010079

[28] Burgos ME. Transitions to the continuum: Three different approaches. Foundations of Physics. 2008;**38**(10): 883-907

[29] Bell M, Gottfried K, Veltman M. John S. Bell on the Foundations of Quantum Mechanics. Word Scientific: Singapore; 2001

[30] Araki H, Yanase MM. Measurement of quantum mechanical operators. Physical Review. 1960;**120**(2):622-626

[31] Caves C. Quantum mechanics of measurements distributed in time. A path-integral formulation. Physical Review D. 1986;**33**:1643-1665

Section 2

Foundational Problems in Quantum Mechanics

Chapter 4

Uncertainty Relations

Kenjiro Yanagi

Abstract

Uncertainty relations are inequalities representing the impossibility of simultaneous measurement in quantum mechanics. The most well-known uncertainty relations were presented by Heisenberg and Schrödinger. In this chapter, we generalize and extend them to produce several types of uncertainty relations.

Keywords: trace inequality, variance, covariance, skew information, metric adjusted skew information, noncommutativity, observable, operator inequality

1. Introduction

Let $M_n(\mathbb{C})$ (resp. $M_{n,sa}(\mathbb{C})$) be the set of all $n \times n$ complex matrices (resp. all $n \times n$ self-adjoint matrices), endowed with the Hilbert-Schmidt scalar product $\langle A, B\rangle = Tr[A^*B]$. Let $M_{n,+}(\mathbb{C})$ be the set of strictly positive elements of $M_n(\mathbb{C})$ and $M_{n,+,1}(\mathbb{C}) \subset M_{n,+}(\mathbb{C})$ be the set of strictly positive density matrices, that is $M_{n,+,1}(\mathbb{C}) = \{\rho \in M_n(\mathbb{C}) | Tr[\rho] = 1, \rho > 0\}$. If not otherwise specified, hereafter, we address the case of faithful states, that is $\rho > 0$. It is known that the expectation of an observable $A \in M_{n,sa}(\mathbb{C})$ in state $\rho \in M_{n,+,1}(\mathbb{C})$ is defined by

$$E_\rho(A) = Tr[\rho A],$$

and the variance of an observable $A \in M_{n,sa}(\mathbb{C})$ in state $\rho \in M_{n,+,1}(\mathbb{C})$ is defined by

$$V_\rho(A) = Tr\left[\rho(A - E_\rho(A)I)^2\right] = Tr\left[\rho A^2\right] - E_\rho(A)^2 = Tr\left[\rho A_0^2\right],$$

where $A_0 = A - E_\rho(A)I$.

In Section 2, we introduce the Heisenberg and Schrödinger uncertainty relations. In Section 3, we present uncertainty relations with respect to the Wigner-Yanase and Wigner-Yanase-Dyson skew information. To represent the degree of noncommutativity between $\rho \in M_{n,+,1}(\mathbb{C})$ and $A \in M_{n,sa}(\mathbb{C})$, the Wigner-Yanase skew information $I_\rho(A)$ is defined by

$$I_\rho(A) = \frac{1}{2}Tr\left[\left(i\left[\rho^{1/2}, A\right]\right)^2\right] = Tr\left[\rho A^2\right] - Tr\left[\rho^{1/2}A\rho^{1/2}A\right],$$

where $[X, Y] = XY - YX$. Furthermore, the Wigner-Yanase-Dyson skew information $I_{\rho,\alpha}(A)$ is defined by

$$I_{\rho,\alpha}(A) = \frac{1}{2} Tr\left[(i[\rho^{\alpha}, A])\left(i\left[\rho^{1-\alpha}, A\right]\right)\right] = Tr\left[\rho A^2\right] - Tr\left[\rho^{\alpha} A \rho^{1-\alpha} A\right], \quad (\alpha \in [0,1]).$$

The convexity of $I_{\rho,\alpha}(A)$ with respect to ρ was famously demonstrated by Lieb [1], and the relationship between the Wigner-Yanase skew information and the uncertainty relation was originally developed by Luo and Zhang [2]. Subsequently, the relationship between the Wigner-Yanase-Dyson skew information and the uncertainty relation was provided by Kosaki [3] and Yanagi-Furuichi-Kuriyama [4]. In Section 4, we discuss the metric adjusted skew information defined by Hansen [5], which is an extension of the Wigner-Yanase-Dyson skew information. The relationship between metric adjusted skew information and the uncertainty relation was provided by Yanagi [6] and generalized by Yanagi-Furuichi-Kuriyama [7] for generalized metric adjusted skew information and the generalized metric adjusted correlation measure. In Sections 5 and 6, we provide non-Hermitian extensions of Heisenberg-type and Schrödinger-type uncertainty relations related to generalized quasi-metric adjusted skew information and the generalized quasi-metric adjusted correlation measure. As a result, we obtain results for non-Hermitian uncertainty relations provided by Dou and Du as corollaries of our results. Finally, in Section 7, we present the sum types of uncertainty relations.

2. Heisenberg and Schrödinger uncertainty relations

Theorem 1.1 (Heisenberg uncertainty relation). For $A, B \in M_{n,sa}(\mathbb{C})$, $\rho \in M_{n,+,1}(\mathbb{C})$,

$$V_\rho(A)V_\rho(B) \geq \frac{1}{4}\left|Tr\left[\rho[A,B]\right]\right|^2, \tag{1}$$

where $[A,B] = AB - BA$ is the commutator.

Theorem 1.2 (Schrödinger uncertainty relation). For $A, B \in M_{n,sa}(\mathbb{C})$, $\rho \in M_{n,+,1}(\mathbb{C})$,

$$V_\rho(A)V_\rho(B) - \left|\mathrm{Re}\,\{Tr[\rho A_0 B_0]\}\right|^2 \geq \frac{1}{4}\left|Tr\left[\rho[A,B]\right]\right|^2.$$

Proof of Theorem 1.2. By the Schwarz inequality

$$\begin{aligned} |Tr[\rho A_0 B_0]|^2 &= \left|Tr\left[\left(\rho^{1/2}B_0\right)^*\left(\rho^{1/2}A_0\right)\right]\right|^2 \\ &\leq Tr\left[\left(\rho^{1/2}B_0\right)^*\left(\rho^{1/2}B_0\right)\right] \cdot Tr\left[\left(\rho^{1/2}A_0\right)^*\left(\rho^{1/2}A_0\right)\right] \\ &= Tr\left[\rho A_0^2\right] \cdot Tr\left[\rho B_0^2\right] = V_\rho(A) \cdot V_\rho(B). \end{aligned}$$

Since

$$\begin{aligned} Tr[\rho[A_0, B_0]] &= Tr[\rho A_0 B_0] - Tr[\rho B_0 A_0] = Tr[\rho A_0 B_0] - \overline{Tr[A_0 B_0 \rho]} \\ &= Tr[\rho A_0 B_0] - \overline{Tr[\rho A_0 B_0]} = 2i\mathrm{Im}\{Tr[\rho A_0 B_0]\}, \end{aligned}$$

we have

$$\begin{aligned} |Tr[\rho A_0 B_0]|^2 &= (\mathrm{Re}\,\{Tr[\rho A_0 B_0]\})^2 + (\mathrm{Im}\{Tr[\rho A_0 B_0]\})^2 \\ &= (\mathrm{Re}\,\{Tr[\rho A_0 B_0]\})^2 + \frac{1}{4}|Tr[\rho[A_0, B_0]]|^2. \end{aligned}$$

Since $Tr[\rho[A_0, B_0]] = Tr[\rho[A, B]]$, we obtain

$$V_\rho(A) \cdot V_\rho(B) - |\operatorname{Re}\{Tr[\rho A_0 B_0]\}|^2 \geq \frac{1}{4} |Tr[\rho[A, B]]|^2. \qquad \square$$

3. Uncertainty relation for Wigner-Yanase-Dyson skew information

3.1 Wigner-Yanase skew information

To represent the degree of noncommutativity between $\rho \in M_{n,+,1}(\mathbb{C})$ and $A \in M_{n,sa}(\mathbb{C})$, the Wigner-Yanase skew information $I_\rho(A)$ and related quantity $J_\rho(A)$ are defined as

$$I_\rho(A) = \frac{1}{2} Tr\left[\left(i\left[\rho^{1/2}, A_0\right]\right)^2\right] = Tr\left[\rho A_0^2\right] - Tr\left[\rho^{1/2} A_0 \rho^{1/2} A_0\right].$$

$$J_\rho(A) = \frac{1}{2} Tr\left[\rho\{A_0, B_0\}^2\right] = Tr\left[\rho A_0^2\right] + Tr\left[\rho^{1/2} A_0 \rho^{1/2} A_0\right],$$

where $\{A, B\} = AB + BA$. The quantity $U_\rho(A)$ representing a quantum uncertainty excluding the classical mixture is defined as

$$U_\rho(A) = \sqrt{I_\rho(A) \cdot J_\rho(A)} = \sqrt{V_\rho(A)^2 - \left(V_\rho(A) - I_\rho(A)\right)^2}.$$

We note the following relation:

$$0 \leq I_\rho(A) \leq U_\rho(A) \leq V_\rho(A). \tag{2}$$

Luo [8] then derived the uncertainty relation of $U_\rho(A)$.
Theorem 1.3. For $A, B \in M_{n,sa}(\mathbb{C})$, $\rho \in M_{n,+,1}(\mathbb{C})$,

$$U_\rho(A) \cdot U_\rho(B) \geq \frac{1}{4} \left|Tr\left[\rho[A, B]\right]\right|^2. \tag{3}$$

Inequality (3) is a refinement of (1) in terms of (2).

3.2 Wigner-Yanase-Dyson skew information

Here, we introduce a one-parameter inequality extended from (3). For $0 \leq \alpha \leq 1$, $A, B \in M_{n,sa}(\mathbb{C})$ and $\rho \in M_{n,+,1}(\mathbb{C})$, we define the Wigner-Yanase-Dyson skew information as follows:

$$I_{\rho,\alpha}(A) = \frac{1}{2} Tr\left[\left(i[\rho^\alpha, A_0]\right)\left(i\left[\rho^{1-\alpha}, A_0\right]\right)\right] = Tr\left[\rho A_0^2\right] - Tr\left[\rho^\alpha A_0 \rho^{1-\alpha} A_0\right].$$

We also define

$$J_{\rho,\alpha}(A) = \frac{1}{2} Tr\left[\{\rho^\alpha, A_0\}\left\{\rho^{1-\alpha}, A_0\right\}\right] = Tr[\rho A_2] + Tr\left[\rho^\alpha A_0 \rho^{1-\alpha} A_0\right]$$

We note that

$$\frac{1}{2} Tr[i[\rho^\alpha, A_0])\left(i\left[\rho^{1-\alpha}, A_0\right]\right)] = \frac{1}{2} Re\left[(i[\rho^\alpha, A])\left(i\left[\rho^{1-\alpha}, A\right]\right)\right];$$

however, we have

$$\frac{1}{2}Tr\left[\{\rho^{\alpha},A_0\}\{\rho^{1-\alpha},A_0\}\right] \neq \frac{1}{2}Tr\left[\{\rho^{\alpha},A\}\{\rho^{1-\alpha},A\}\right].$$

We then have the following inequalities:

$$I_{\rho,\alpha}(A) \leq I_{\rho}(A) \leq J_{\rho}(A) \leq J_{\rho,\alpha}(A), \tag{4}$$

because $Tr\left[\rho^{1/2}A\rho^{1/2}A\right] \leq Tr[\rho^{\alpha}A\rho^{1-\alpha}A]$. We define

$$U_{\rho,\alpha}(A) = \sqrt{I_{\rho,\alpha}(A)\cdot J_{\rho,\alpha}(A)} = \sqrt{V_{\rho}(A)^2 - (V_{\rho}(A) - I_{\rho,\alpha}(A)}. \tag{5}$$

From (2), (4), and (5), we have

$$0 \leq I_{\rho,\alpha}(A) \leq I_{\rho}(A) \leq U_{\rho}(A)$$

and

$$0 \leq I_{\rho,\alpha}(A) \leq U_{\rho,\alpha}(A) \leq U_{\rho}(A).$$

We provide the following uncertainty relation with respect to $U_{\rho,\alpha}(A)$ as a direct generalization of (3).

Theorem 1.4 ([9]). For $A, B \in M_{n,sa}(\mathbb{C})$, $\rho \in M_{n,+,1}(\mathbb{C})$,

$$U_{\rho,\alpha}(A)\cdot U_{\rho,\alpha}(B) \geq \alpha(1-\alpha)\left|Tr\left[\rho[A,B]\right]\right|^2. \tag{6}$$

Proof of Theorem 1.4. By spectral decomposition, there exists an orthonormal basis $\{|\phi_1\rangle, |\phi_2\rangle, \ldots, |\phi_n\rangle\}$ consisting of eigenvectors of ρ. Let $\lambda_1, \lambda_2, \ldots, \lambda_n$ be the corresponding eigenvalues, where $\sum_{i=1}^{n}\lambda_i = 1$ and $\lambda_i \geq 0$. Thus ρ has a spectral representation $\rho = \sum_{i=1}^{n}\lambda_i|\phi_i\rangle\langle\phi_i|$. We can obtain the following representations of $I_{\rho,\alpha}(A)$ and $J_{\rho,\alpha}(A)$:

$$I_{\rho,\alpha}(A) = \sum_{i<j}\left(\lambda_i + \lambda_j - \lambda_i^{\alpha}\lambda_j^{1-\alpha} - \lambda_i^{1-\alpha}\lambda_j^{\alpha}\right)|\langle\phi_i|A_0\left|\phi_j\rangle\right|^2.$$

$$J_{\rho,\alpha}(A) \geq \sum_{i<j}\left(\lambda_i + \lambda_j + \lambda_i^{\alpha}\lambda_j^{1-\alpha} + \lambda_i^{1-\alpha}\lambda_j^{\alpha}\right)|\langle\phi_i|A_0\left|\phi_j\rangle\right|^2.$$

Since $(1-2\alpha)^2(t-1)^2 - (t^{\alpha} - t^{1-\alpha})^2 \geq 0$ for any $t > 0$ and $0 \leq \alpha \leq 1$, we define $t = \frac{\lambda_i}{\lambda_j}$ and have

$$(1-2\alpha)^2\left(\frac{\lambda_i}{\lambda_j} - 1\right)^2 - \left(\left(\frac{\lambda_i}{\lambda_j}\right)^{\alpha} - \left(\frac{\lambda_i}{\lambda_j}\right)^{1-\alpha}\right)^2 \geq 0.$$

Then,

$$(\lambda_i + \lambda_j)^2 - \left(\lambda_i^{\alpha}\lambda_j^{1-\alpha} + \lambda_i^{1-\alpha}\lambda_j^{\alpha}\right)^2 \geq 4\alpha(1-\alpha)(\lambda_i - \lambda_j)^2. \tag{7}$$

Since

$$\begin{aligned}Tr[\rho[A,B]] &= Tr[\rho[A_0,B_0]] = 2iImTr[\rho A_0 B_0] = 2iIm\sum_{i<j}(\lambda_i-\lambda_j)\left\langle\phi_i|A_0|\phi_j\right\rangle\left\langle\phi_j|B_0|\phi_i\right\rangle\\ &= 2i\sum_{i<j}(\lambda_i-\lambda_j)Im\left\langle\phi_i|A_0|\phi_j\right\rangle\left\langle\phi_j|B_0|\phi_i\right\rangle,\end{aligned}$$

$$\begin{aligned}|Tr[\rho[A,B]]| = 2|\sum_{i<j}(\lambda_i-\lambda_j)Im\left\langle\phi_i|A_0|\phi_j\right\rangle\left\langle\phi_j|B_0|\phi_i\right\rangle| \le 2\sum_{i<j}|\lambda_i\\ -\lambda_j||Im\left\langle\phi_i|A_0|\phi_j\right\rangle\left\langle\phi_j|B_0|\phi_i\right\rangle|.\end{aligned}$$

We then have

$$|Tr[\rho[A,B]]|^2 \le 4\left\{\sum_{i<j}|\lambda_i-\lambda_j||Im\left\langle\phi_i|A_0|\phi_j\right\rangle\langle\phi_j|B_0|\phi_i\rangle|\right\}^2.$$

By (7) and the Schwarz inequality,

$$\begin{aligned}\alpha(1-\alpha)|Tr[\rho[A,B]]|^2 &\le 4\alpha(1-\alpha)\left\{\sum_{i<j}|\lambda_i-\lambda_j||Im\left\langle\phi_i|A_0|\phi_j\right\rangle\langle\phi_j|B_0|\phi_i\rangle|\right\}^2\\ &= \left\{\sum_{i<j}2\sqrt{\alpha(1-\alpha)}|\lambda_i-\lambda_j||Im\left\langle\phi_i|A_0|\phi_j\right\rangle\langle\phi_j|B_0|\phi_i\rangle\right\}^2\\ &\le \left\{\sum_{i<j}2\sqrt{\alpha(1-\alpha)}|\lambda_i-\lambda_j||\left\langle\phi_i|A_0|\phi_j\right\rangle\langle\phi_j|B_0|\phi_i\rangle|\right\}^2\\ &\le \left\{\sum_{i<j}\left\{(\lambda_i+\lambda_j)^2-\left(\lambda_i^\alpha\lambda_j^{1-\alpha}+\lambda_i^{1-\alpha}\lambda_j^\alpha\right)^2\right\}^{1/2}|\langle\phi_i|A_0|\phi_j\rangle||\langle\varphi_j|B_0|\phi_i\rangle|\right\}^2\\ &\le \sum_{i<j}\left(\lambda_i+\lambda_j-\lambda_i^\alpha\lambda_j^{1-\alpha}-\lambda_i^{1-\alpha}\lambda_j^\alpha\right)|\langle\phi_i|A_0|\phi_j\rangle|^2\\ &\times\sum_{i<j}\left(\lambda_i+\lambda_j+\lambda_i^\alpha\lambda_j^{1-\alpha}+\lambda_i^{1-\alpha}\lambda_j^\alpha\right)|\langle\phi_i|B_0|\phi_j\rangle|^2.\end{aligned}$$

Then, we have

$$I_{\rho,\alpha}(A)J_{\rho,\alpha}(B) \ge \alpha(1-\alpha)|Tr[\rho[A,B]]|^2.$$

We also have

$$I_{\rho,\alpha}(B)J_{\rho,\alpha}(A) \ge \alpha(1-\alpha)|Tr[\rho[A,B]]|^2.$$

Thus, we have the final result, (6). □

When $\alpha = \frac{1}{2}$, we obtain the result in Theorem 1.3.

4. Metric adjusted skew information and metric adjusted correlation measure

4.1 Operator monotone function

A function $f : (0, +\infty) \to \mathbb{R}$ is considered operator monotone if, for any $n \in \mathbb{N}$, and $A, B \in M_n$ such that $0 \le A \le B$, the inequalities $0 \le f(A) \le f(B)$ hold. An operator monotone function is said to be symmetric if $f(x) = xf(x^{-1})$ and normalized if $f(1) = 1$.

Definition 1 $\mathcal{F}_{op}$ is the class of functions $f : (0, +\infty) \to (0, +\infty)$ such that:

1. $f(1) = 1$.

2. $tf(t^{-1}) = f(t)$.

3. f is operator monotone.

Example 1. Examples of elements of $\mathcal{F}_{op}$ are given by the following:

$$f_{RLD}(x) = \frac{2x}{x+1}, \quad f_{WY}(x) = \left(\frac{\sqrt{x}+1}{2}\right)^2, \quad f_{BKM}(x) = \frac{x-1}{\log x},$$

$$f_{SLD}(x) = \frac{x+1}{2}, \quad f_{WYD}(x) = \alpha(1-\alpha)\frac{(x-1)^2}{(x^\alpha - 1)(x^{1-\alpha} - 1)}, \quad \alpha \in (0,1).$$

Remark 1. Any $f \in \mathcal{F}_{op}$ satisfies

$$\frac{2x}{x+1} \leq f(x) \leq \frac{x+1}{2}, \quad x > 0.$$

For $f \in \mathcal{F}_{op}$, we define $f(0) = \lim_{x\to 0} f(x)$. We introduce the sets of regular and non-regular functions

$$\mathcal{F}_{op}^r = \{ f \in \mathcal{F}_{op} | f(0) \neq 0 \}, \quad \mathcal{F}_{op}^n \{ f \in \mathcal{F}_{op} | f(0) = 0 \}$$

and notice that trivially $\mathcal{F}_{op} = \mathcal{F}_{op}^r \cup \mathcal{F}_{op}^n$.

Definition 2. For $f \in \mathcal{F}_{op}^r$, we set

$$\tilde{f}(x) = \frac{1}{2}\left[(x+1) - (x-1)^2 \frac{f(0)}{f(x)}\right], \quad x > 0.$$

Theorem 1.5 ([10]). The correspondence $f \to \tilde{f}$ is a bijection between $\mathcal{F}_{op}^r$ and $\mathcal{F}_{op}^n$.

4.2 Metric adjusted skew information

In the Kubo-Ando theory [11] of matrix means, a mean is associated with each operator monotone function $f \in \mathcal{F}_{op}$ by the following formula:

$$m_f(A, B) = A^{1/2} f\left(A^{-1/2} B A^{-1/2}\right) A^{1/2},$$

where $A, B \in M_{n,+}(\mathbb{C})$. Using the notion of matrix means, the class of monotone metrics can be defined by the following formula:

$$\langle A, B \rangle_{\rho, f} = Tr\left[A \cdot m_f(L_\rho, R_\rho)^{-1}(B)\right],$$

where $L_\rho(A) = \rho A, R_\rho(A) = A\rho$.

Definition 3. For $A \in M_{n,sa}(\mathbb{C})$, we define as follows:

$$I_\rho^f(A) = \frac{f(0)}{2}\langle i[\rho, A], i[\rho, A]\rangle_{\rho,f},$$
$$C_\rho^f(A) = Tr\left[m_f(L_\rho, R_\rho)(A)\cdot A\right],$$
$$U_\rho^f(A) = \sqrt{V_\rho(A)^2 - \left(V_\rho(A) - I_\rho^f(A)\right)^2}.$$

Quantity $I_\rho^f(A)$ is referred to as the metric adjusted skew information, and $\langle A, B\rangle_{\rho,f}$ is referred to as the metric adjusted correlation measure.

Proposition 1. The following holds:

1. $I_\rho^f(A) = I_\rho^f(A_0) = Tr(\rho A_0^2) - Tr\left(m_{\tilde{f}}(L_\rho, R_\rho)(A_0)\cdot A_0\right) = V_\rho(A) - C_\rho^{\tilde{f}}(A_0).$

2. $J_\rho^f(A) = Tr(\rho A_0^2) + Tr\left(m_{\tilde{f}}(L_\rho, R_\rho)(A_0)\cdot A_0\right) = V_\rho(A) + C_\rho^{\tilde{f}}(A_0).$

3. $0 \leq I_\rho^f(A) \leq U_\rho^f(A) \leq V_\rho(A).$

4. $U_\rho^f(A) = \sqrt{I_\rho^f(A)\cdot J_\rho^f(A)}.$

Theorem 1.6 ([6]). For $f \in F_{op}^r$, if

$$\frac{x+1}{2} + \tilde{f}(x) \geq 2f(x), \tag{8}$$

then it holds that

$$U_\rho^f(A)\cdot U_\rho^f(B) \geq f(0)\left|Tr(\rho[A,B])\right|^2, \tag{9}$$

where $A, B \in M_{n,sa}(\mathbb{C})$.

To prove Theorem 1.6, several lemmas are used.

Lemma 1. If (8) holds, then the following inequality is satisfied:

$$\left(\frac{x+y}{2}\right)^2 - m_{\tilde{f}}(x,y)^2 \geq f(0)(x-y)^2.$$

Proof of Lemma 1. By (8), we have

$$\frac{x+y}{2} + m_{\tilde{f}}(x,y) \geq 2m_f(x,y). \tag{10}$$

Since

$$m_{\tilde{f}}(x,y) = y\tilde{f}\left(\frac{x}{y}\right) = \frac{y}{2}\left\{\frac{x}{y} + 1 - \left(\frac{x}{y} - 1\right)^2 \frac{f(0)}{f(x/y)}\right\} = \frac{x+y}{2} - \frac{f(0)(x-y)^2}{2m_f(x,y)},$$

we have

$$\left(\frac{x+y}{2}\right)^2 - m_{\tilde{f}}(x,y)^2 = \left\{\frac{x+y}{2} - m_{\tilde{f}}(x,y)\right\}\left\{\frac{x+y}{2} + m_{\tilde{f}}(x,y)\right\}$$
$$= \frac{f(0)(x-y)^2}{2m_f(x,y)}\left\{\frac{x+y}{2} + m_{\tilde{f}}(x,y)\right\} \geq f(0)(x-y)^2. \quad (\text{by } (10))$$

Lemma 2. Let $\{|\phi_1\rangle, |\phi_2\rangle, \cdots, |\phi_n\rangle\}$ be a basis of eigenvectors of ρ, corresponding to the eigenvalues $\{\lambda_1, \lambda_2, \cdots, \lambda_n\}$. We set $a_{jk} = \left\langle \phi_j|A_0|\phi_k \right\rangle, b_{jk} = \left\langle \phi_j|B_0|\phi_k \right\rangle$. Then, we have

$$I_\rho^f(A) = \frac{1}{2}\sum_{j,k}(\lambda_j + \lambda_k)a_{jk}a_{kj} - \sum_{j,k} m_{\tilde{f}}(\lambda_j, \lambda_k)a_{jk}a_{kj},$$

$$J_\rho^f(A) = \frac{1}{2}\sum_{j,k}(\lambda_j + \lambda_k)a_{jk}a_{kj} + \sum_{j,k} m_{\tilde{f}}(\lambda_j, \lambda_k)a_{jk}a_{kj},$$

$$\left(U_\rho^f(A)\right)^2 = \frac{1}{4}\left(\sum_{j,k}(\lambda_j + \lambda_k)|a_{jk}|^2\right)^2 - \left(\sum_{j,k} m_{\tilde{f}}(\lambda_j, \lambda_k)|a_{jk}|^2\right)^2.$$

Proof of Theorem 1.6. Since

$$Tr(\rho[A,B]) = Tr(\rho[A_0,B_0]) = \sum_{j,k}(\lambda_j - \lambda_k)a_{jk}b_{kj},$$

we have

$$\begin{aligned}
f(0)|Tr(\rho[A,B])|^2 &\leq \left(\sum_{j,k} f(0)^{1/2}|\lambda_j - \lambda_k||a_{jk}||b_{kj}|\right)^2 \\
&\leq \left(\sum_{j,k}\left\{\left(\frac{\lambda_j + \lambda_k}{2}\right)^2 - m_{\tilde{f}}(\lambda_j, \lambda_k)^2\right\}^{1/2}|a_{jk}||b_{kj}|\right)^2 \\
&\leq \left(\sum_{j,k}\left\{\frac{\lambda_j + \lambda_k}{2} - m_{\tilde{f}}(\lambda_j, \lambda_k)\right\}|a_{jk}|^2\right) \\
&\quad \times \left(\sum_{j,k}\left\{\frac{\lambda_j + \lambda_k}{2} + m_{\tilde{f}}(\lambda_j, \lambda_k)\right\}|b_{kj}|^2\right) = I_\rho^f(A)J_\rho^f(B).
\end{aligned}$$

We also have

$$I_\rho^f(B)J_\rho^f(A) \geq f(0)|Tr(\rho[A,B])|^2.$$

Thus, we have the final result (9). □

5. Generalized metric adjusted skew information

We assume that $f \in \mathcal{F}_{op}^r$ satisfies the following condition (A):

$$g(x) \geq k\frac{(x-1)^2}{f(x)}, \text{ for some } k > 0.$$

Let

$$\Delta_g^f(x) = g(x) - k\frac{(x-1)^2}{f(x)} \in F_{op}.$$

Definition 4. For $A, B \in M_{n,sa}(\mathbb{C}), \rho \in M_{n,+,1}(\mathbb{C})$ we define the following:

$$Corr_\rho^{(g,f)}(A,B) = k\langle i[\rho, A_0], i[\rho, B_0]\rangle_f$$
$$= Tr[A_0 m_g(L_\rho, R_\rho)B_0] - Tr\left[A_0 m_{\Delta_g^f}(L_\rho, R_\rho)B_0\right].$$

$$I_\rho^{(g,f)}(A) = Corr_\rho^{(g,f)}(A,A)$$

$$= Tr[A_0 m_g(L_\rho, R_\rho)A_0] - Tr\left[A_0 m_{\Delta_g^f}(L_\rho, R_\rho)A_0\right] - Tr\left[A_0 m_{\Delta_g^f}(L_\rho, R_\rho)A_0\right].$$

$$J_\rho^{(g,f)}(A) = Tr[A_0 m_g(L_\rho, R_\rho)A_0] - Tr\left[A_0 m_{\Delta_g^f}(L_\rho, R_\rho)A_0\right] + Tr\left[A_0 m_{\Delta_g^f}(L_\rho, R_\rho)A_0\right].$$

$$U_\rho^{(g,f)}(A) = \sqrt{I_\rho^{(g,f)}(A) \cdot J_\rho^{(g,f)}(A)}.$$

$I_\rho^{(g,f)}(A)$ is referred to as the generalized metric adjusted skew information, and $Corr_\rho^{(g,f)}(A,B)$ is referred to as the generalized metric adjusted correlation measure.

Theorem 1.7 ([7]). Under condition (A), the following holds:

1. (Schrödinger type) For $A, B \in M_{n,sa}(\mathbb{C}), \rho \in M_{n,+,1}(\mathbb{C})$,

$$I_\rho^{(g,f)}(A) \cdot I_\rho^{(g,f)}(B) \geq \left|Corr_\rho^{(g,f)}(A,B)\right|^2.$$

2. (Heisenberg type) For $A, B \in M_{n,sa}(\mathbb{C}), \rho \in M_{n,+,1}(\mathbb{C})$, we assume the following condition (B):

$$g(x) + \Delta_g^f(x) \geq \ell f(x) \text{ for some } \ell > 0.$$

Then,

$$U_\rho^{(g,f)}(A) \cdot U_\rho^{(g,f)}(B) \geq k\ell \left|Tr\left[\rho[A,B]\right]\right|^2.$$

6. Generalized quasi-metric adjusted skew information

In this section, we present general uncertainty relations for non-Hermitian observables $X, Y \in M_n(\mathbb{C})$.

Definition 5. For $X, Y \in M_n(\mathbb{C}), A, B \in M_{n,+}(\mathbb{C})$ we define the following:

$$\Gamma_{A,B}^{(g,f)}(X,Y) = k\langle (L_A - R_B)X, (L_A - R_B)Y\rangle_f$$
$$= kTr\left[X^*(L_A - R_B)m_f(L_A, R_B)^{-1}(L_A - R_B)Y\right]$$
$$= Tr[X^* m_g(L_A, R_B)Y] - Tr\left[X^* m_{\Delta_b^f}(L_A, R_B)Y\right],$$

$$\Psi_{A,B}^{(g,f)}(X,Y) = Tr[X^* m_g(L_A, R_B)Y] + Tr\left[X^* m_{\Delta_g^f}(L_A, R_B)Y\right],$$

$$I_{A,B}^{(g,f)}(X) = \Gamma_{A,B}^{(g,f)}(X,X), \quad J_{A,B}^{(g,f)}(X) = \Psi_{A,B}^{(g,f)}(X,X), \quad U_{A,B}^{(g,f)}(X) = \sqrt{I_{A,B}^{(g,f)}(X) \cdot J_{A,B}^{(g,f)}(X)}.$$

$I_{A,B}^{(g,f)}(X)$ is referred to as the generalized quasi-metric adjusted skew information, and $\Gamma_{A,B}^{(g,f)}(X,Y)$ is referred to as the generalized quasi-metric adjusted correlation measure.

Theorem 1.8 ([12]). Under condition (A), the following holds:

1. (Schrödinger type) For $X, Y \in M_n(\mathbb{C}), A, B \in M_{n,+}(\mathbb{C})$,

$$I_{A,B}^{(g,f)}(X) \cdot I_{A,B}^{(g,f)}(Y) \geq \left|\Gamma_{A,B}^{(g,f)}(X,Y)\right|^2 \geq \frac{1}{16}\left(I_{A,B}^{(g,f)}(X+Y) - I_{A,B}^{(g,f)}(X-Y)\right)^2.$$

2. (Heisenberg type) For $X, Y \in M_n(\mathbb{C}), A, B \in M_{n,+}(\mathbb{C})$, we assume condition (B). Then,

$$U_{A,B}^{(g,f)}(X) \cdot U_{A,B}^{(g,f)}(Y) \geq k\ell \left|Tr\left[X^* |L_A - R_B| Y\right]\right|^2.$$

In particular,

$$k\ell \left|Tr[X^* |L_A - R_B| X]\right|^2 \leq Tr\left[X^*\left(m_g(L_A, R_B) - m_{\Delta_g^f}(L_A, R_B)\right)X\right] \times Tr\left[X^*\left(m_g(L_A, R_B) + m_{\Delta_g^f}(L_A, R_B)\right)X\right], \tag{11}$$

where $X \in M_n(\mathbb{C})$ and $A, B \in M_{n,+}(\mathbb{C})$.
Proof of 1 in Theorem 1.8. By the Schwarz inequality, we have

$$I_{A,B}^{(g,f)}(X) \cdot I_{A,B}^{(g,f)}(Y) = \Gamma_{A,B}^{(g,f)}(X,X) \cdot \Gamma_{A,B}^{(g,f)}(Y,Y) \geq \left|\Gamma_{A,B}^{(g,f)}(X,Y)\right|^2.$$

Now, we prove the second inequality. Since

$$\begin{aligned} I_{A,B}^{(g,f)}(X+Y) &= Tr\left[(X^* + Y^*) m_g(L_A, R_B)(X+Y)\right] \\ &\quad - Tr\left[(X^* + Y^*) m_{\Delta_g^f}(L_A, R_B)(X+Y)\right], \\ I_{A,B}^{(g,f)}(X-Y) &= Tr\left[(X^* - Y^*) m_g(L_A, R_B)(X-Y)\right] \\ &\quad - Tr\left[(X^* - Y^*) m_{\Delta_g^f}(L_A, R_B)(X-Y)\right], \end{aligned}$$

we have

$$\begin{aligned} &I_{A,B}^{(g,f)}(X+Y) - I_{A,B}^{(g,f)}(X-Y) \\ &= 2Tr\left[X^* m_g(L_A, R_B) Y\right] + 2Tr Y^* m_g(L_A, R_B) X] - 2Tr\left[X^* m_{\Delta_g^f}(L_A, R_B) Y\right] \\ &-2Tr\left[Y^* m_{\Delta_g^f}(L_A, R_B) X\right] = 2\Gamma_{A,B}^{(g,f)}(X,Y) + 2\Gamma_{A,B}^{(g,f)}(Y,X) = 4\,Re\left\{\Gamma_{A,B}^{(g,f)}(X,Y)\right\}. \end{aligned}$$

Similarly, we have

$$I_{A,B}^{(g,f)}(X+Y) + I_{A,B}^{(g,f)}(X-Y) = 2\left(I_{A,B}^{(g,f)}(X) + I_{A,B}^{(g,f)}(Y)\right).$$

Then,

$$\Gamma_{A,B}^{(g,f)}(X,Y) = Re\left\{\Gamma_{A,B}^{(g,f)}(X,Y)\right\} + iIm\left\{\Gamma_{A,B}^{(g,f)}(X,Y)\right\}$$

$$= \frac{1}{4}\left(I_{A,B}^{(g,f)}(X+Y) - I_{A,B}^{(g,f)}(X-Y)\right) + iIm\{\Gamma_{A,B}^{(g,f)}(X,Y).$$

Thus,

$$\left|\Gamma_{A,B}^{(g,f)}(X,Y)\right|^2 = \frac{1}{16}\left(I_{A,B}^{(g,f)}(X+Y) - I_{A,B}^{(g,f)}(X-Y)\right)^2 + \left(Im\left\{\Gamma_{A,B}^{(g,f)}(X,Y)\right\}\right)^2$$
$$\geq \frac{1}{16}\left(I_{A,B}^{(g,f)}(X+Y) - I_{A,B}^{(g,f)}(X-Y)\right)^2.$$

□

We use the following lemma to prove 2:
Lemma 3

$$m_g(x,y)^2 - m_{\Delta_g^f}(x,y)^2 \geq k\ell(x-y)^2.$$

Proof of Lemma 3. By conditions (A) and (B), we have

$$m_{\Delta_g^f}(x,y) = m_g(x,y) - k\frac{(x-y)^2}{m_f(x,y)},$$
$$m_g(x,y) + m_{\Delta_g f}(x,y) \geq \ell m_f(x,y).$$

We then have

$$m_g(x,y)^2 - m_{\Delta_g^f}(x,y)^2 = \left\{m_g(x,y) - m_{\Delta_g^f}(x,y)\right\}\left\{m_g(x,y) + m_{\Delta_g^f}(x,y)\right\}$$
$$\geq \frac{k(x-y)^2}{m_f(x,y)}\ell m_f(x,y) = k\ell(x-y)^2.$$

□

Proof of 2 in Theorem 1.8. Let

$$A = \sum_{i=1}^{n} \lambda_i|\phi_i\rangle\left\langle \phi_i\right|, \quad B = \sum_{i=1}^{n} \mu_i|\psi_i\left.\right\rangle\langle\psi_i|$$

be the spectral decompositions of A and B, respectively. Then, we have

$$I_{A,B}^{(g,f)}(X) = \sum_{i,j}\left\{m_g\left(\lambda_i,\mu_j\right) - m_{\Delta_g^f}\left(\lambda_i,\mu_j\right)\right\}|\langle\phi_i|X\left|\psi_j\rangle\right|^2,$$
$$J_{A,B}^{(g,f)}(Y) = \sum_{i,j}\left\{m_g\left(\lambda_i,\mu_j\right) + m_{\Delta_g^f}\left(\lambda_i,\mu_j\right)\right\}|\langle\phi_i|Y\left|\psi_j\rangle\right|^2,$$

Since

$$|L_A - R_B| = \sum_{i=1}^{n}\sum_{j=1}^{n} |\lambda_i - \mu_j| L_{|\phi_i\rangle\langle\phi_i|} R_{|\psi_j\rangle\langle\psi_j|},$$

we have

$$Tr[X^*|L_A - R_B|Y] = \sum_{i=1}^{n}\sum_{j=1}^{n}|\lambda_i - \mu_j|\overline{\langle\phi_i|X|\psi_j\rangle}\langle\phi_i|Y|\psi_j\rangle.$$

Then, by Lemma 3, we have

$$k\ell|Tr[X^*|L_A - R_B|Y]|^2 \le \left\{\sum_{i=1}^{n}\sum_{j=1}^{n}\sqrt{k\ell}|\lambda_i - \mu_j||\langle\phi_i|X|\psi_j\rangle||\langle\phi_j|Y|\psi_i\rangle|\right\}^2$$

$$\le \left\{\sum_{i=1}^{n}\sum_{j=1}^{n}(m_g\left(\lambda_i,\mu_j\right)^2 - m_{\Delta_g^f}(\lambda_i,\mu_j)^2)|\langle\phi_i|X|\psi_j\rangle||\langle\phi_j|Y|\psi_i\rangle|\right\}^2$$

$$\le \left\{\sum_{i=1}^{n}\sum_{j=1}^{n}(m_g\left(\lambda_i,\mu_j\right) - m_{\Delta_g^f}(\lambda_i,\mu_j))|\langle\phi_i|X\left|\phi_j\rangle\right|^2\right\}$$

$$\left\{\sum_{i=1}^{n}\sum_{j=1}^{n}(m_g\left(\lambda_i,\mu_j\right) + m_{\Delta_g^f}(\lambda_i,\mu_j))|\langle\phi_j|Y|\psi_i\rangle|^2\right\}$$

$$= I_{A,B}^{(g,f)}(X)\cdot J_{A,B}^{(g,f)}(Y).$$

Similarly, we have $k\ell|Tr[X^*|L_A - R_B|Y]|^2 \le I_{A,B}^{(g,f)}(Y)\cdot J_{A,B}^{(g,f)}(X)$. Therefore,

$$U_{A,B}^{(g,f)}(X)\cdot U_{A,B}^{(g,f)}(Y) \ge k\ell|Tr\left[X^*|L_A - R_B|Y\right]|^2. \qquad \square$$

When $A = B = \rho \in M_{n,+,1}(\mathbb{C}), X = A \in M_n(\mathbb{C})$), and $Y = B \in M_n(\mathbb{C})$, we obtain the result in Theorem 1.7.

We assume that

$$g(x) = \frac{x+1}{2},\; f(x) = \alpha(1-\alpha)\frac{(x-1)^2}{(x^\alpha - 1)(x^{1-\alpha} - 1)},\quad k = \frac{f(0)}{2},\;\ell = 2.$$

We then obtain the following trace inequality by substituting $X = I$ in (11).

$$\alpha(1-\alpha)(Tr[|L_A - R_B|I])^2 \le \left(\frac{1}{2}Tr[A+B]\right)^2 - \left(\frac{1}{2}Tr\left[A^\alpha B^{1-\alpha} + A^{1-\alpha}B^\alpha\right]\right)^2. \quad (12)$$

This is a generalization of the trace inequality provided in [13]. In addition, we produce the following new inequality by combining a Chernoff-type inequality with Theorem 1.8.

Theorem 1.9 ([14]). We have the following:

$$\frac{1}{2}Tr[A + B - |L_A - R_B|I] \le \inf_{0\le\alpha\le1} Tr\left[A^{1-\alpha}B^\alpha\right] \le Tr\left[A^{1/2}B^{1/2}\right]$$

$$\le \frac{1}{2}Tr\left[A^\alpha B^{1-\alpha} + A^{1-\alpha}B^\alpha\right] \le \sqrt{\left(\frac{1}{2}Tr[A+B]\right)^2 - \alpha(1-\alpha)(Tr[|L_A - R_B|I)^2}.$$

The following lemma is necessary to prove Theorem 1.9.

Lemma 4. Let $f(s) = Tr\left[A^{1-s}B^s\right]$ for $A, B \in M_n(\mathbb{C})$ and $0 \le s \le 1$. Then $f(s)$ is convex in s.

Proof of Lemma 4. $f'(s) = Tr\left[-A^{1-s}\log AB^s + A^{1-s}B^s\log B\right]$. And then

$$f''(s) = Tr\left[A^{1-s}(\log A)^2 B^s - A^{1-s}\log AB^s \log B\right] - Tr\left[A^{1^s}\log AB^s \log B - A^{1-s}B^s(\log B)^2\right]$$

$$= Tr\left[A^{1-s}(\log A)^2 B^s\right] - Tr\left[A^{1-s}\log A\log BB^s\right] - Tr\left[\log B\log AA^{1-s}B^s\right] + Tr\left[A^{1-s}(\log B)^2 B^s\right]$$

$$= Tr\left[A^{1-s}\log A(\log A - \log B)B^s\right] - Tr\left[A^{1-s}(\log A - \log B)\log BB^s\right]$$

$$= Tr\left[A^{1-s}(\log A - \log B)B^s \log A\right] - Tr\left[A^{1-s}(\log A - \log B)\log BB^s\right]$$

$$= Tr\left[A^{1-s}(\log A - \log B)B^s(\log A - \log B)\right]$$

$$= Tr\left[A^{(1-s)/2}(\log A - \log B)B^s(\log A - \log B)A^{(1-s)/2}\right] \geq 0.$$

$f(s)$ is convex in s. □

Proof of Theorem 1.9. The third and fourth inequalities follow from Lemma 4 and (12), respectively. Thus, we only prove the following inequality:

$$Tr[A + B - |L_A - R_B|I] \leq 2Tr\left[A^{1-\alpha}B^{\alpha}\right] \quad (0 \leq \alpha \leq 1).$$

Let

$$A = \sum_i \lambda_i |\phi_i\rangle \left\langle \phi_i \right| = \sum_{i,j} \lambda_i |\phi_i \left\rangle \right\langle \phi_i | \psi_j \left\rangle \right\langle \psi_j |,$$

$$B = \sum_j \mu_j |\psi_j\rangle \left\langle \psi_j \right| = \sum_{i,j} \mu_j |\phi_i \left\rangle \right\langle \phi_i | \psi_j \left\rangle \right\langle \psi_j |.$$

Then, we have

$$Tr[A] = \sum_{i,j} \lambda_i |\langle \phi_i | \psi_j \rangle|^2, \quad Tr[B] = \sum_{i,j} \mu_j |\langle \phi_i | \psi_j \rangle|^2.$$

And since

$$|L_A - R_B| = \sum_{i,j} |\lambda_i - \mu_j| L_{|\phi_i\rangle\langle\phi_i|} R_{|\psi_j\rangle\langle\psi_j|},$$

we have

$$|L_A - R_B| I = \sum_{i,j} |\lambda_i - \mu_j| |\phi_i\rangle \left\langle \phi_i | \psi_j \right\rangle \langle \psi_j |.$$

Then, we have

$$Tr[|L_A - R_B| I] = \sum_{i,j} |\lambda_i - \mu_j| |\langle \phi_i | \psi_j \rangle|^2.$$

Therefore,

$$Tr[A + B - |L_A - R_B| I] = \sum_{i,j} \left(\lambda_i + \mu_j - |\lambda_i - \mu_j|\right) |\langle \phi_i | \psi_j \rangle|^2.$$

However, since we have

$$A^{\alpha} = \sum_i \lambda_i^{\alpha} |\phi_i\rangle\Big\langle \phi_i| = \sum_{i,j} \lambda_i^{\alpha} |\phi_i\Big\rangle\Big\langle \phi_i|\psi_j\Big\rangle\langle\psi_j|,$$

$$B^{1-\alpha} = \sum_j \mu_j^{1-\alpha} |\psi_j\rangle\Big\langle \psi_j| = \sum_{i,j} \mu_j^{1-\alpha} |\phi_i\Big\rangle\Big\langle \phi_i|\psi_j\Big\rangle\langle\psi_j|,$$

$$A^{\alpha}B^{1-\alpha} = \sum_{i,j} \lambda_i^{\alpha}\mu_j^{1-\alpha} |\phi_i\rangle\Big\langle \phi_i|\psi_j\Big\rangle\langle\psi_j|.$$

Then,

$$Tr\left[A^{\alpha}B^{1-\alpha}\right] = \sum_{i,j} \lambda_i^{\alpha}\mu_j^{1-\alpha} |\langle\phi_i|\psi_j\rangle|^2.$$

Thus,

$$2Tr\left[A^{\alpha}B^{1-\alpha}\right] - Tr[A+B-|L_A - R_B|I] = \sum_{i,j}\left\{2\lambda_i^{\alpha}\mu_j^{1-\alpha} - \left(\lambda_i + \mu_j - |\lambda_i - \mu_j|\right)\right\}|\langle\phi_i|\psi_j\rangle|^2.$$

Since $2x^{\alpha}y^{1-\alpha} - (x+y-|x-y|) \geq 0$ for $x, y > 0, 0 \leq \alpha \leq 1$ in general, we can obtain Theorem 1.9. □

Remark 2. We note the following 1, 2:

1. $\frac{1}{2}Tr[A+B-|A-B|] \leq \inf_{0\leq\alpha\leq 1} Tr\left[A^{1-\alpha}B^{\alpha}\right] \leq Tr\left[A^{1/2}B^{1/2}\right]$

$$\leq \sqrt{\left(\frac{1}{2}Tr[A+B]\right)^2 - \frac{1}{4}(Tr[|A-B|])^2}.$$

2. There is no relationship between $Tr[|L_A - R_B|I]$ and $Tr[|A-B|]$. When

$$A = \begin{pmatrix} \frac{3}{2} & \frac{1}{2} \\ \frac{1}{2} & \frac{3}{2} \end{pmatrix}, \quad B = \begin{pmatrix} 4 & 0 \\ 0 & 1 \end{pmatrix},$$

we have $Tr[|L_A - R_B|I] = 3, \quad Tr[|A-B|] = \sqrt{10}$. When

$$A = \begin{pmatrix} \frac{13}{2} & \frac{7}{2} \\ \frac{7}{2} & \frac{13}{2} \end{pmatrix}, \quad B = \begin{pmatrix} 2 & 0 \\ 0 & 5 \end{pmatrix},$$

we have $Tr[|L_A - R_B|I] = 8, \quad Tr[|A-B|] = \sqrt{58}$.

7. Sum type of uncertainty relations

Let $A, B \in M_{n,sa}(\mathbb{C})$ have the following spectral decompositions:

$$A = \sum_{i=1}^{n} \lambda_i |\phi_i\rangle\Big\langle \phi_i|, \quad B = \sum_{i=1}^{n} \mu_i |\psi_i\Big\rangle\langle\psi_i|.$$

For any quantum state $|\phi\rangle$, we define the two probability distributions

$$P = (p_1, p_2, \cdots, p_n), \quad Q = (q_i, q_2, \cdots, q_n),$$

where $p_i = |\langle \phi_i | \phi \rangle|^2, \quad q_j = |\langle \psi_j | \phi \rangle|^2$. Let

$$H(P) = -\sum_{i=1}^{n} p_i \log p_i, \quad H(Q) = -\sum_{j=1}^{n} q_j \log q_j$$

be the Shannon entropies of P and Q, respectively.
Theorem 1.10. The following uncertainty relation holds:

$$H(P) + H(Q) \geq -2 \log c,$$

where $c = \max_{i,j} |\left\langle \phi_i | \psi_j \right\rangle|$.
For details, see [15, 16].
Definition 6. The Fourier transformation of $\psi \in L^2(\mathbb{R})$ is defined as

$$\hat{\psi}(\omega) = \int_{-\infty}^{\infty} \psi(t) e^{-2\pi i \omega t} dt.$$

We also define

$$Q(\mathbb{R}) = \left\{ f \in L^2(\mathbb{R}); \int_{-\infty}^{\infty} t^2 |f(t)|^2 dt < \infty \right\}.$$

Proposition 2. If $\psi \in L^2(\mathbb{R})$, $\|\psi\|^2 = 1$ satisfies $\psi, \hat{\psi} \in Q(\mathbb{R})$, then

$$S(\psi) + S(\hat{\psi}) \geq \log \frac{e}{2},$$

where

$$S(\psi) = -\int_{-\infty}^{\infty} |\psi(t)|^2 \log |\psi(t)|^2 dt, \quad S(\hat{\psi}) = -\int_{-\infty}^{\infty} |\hat{\psi}(t)|^2 \log |\hat{\psi}(t)|^2 dt.$$

For details, see [17].
Theorem 1.11 ([18]). For any $X, Y \in M_n(\mathbb{C}), A, B \in M_{n,+}(\mathbb{C})$, the following holds:

1. $I_{A,B}^{(g,f)}(X, Y) + I_{A,B}^{(g,f)}(Y) \geq \frac{1}{2} \max \left\{ I_{A,B}^{(g,f)}(X+Y), I_{A,B}^{(g,f)}(X-Y) \right\}$.

2. $\sqrt{I_{A,B}^{(g,f)}(X)} + \sqrt{I_{A,B}^{(g,f)}(Y)} \geq \max \left\{ \sqrt{I_{A,B}^{(g,f)}(X+Y)}, \sqrt{I_{A,B}^{(g,f)}(X-Y)} \right\}$.

3. $\sqrt{I_{A,B}^{(g,f)}(X)} + \sqrt{I_{A,B}^{(g,f)}(Y)} \leq 2 \max \left\{ \sqrt{I_{A,B}^{(g,f)}(X+Y)}, \sqrt{I_{A.B}^{(g,f)}(X-Y)} \right\}$.

Proof 1. The Hilbert-Schmidt norm $\|\cdot\|$ satisfies

$$\|X\|^2 + \|Y\|^2 = \frac{1}{2}\left(\|X+Y\|^2 + \|X-Y\|^2\right) \geq \frac{1}{2} \max\left\{\|X+Y\|^2, \|X-Y\|^2\right\}. \quad (13)$$

Since $I_{A,B}^{(g,f)}(X,X)$ is the second power of the Hilbert-Schmidt norm, $\|X\| = \sqrt{I_{A,B}^{(g,f)}(X)}$. We then obtain the result by substituting (13),

2. By the triangle inequality of a general norm, we apply the triangle inequality for $\|X\| = \sqrt{I_{A,B}^{(g,f)}(X)}$.

3. We prove the following norm inequality:

$$\|X\| + \|Y\| \leq \|X+Y\| + \|X-Y\|. \tag{14}$$

Since

$$\|X\| = \|\frac{1}{2}(X+Y) + \frac{1}{2}(X-Y)\| \leq \frac{1}{2}\|X+Y\| + \frac{1}{2}\|X-Y\|$$

and

$$\|Y\| = \|\frac{1}{2}(Y+X) + \frac{1}{2}(Y-X)\| \leq \frac{1}{2}\|Y+X\| + \frac{1}{2}\|Y-X\|,$$

we add two inequalities and obtain (14). □

Author details

Kenjiro Yanagi[1,2]

1 Faculty of Science, Department of Mathematics, Josai University, Japan

2 Yamaguchi University, Japan

*Address all correspondence to: yanagi@yamaguchi-u.ac.jp

References

[1] Lieb EH. Convex trace functions and the Wigner-Yanase-Dyson conjecture. Advances in Mathematics. 1973;**11**: 267-288

[2] Luo S, Zhang Q. On skew information. IEEE Transactions of Information Theory. 2004;**50**:1778-1782, and Correction to "On skew information", IEEE Transactions of Information Theory. 2005;**51**:4432

[3] Kosaki H. Matrix trace inequality related to uncertainty principle. International Journal of Mathematics. 2005;**16**:629-646

[4] Yanagi K, Furuichi S, Kuriyama K. A generalized skew information and uncertainty relations. IEEE Transactions of Information Theory. 2005;**IT-51**(12): 4401-4404

[5] Hansen F. Metric adjusted skew information. Proceedings of the National Academy of Sciences of the United States of America. 2008;**105**: 9909-9916

[6] Yanagi K. Metric adjusted skew information and uncertainty relation. Journal of Mathematical Analysis and Applications. 2011;**380**(2):888-892

[7] Yanagi K, Furuichi S, Kuriyama K. Uncertainty relations for generalized metric adjusted skew information and generalized metric adjusted correlation measure. Journal of Uncertainty Analysis and Applications. 2013;**1**(12): 1-14

[8] Luo S. Heisenberg uncertainty relation for mixed states. Physical Review A. 2005;**72**:042110

[9] Yanagi K. Uncertainty relation on Wigner-Yanase-Dyson skew information. Journal of Mathematical Analysis and Applications. 2010;**365**: 12-18

[10] Gibilisco P, Hansen F, Isola T. On a correspondence between regular and non-regular operator monotone functions. Linear Algebra and its Applications. 2009;**430**:2225-2232

[11] Kubo F, Ando T. Means of positive linear operators. Mathematische Annalen. 1980;**246**:205-224

[12] Yanagi K. On the trace inequalities related to left-right multiplication operators and their applications. Linear and Nonlinear Analysis. 2018;**4**(3): 361-370

[13] Audenaert KMR, Calsamiglia J, Masancs LI, Munnoz-Tapia R, Acin A, Bagan E, et al. The quantum Chernoff bound. Review Letters. 2007;**98**: 160501-1-160501-4

[14] Yanagi K. Generalized trace inequalities related to fidelity and trace distance. Linear and Nonlinear Analysis. 2016;**2**(2):263-270

[15] Maassen H, Uffink JBM. Generalized entropic uncertainty relations. Physical Review Letters. 1988; **60**(12):1103-1106

[16] Nielsen MA, Chuang IL. Quantum Computation and Quantum Information. Cambridge University Press; 2000

[17] Hirschman II Jr. A note on entropy. American Journal of Mathematics. 1957; **79**:152-156

[18] Yanagi K. Sum types of uncertainty relations for generalized quasi-metric adjusted skew informations. International Journal of Mathematical Analysis and Applications. 2018;**4**(4): 85-94

Chapter 5

Complex Space Nature of the Quantum World: Return Causality to Quantum Mechanics

Ciann-Dong Yang and Shiang-Yi Han

Abstract

As one chapter, we about to begin a journey with exploring the limitation of the causality that rules the whole universe. Quantum mechanics is established on the basis of the phenomenology and the lack of ontology builds the wall which blocks the causality. It is very difficult to reconcile the probability and the causality in such a platform. A higher dimension consideration may leverage this dilemma by expanding the vision. Information may seem to be discontinuous or even so weird if only be viewed from a part of the degree of freedoms. Based on this premise, we reexamined the microscopic world within a complex space. Significantly, some knowledge beyond the empirical findings is revealed and paves the way for a more detailed exploration of the quantum world. The random quantum motion is essential for atomic particle and exhibits a wave-related property with a bulk of trajectories. It seems we can break down the wall which forbids the causality entering the quantum kingdom and connect quantum mechanics with classical mechanics. The causality returns to the quantum world without any assumption in terms of the quantum random motion under the optimal guidance law in complex space. Thereby hangs a tale, we briefly introduce this new formulation from the fundamental theoretical description to the practical technology applications.

Keywords: random quantum trajectory, optimal guidance law, complex space

1. Introduction

It took scientists nearly two centuries from first observation of flower powder's Brownian motion to propose a mathematical qualitative description [1]. Time is an arrow launched from the past to the future, every event happens for a reason. "The world is woven from billions of lives, every strand crossing every other. What we call premonition is just movement of the web. If you could attenuate to every strand of quivering data, the future would be entirely calculable. As inevitable as mathematics [2]." All physical phenomena are connected to the same web. As long as we can see through the quivering data and cut into the very core, we can glimpse the most elegant beauty of nature. As precise as physics.

It took nearly 30 years for physicists to establish quantum mechanics but nearly 100 years to seek for its essence. Quantum mechanics is the most precise theory to describe the microscopic world but also is the most obscure one among all theories. It collects lots data but not all. Just like what we can observed is the shadow on the ground not the actual object in the air. It is impossible to see the whole appearance of the object by observing its shadow. The development of the quantum era seems started in such circumstances and missed something we call the essence of nature. In this chapter, we hope to recover the missing part by considering a higher dimension to capture the actual appearance of nature. At the end, we will find out that nature dominates the web where we live as well as the theories we develop. Everything should follow the law of the nature, and there is no exception.

Trajectory is a typical classical feature of the macroscopic object solved by the equation of motion. The trajectory of the microscopic particle is supposed to be observed if the law of nature remains consistent all the way down to the atomic scale. However, such an observation cannot be made till 2011. Kocsis and his coworkers propose an observation of the average trajectories of single photons in a two-slit interferometer on the basis of weak measurement [3]. Since then quantum trajectories are observed for many quantum systems, such as superconducting quantum bit, mechanical resonator, and so on [4–6]. Weak measurement provides the weak value which is a measurable quantity definable to any quantum observable under the weak coupling between the system and the measurement apparatus [7]. The significant characteristic of the weak value does not lie within the range of eigenvalues and is complex. It is pointed out that the real part of the complex weak value represents the average quantum value [8], and the imaginary part is related to the rate of variation in the interference observation [9].

The trajectory interpretation of quantum mechanics is developed on the basis of de Broglie's matter wave and Bohm's guidance law. In recent years, the importance of the quantum trajectory in theoretical treatment and experimental test has been discussed in complex space [10–21]. All these research indirectly or directly show that the complex space extension is more than a mathematical tool, it implies a causal essence of the quantum world.

On the other hand, it is found out that the real part of momentum's weak value is the Bohmian momentum representing the average momentum conditioned on a position detection; while its imaginary part is proportional to the osmotic velocity that describes the logarithmic derivative of the probability density for measuring the particular position directed along the flow generated by the momentum [22]. This not only implies the existence of randomness in a quantum system, but also discloses that the random motion occurs in complex space. Numerous studies with the complex initial condition and the random property have been discussed [23–25]. A stochastic interpretation of quantum mechanics is proposed which regards the random motion as a nature property of the quantum world not the interference made by the measurement devices [26, 27]. These investigations suggest that a complex space and the random motion are two important features of the quantum world.

Based on the complex space structure, we propose a new perspective of quantum mechanics that allows one to reexamine quantum phenomena in a classical way. We will see in this chapter how the quantum motion can provide the classical description for the quantum kingdom and is in line with the probability distribution. One thing particular needed to be emphasized is that the stochastic Hamilton Jacobi Bellman equation can reduce to the Schrödinger equation under some specific conditions. In other words, the Schrödinger equation is one special case of all kinds of random motions in complex space. A further discussion of the relationship between the trajectory interpretation and probability interpretation is presented in Section 2. In particular, the solvable nodal issue is put into discussion, and the continuity equation for the complex probability density function is proposed. In Section 3, we demonstrate how the quantum force could play the crucial role in the force balanced condition within the hydrogen atom and how the quantum potential forms the shell structure where the orbits are quantized. A practical application to the Nano-scale is demonstrated in Section 4. We consider the quantum potential relation to the electronic channel in a 2D Nano-structure. In addition, the conductance quantization is realized in terms of the quantum potential which shows that the lower potential region is where the most electrons pass through the channel. And then, concluding remarks are presented in Section 5.

2. Random quantum motion in the complex plane

In the macroscopic world, it is natural to see an object moving along with a specific path which is determined by the resultant optimal action function. However, in the microscopic world, we cannot repetitively carry out this observation since there is no definition of the trajectory for a quantum particle. With the limit on the observation, only a part of trajectory, more precisely, the trajectory in the real part of complex space can be detected. As particle passing or staying in the imaginary part of complex space it disappears from our visible world and becomes untraceable. The particle randomly transits in and out of the real part and imaginary part of complex space, causes a discontinuous trajectory viewed from the observable space. Therefore, it can only be empirically described by the probability in quantum mechanics.

In this section, we briefly introduce how particle's motion can be fully described by the optimal guidance law in the complex plane [28]. Then we will discuss under what condition the statistical distribution of an ensemble of trajectories in the complex plane will be compatible with the quantum mechanical and classical results. In the following, we consider a complex plane for the purpose of simplicity; however, there should be no problem to implement the optimal guidance law in complex space. Let us consider a particle with random motion in the complex plane whose dynamic evolution reads

$$dx = f(t,x,u)dt + g(x,u)dw, \quad x = x_R + ix_I \in \mathbb{C}, \tag{1}$$

where x represents a vector, u is the guidance law needed to be determined, w is Wiener process with properties $\langle dw \rangle = 0$ and $\langle dw^2 \rangle = dt$, $f(t,x,u)$ is the drift velocity, and $g(x,u)$ is the diffusion velocity. The cost function for $x(t)$ with randomness property reads

$$J(t,x,u) = E_{t,x}\left[\int_t^{t_f} L(\tau, x(\tau), u(\tau))d\tau\right], \tag{2}$$

where $E_{t,x}$ represents the expectation of the cost function over all infinite trajectories launched from the single initial condition, $x(t) = x$ in time interval $[t, t_f]$. To find the minimum cost function, we define the value function,

$$V(t,x) = \min_{u[t,t_f]} J(t,x,u). \tag{3}$$

Instead of using the variational method, we apply the dynamic programming method to Eq. (3) for the random motion. We then have the following expression after having the Taylor expansion:

$$-\frac{\partial V(t,x)}{\partial t} = \min_{u[t,t_f]}\left\{L + \frac{\partial V(t,x)}{\partial x}f + \frac{1}{2}tr\left[g^T(x,u)\frac{\partial^2 V(t,x)}{\partial x^2}g(x,u)\right]\right\}, \tag{4}$$

which is recognized as the Hamilton-Jacobi-Bellman (HJB) equation and $\partial^2 V(t,x)/\partial x^2$ is Jacobi matrix. Finding the minimum of the cost function leads to the momentum for the optimal path,

$$p = \frac{\partial L(t,x,u)}{\partial u} = \frac{\partial L(t,x,\dot{x})}{\partial \dot{x}} = -\nabla V(t,x), \tag{5}$$

and determines the optimal guidance law,

$$u = u(t,x,p)|_{p=-\nabla V}. \tag{6}$$

If one replaces Lagrange L by Hamiltonian $H(t,x,p) = p^T u - L(t,x,u)$, defines the action function as $S(t,x) = -V(t,x)$ and let $g(x,u) = \sqrt{-i\hbar/m}$, Eq. (4) can be transferred to the quantum Hamilton-Jacobi (HJ) equation,

$$\frac{\partial S}{\partial t} + H(t,x,p)|_{p=\nabla S} + \frac{i\hbar}{2m}\nabla^2 S = 0. \tag{7}$$

Please notice that the last term in Eq. (7) is what makes the quantum HJ equation differs from its classical counterpart. It is called the quantum potential,

$$Q = \frac{i\hbar}{2m}\nabla^2 S \tag{8}$$

in dBB theory, Bohmian mechanics, and quantum Hamilton mechanics [29–33]. Even the quantum potential we derive here has the same expression appeared in Bohmian mechanics, its relation to the random motion should be noticed. However, it is not yet suitable to claim that the random motion attributes to the quantum potential or vice versa. It is worthwhile to bring into discussion. Before inspecting this question more deeply, we still can take advantage of the quantum potential to describe or even explain some quantum phenomena.

We can transfer the quantum HJ equation (7) to the Schrödinger equation,

$$i\hbar\frac{\partial \Psi(t,x)}{\partial t} = -\frac{\hbar^2}{2m}\nabla^2\Psi(t,x) + U\Psi(t,x) \tag{9}$$

via the relation between the action function and wave function,

$$S(t,x) = -i\hbar \ln \Psi(t,x), \tag{10}$$

where U represents the external potential. This simple relation reveals a connection between the trajectory and the wave description. In classical mechanics, a particle follows the principle of least action; while the wave picture took place in quantum mechanics. Eq. (10) implies that if we collect all action functions determined by different initial conditions which satisfy the initial probability distribution, a collection of corresponding wave patterns arise and eventually forms the solution wave function of the Schrödinger equation. This process is the same as what Schrödinger attempted to cope with the observable wave and tried to deduce the suitable wave equation based on the classical wave theory. The only difference is that Schrödinger started his deduction from the wave perspective; however, we start from the particle perspective. Even the wave-particle duality troubles physicists to inspect advanced about the essence of nature, the recent experiment confirms relation (10) by observing an ensemble of quantum trajectories [3].

This becomes a solid evidence to support the deduction that the matter wave is formed by a huge number of trajectories.

To fully understand the property of these trajectories under the influence of the guidance law, we consider a particle experiencing a randomness,

$$dx = u(t, x, p)dt + \sqrt{\frac{-i\hbar}{m}}dw, \tag{11}$$

where we have replaced $f(t, x, u)$ by the optimal guidance law $u(t, x, p)$, and assigned $g(x, u) = \sqrt{-i\hbar/m}$ into Eq. (1). Combining Eqs. (6) and (10), the optimal guidance law can be expressed in terms of the wave function,

$$u(t, x, p) = \frac{-i\hbar}{m}\frac{\nabla\Psi(t, x)}{\Psi(t, x)}. \tag{12}$$

Therefore, Eq. (11) can be recast into the following expression:

$$dx = \frac{-i\hbar}{m}\frac{\nabla\Psi(t, x)}{\Psi(t, x)}dt + \sqrt{\frac{-i\hbar}{m}}dw. \tag{13}$$

Eq. (13) will reduce to the equation of motion given by the quantum HJ equation (7) if we take the average of both sides,

$$\dot{x} = \frac{-i\hbar}{m}\frac{\nabla\Psi(t, x)}{\Psi(t, x)}, \tag{14}$$

since the random motion in Eq. (13) has zero mean. This result shows that the quantum HJ equation represents the mean motion of the particle. The trajectory in the complex plane solved from Eq. (13) is random and will become the mean trajectory solved from Eq. (14) after being averaged out. **Figure 1** illustrates this property by demonstrating the quantum motion of the Gaussian wave packet [28].

The first question we would like to answer by the complex random trajectory (CRT) interpretation is its connection to the probability interpretation. In quantum mechanics, the amplitude square of the wave function gives the probability density of physical quantities as shown in **Figure 2(a)**, in which the solid line stands for the quantum harmonic oscillator in $n = 1$ state. The trajectory interpretation is supported by the excellent agreement of the statistical spatial distribution made by collecting all crossovers on the real axis of an ensemble of CRTs as the dots displayed in **Figure 2(a)**. It shows a good agreement of the statistical spatial distribution and the quantum mechanical probability distribution [36].

In most text book of quantum mechanics, the nodes of the probability of harmonic oscillator either be ignored or be regarded as the quantum characteristic. Only the classical-like curve of the averaged probability has been mentioned. The other significant finding brought out by the CRT interpretation is the nodal vanished condition given by the statistical distribution of the collection of all pointes be projected onto the real axis as **Figure 2(b)** shows. It starts to approach the classical probability distribution for high quantum number as **Figure 2(c)** presents. The leverage of complex space structure deals with the probability nodes, and even further to reach the classical region dominated by Newtonian mechanics (more detail refers to [36]). After the matter wave can be interpreted by an ensemble of trajectories in both theoretical and experimental results [3, 18, 34, 35], the CRT interpretation shows both quantum mechanical and classical compatible

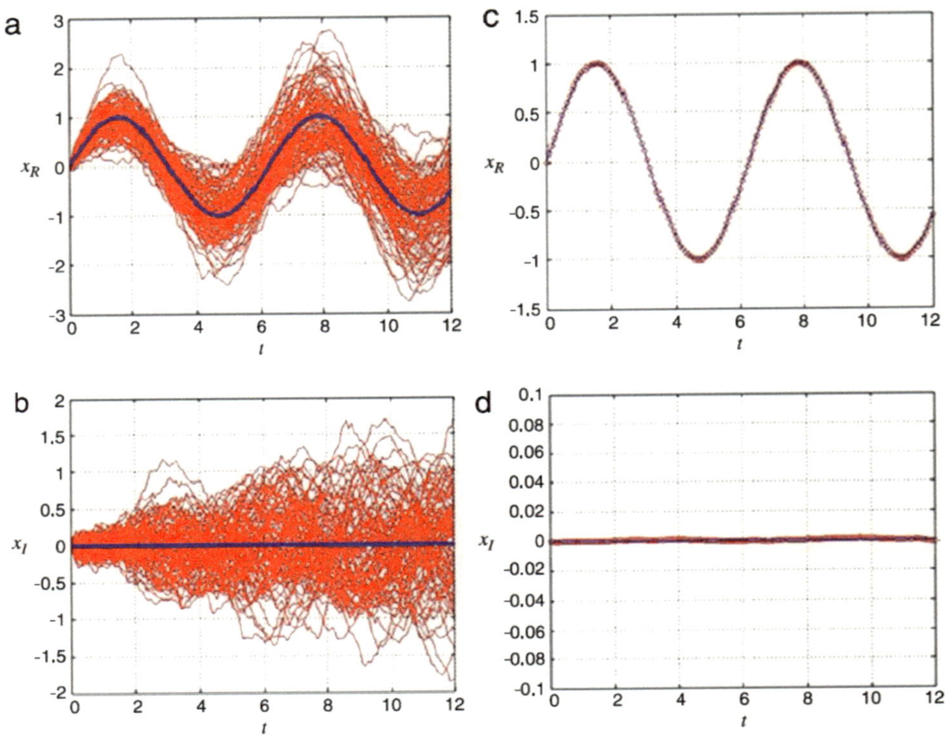

Figure 1.
100,000 trajectories solved from Eq. (13) with the same initial condition of the Gaussian wave packet in the complex plane: (a) the time evolution on the real axis for which the mean is denoted by the blue line; (b) the time evolution on the imaginary axis with zero mean represented by the blue line. The complex trajectory solved from Eq. (14) with one initial condition: (c) the time evolution on the real axis; (d) the time imaginary part of the motion. This figure reveals that the mean of the CRT is the trajectory solved from the quantum Hamilton equations of motion [28].

results under two kinds of point collections. In other words, Bohr's correspondence principle can be interpreted by the CRT interpretation without loss of generality [36].

The second question we try to cope with by means of the CRT interpretation is the conservation of the complex probability. In quantum mechanics, the continuity equation for the probability density function is given by Bohr's law $\rho_{QM} = |\Psi|^2$, and the current density J,

$$\frac{\partial \rho_{QM}}{\partial t} = -\nabla \cdot J. \tag{15}$$

The probability density function of the CRT interpretation satisfies the Fokker-Planck equation,

$$\frac{\partial \rho(t,x)}{\partial t} = -\nabla \cdot \left(\dot{\overline{x}}(t,x)\rho(t,x)\right) - \frac{i\hbar}{2m}\nabla^2\rho(t,x), \tag{16}$$

and has the complex value. Multiplying Eq. (16) and its complex conjugate then dividing by 2, we obtain the continuity equation for complex probability density,

$$\frac{\partial \rho(t,\overline{x})}{\partial t} = -\nabla \cdot \left(\dot{\overline{x}}\rho(t,\overline{x})\right), \tag{17}$$

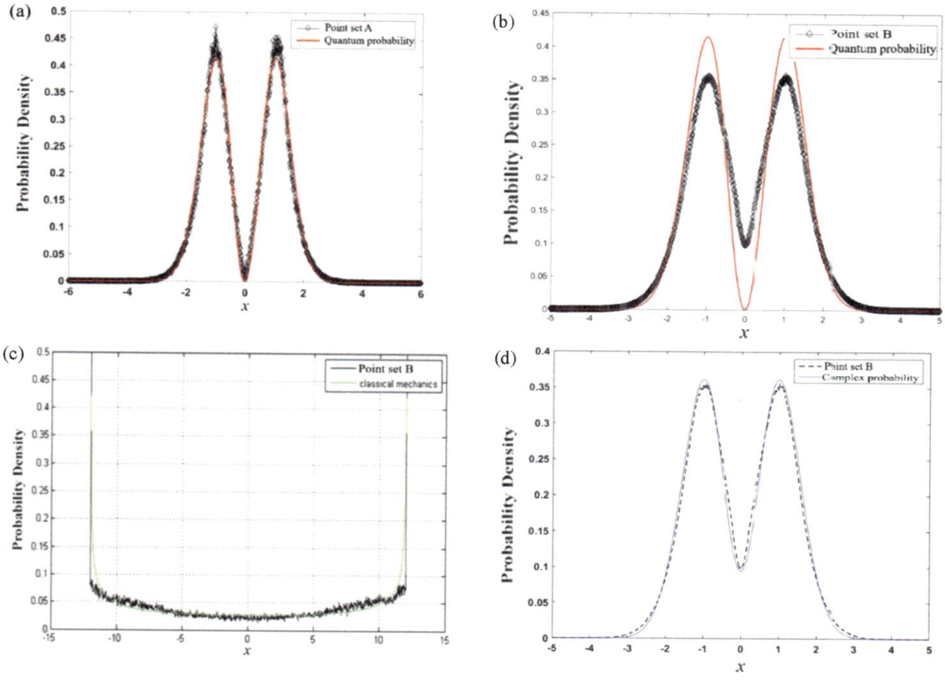

Figure 2.
(a) The quantum mechanical compatible outcome proposed by point collections of an ensemble of CRTs crossing the real axis for quantum harmonic oscillator in n = 1 *state with coefficient correlation,* Γ = 0.995. *(b) The dismissed nodal condition is given by the same trajectory ensemble but is composed of all projected points onto the real axis. (c) The classical-like probability distribution is presented by collecting all projection points on the real axis for* n = 70 *state with coefficient correlation,* Γ = 0.9412. *(d) The analytical solution of the complex probability density function solved from the Fokker-Planck equation shows good agreement with the spatial distribution composed of all projection points on the real axis with coefficient correlation,* Γ = 0.9975 *[36].*

where $\overline{x}$ denotes the mean of valuable x. From Eq. (17) we can see that the complex probability density is conserved in the complex plane, neither on the real axis nor imaginary axis. **Figure 2(d)** illustrates the good agreement between the solution solved from Eq. (17) (blue dotted line) and the statistical spatial distribution (black solid line) contributed by all points collected by the projections onto the real axis. This result verifies that the analytical solution coheres with the statistical distribution made by CRT. It shows that the same results obtained from two different ways stand from the equal footing of the classical concept.

3. Shell structure in hydrogen atom

In quantum mechanics, the quantized orbits of the electron in the hydrogen atom is determined by solving the Schrödinger equation for different eigen states. There is no further description of these orbits, especially no explanation about the force balanced condition under the influence of the Coulomb force. Less study reports the role that the quantum potential plays in atomic analysis. In this section, a quest for describing the hydrogen atom is stretching underlying the quantum potential in complex space. We show our most equations in dimensionless form for the purposes of simplifying the question.

Let us consider the quantum Hamiltonian with Coulomb potential in complex space [37],

$$H = \frac{1}{2m}\left[\left(\frac{\partial S}{\partial r}\right)^2 + \frac{\hbar}{i}\left(\frac{2}{r}\frac{\partial S}{\partial r} + \frac{\partial^2 S}{\partial r^2}\right)\right] + \frac{1}{2mr^2}\left[\left(\frac{\partial S}{\partial \theta}\right)^2 + \frac{\hbar}{i}\left(\cot\theta\frac{\partial S}{\partial \theta} + \frac{\partial^2 S}{\partial \theta^2}\right)\right.$$
$$\left. + \frac{1}{\sin^2\theta}\left(\left(\frac{\partial S}{\partial \phi}\right)^2 + \frac{\hbar}{i}\frac{\partial^2 S}{\partial \phi^2}\right)\right] + \frac{-Ze^2}{4\pi\epsilon_0 r}, \qquad (18)$$

where S is the action function. Hamiltonian (18) is state dependent if we apply the simple relation (9) to it. We can therefore have the dimensionless total potential in terms of the wave function,

$$V_{nlm_l} = -\frac{2}{r} + \left[\frac{1}{4r^2}\left(4 + \cot^2\theta\right) - \frac{d^2 \ln R_{nl}(r)}{dr^2} - \frac{1}{r^2}\frac{d^2 \ln \Theta_{lm_l}(\theta)}{d\theta^2}\right], \qquad (19)$$

where n, l, and m_l denote the principle quantum number, azimuthal quantum number, and magnetic quantum number, respectively. The first term in Eq. (19) is recognized as the Coulomb potential; while the remaining terms are the components of the quantum potential. **Figure 3(a)** illustrates the three potentials varying in radial direction of $(n, l, m_l) = (1, 0, 0)$ state; they are the total potential, Coulomb potential, and quantum potential. The quantum potential yields the opposite spatial distribution to the Coulomb potential, therefore, the total potential performs a neutral situation. When the electron is too close (less than the Bohr radius) to the nucleus, the total potential forms a solid wall that forbids the electron getting closer. The total potential holds an appropriate distribution such that the electron is subject to an attractive force when it is too far away from the nucleus. From the perspective of the electron, it is quantum potential maintains the orbit stable and stop the disaster of crashing on the nucleus.

From Eq. (19) we can obtain the total forces for $(n, l, m_l) = (1, 0, 0)$ state:

$$f^r_{100} = -\frac{2}{r^2} + \frac{1}{2r^3}\left(4 + \cot^2\theta\right), \quad f^\theta_{100} = \frac{1}{2r^2}\frac{\cos\theta}{\sin^3\theta}, \quad f^\phi_{100} = 0. \qquad (20)$$

Under a specific condition $f^r_{100} = f^\theta_{100} = 0$, the electron stays stationary at the equilibrium position $(r, \theta) = (1, \pi/2)$ for which $r = 1$ corresponds to the Bohr radius. The motion of electron at the equilibrium point is determined by

$$f^r_{100}(r, \pi/2) = f^r_Q + f^r_V = \frac{2}{r^3} - \frac{2}{r^2}, \qquad (21)$$

where the first and the second term represent the repulsive quantum force and the attractive Coulomb force with lower label Q and V, respectively. As the distance between the electron and the nucleus changes, the two forces take the lead in turn as **Figure 3(b)** illustrates. It is clear to see that the zero force location happens at $r = 1$ (Bohr radius) owing to the force balancing formed by the Coulomb force and quantum force.

In quantum mechanics, the maximum probability of finding the electron is at the Bohr radius according to

$$\frac{d}{dr}P_{10}(r) = \frac{d}{dr}\left(4\pi r^2 e^{-2r}\right) = 0. \qquad (22)$$

The balanced force and the probability are totally different concepts; however, present the same description of the hydrogen atom. This may reflect the equivalent

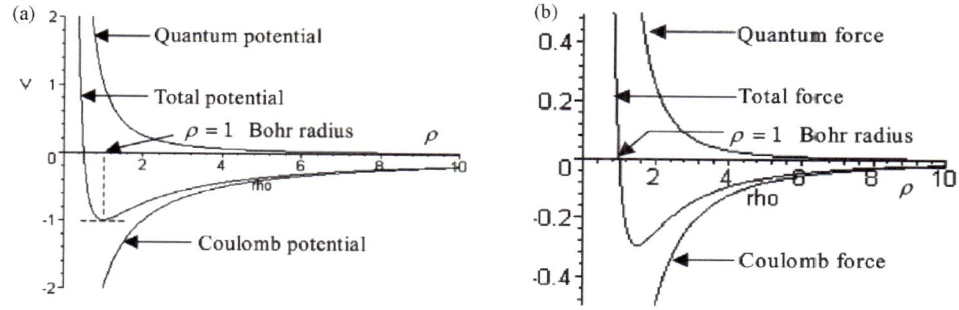

Figure 3.
(a) The variations of three potentials in radial direction for the ground state. (b) The total radial force in the ground state which is composed of the coulomb force and quantum force with zero value at the Bohr radius [37].

meaning between the classical shell layers and the quantum probability. Furthermore, it may help us to realize the probabilistic electron cloud in a classical standpoint.

Let us consider $(n, l, m_l) = (2, 0, 0)$ state, which has the total potential as

$$V_{200} = V + Q = -\frac{2}{r} + \left[\frac{1}{(2-r)^2} + \frac{1}{4r^2}\left(4 + \cot^2\theta\right)\right], \tag{23}$$

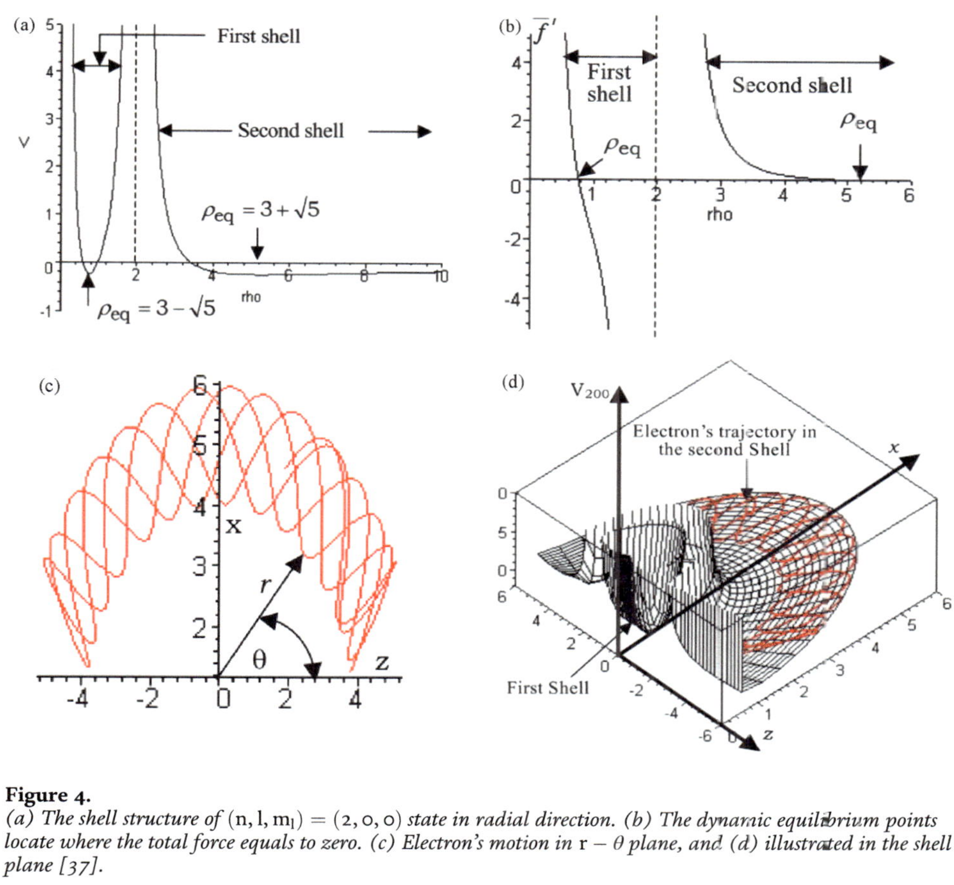

Figure 4.
(a) The shell structure of $(n, l, m_l) = (2, 0, 0)$ *state in radial direction. (b) The dynamic equilibrium points locate where the total force equals to zero. (c) Electron's motion in* $r - \theta$ *plane, and (d) illustrated in the shell plane [37].*

and the force distributions in three directions:

$$f^r_{200} = -\frac{2}{r^2} + \left[-\frac{1}{(2-r)^3} + \frac{1}{2r^3}\left(4 + \cot^2\theta\right) \right], f^\theta_{200} = \frac{1}{2r^2}\frac{\cos\theta}{\sin^3\theta}, f^\phi_{200} = 0, \quad (24)$$

which indicates the same equilibrium point location $(r_{eq}, \theta_{eq}) = (3 \pm \sqrt{5}, \pi/2)$ given by the equations of motion from Eq. (14):

$$\frac{dr}{dt} = 4i\frac{r^2 - 6r + 4}{r(r-2)}, \frac{d(\cos\theta)}{dt} = i\frac{\cos\theta}{r^2}, \frac{d\phi}{dt} = 0, \quad (25)$$

under the zero resultant force condition and the electron dynamic equilibrium condition. **Figure 4(a)** presents the shell structures in radial direction according to Eq. (24). The range of the layers are constrained by the total potential and divided into two different parts. The two equilibrium points individually correspond to the zero force locations in the two shells as **Figure 4(b)** indicates. Eq. (25) offers how electron move in this state. **Figure 4(c)** illustrates electron's trajectory in the $r - \theta$ plane; while **Figure 4(d)** embodies trajectory in the shell structure.

4. Channelized quantum potential and conductance quantization in 2D Nano-channels

The practical technology usage of the proposed formalism is applied to 2D Nano-channels in this section. Instead of the probability density function offered by the conventional quantum mechanics, we stay in line with causalism to perceive what role played by the quantum potential. Consider a 2D straight channel made by GaAs-GaAlAs and is surrounded by infinite potential barrier except the two reservoirs and the channel. The schematic plot of the channel refers to **Figure 5**. The dynamic evolution of the wave function $\psi(x, y)$ in the channel is described by the Schrödinger equation,

$$-\frac{\hbar^2}{2m^*}\left(\frac{\partial^2}{\partial x^2} + \frac{\partial^2}{\partial y^2}\right)\psi(x,y) = E\psi(x,y), \quad (26)$$

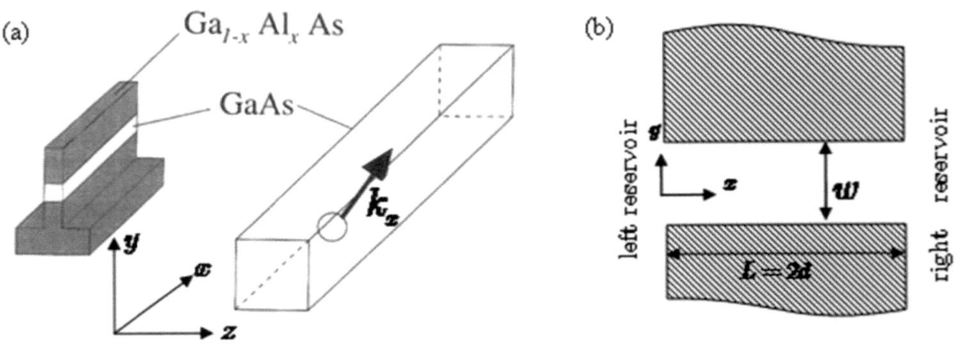

Figure 5.
(a) A single quantum wire and an expanded view showing schematically the single degree of freedom in the x direction. (b) 2D straight channel made up of quantum wire with length 2d *and width* w *connects the left reservoir to the right reservoir.*

where $m^* = 0.067m_e$ is the effective mass of the electron, and E is the total energy of the incident electron. The general solution of Eq. (26) has the form as

$$\psi_k^C(x,y) = \sum_{n=1}^{N} \left(B_n e^{ik_n x} + C_n e^{-ik_n x}\right)\phi_n(y), \phi_n(y) = \sin\left[\frac{n\pi}{w}\left(y + \frac{w}{2}\right)\right], \tag{27}$$

where N is the number of mode, w is the width of the channel, and k_n is the wave number which satisfies the energy conservation law:

$$E_x + E_y = \frac{(k_n\hbar)^2}{2m^*} + E_n = E, \tag{28}$$

in which $E_x = p_x^2/(2m^*) = (k_n\hbar)^2/(2m^*)$ is the free particle energy in the x direction, and $E_y = E_n = n^2\hbar^2\pi^2/(2m^* w^2)$, $n = 1, 2, \cdots$, is quantized energy in the y direction due to the presence of the infinite square well. From Eq. (28), we have the wave number read

$$k_n = \sqrt{2m^*(E - E_n)/\hbar^2}. \tag{29}$$

The function $B_n e^{ik_n x} + C_n e^{-ik_n x}$ in Eq. (27) is the free-particle wave function in the x direction, and $\phi_n(y)$ is an eigen function for the infinite well in the y direction satisfying the boundary condition $\phi_n(y)(w/2) = \phi_n(y)(-w/2) = 0$. The coefficients B_n and C_n are uniquely determined by the incident energy E and incident angle ϕ. (More detail refers to [38].) The quantum potential in the channel can be obtained by combing Eqs. (8), (10) and the wave function (27) (in dimensionless form),

$$Q(x,y) = -\left(\frac{\partial^2}{\partial x^2} + \frac{\partial^2}{\partial y^2}\right) \ln \psi_k^C(x,y). \tag{30}$$

The quantum potential provides fully information of electron's motion, its characteristic of inverse proportional to the probability density displays more knowledge in the channel. The inverse proportional relation reads

$$|Q(x,y)| = \frac{1}{P(x,y)}\left[\left(\frac{\partial \psi_k^C}{\partial x}\right)^2 + \left(\frac{\partial \psi_k^C}{\partial y}\right)^2\right], \tag{31}$$

which represents that the high quantum potential region corresponds to the low probability of electrons passing through as **Figure 6** displays; and **Figure 7** illustrates how the quantum potential gradually form the quantized channels as the incident angle increases, which shows the state dependent characteristic of the quantum potential.

The other quantum feature originating from the quantum potential is the quantization of conductance in the channel as **Figure 8** presents. We will show that the high conductance region is where the most electrons gather. To simplify the system, we firstly replace the motion in 2D channel by a motion in 1D square barriers [39]. Therefore, we consider the wave function $\psi_n(x)$ satisfying the following Schrödinger equation,

$$\frac{d^2\psi_n(x)}{dx^2} + \frac{2m^*}{\hbar^2}(E - V_n)\psi_n(x) = 0, \tag{32}$$

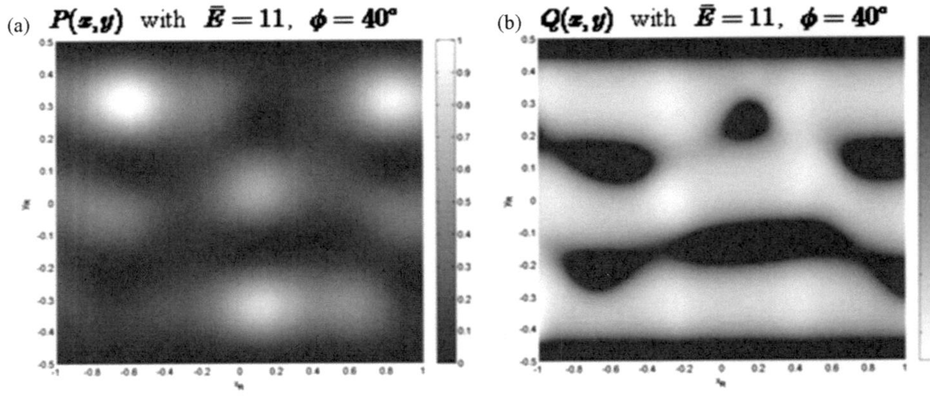

Figure 6.
The incident energy E = 11 *and the incident angle* $\phi = 40^{\circ}$ *for: (a) the probability density function; (b) the corresponding quantum potential of the cross-section in the channel. The bright regions of the quantum potential in (b) represent the lower potential barriers which are in accord with the bright regions in (a) where are the locations with higher probability of finding electrons [38].*

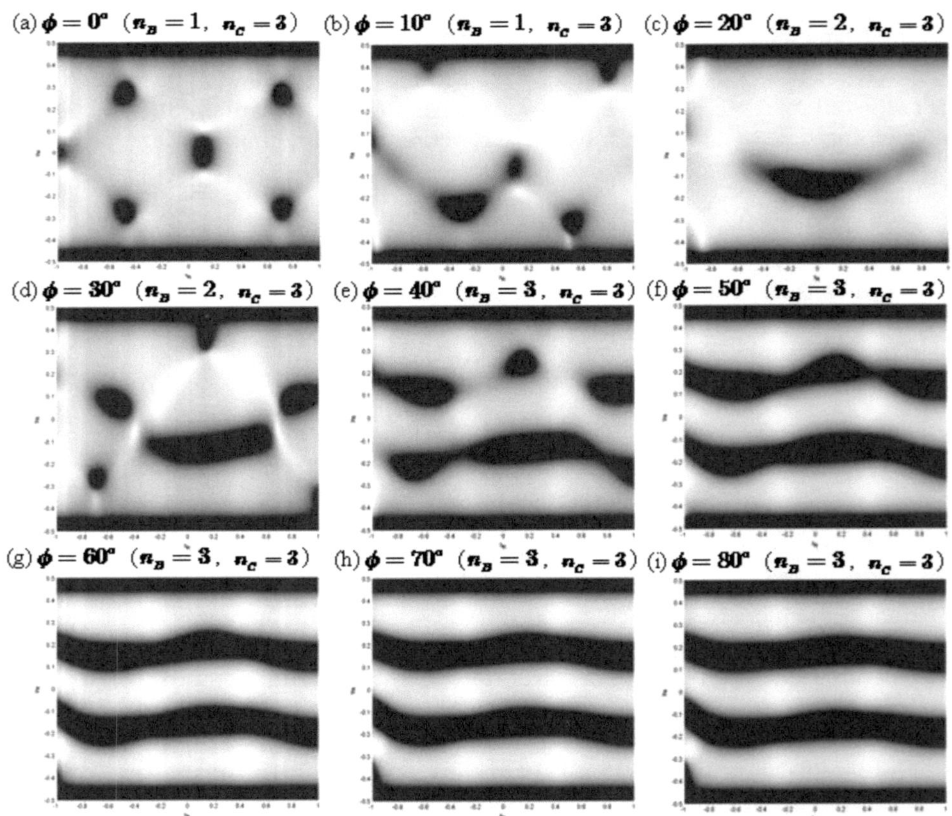

Figure 7.
The variation of the quantum potential with respect to the incident angle ϕ *for a fixed incident energy* E = 11. *It is seen that the channelized structure becomes more and more apparent with the increasing incident angle* ϕ *[38].*

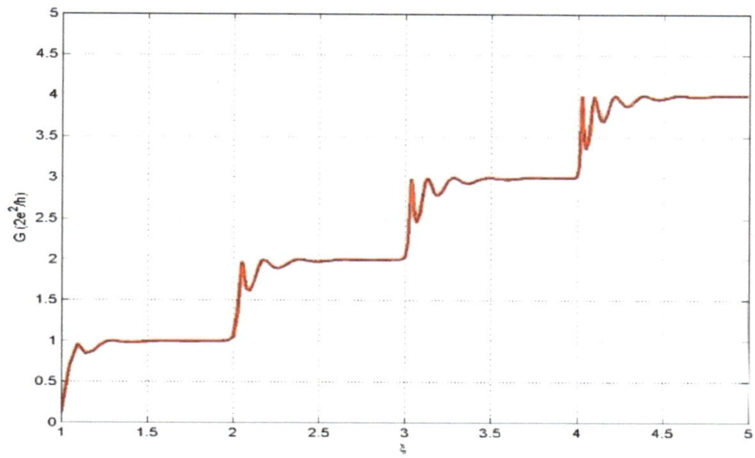

Figure 8.
The conductance G of a narrow channel shows plateaus at integer multiples of $2e^2/h$ *as the electron's energy* $\xi = \sqrt{E}$ *increases [39].*

where V_n is the equivalent square barrier,

$$V_n = \begin{cases} \dfrac{n^2\hbar^2\pi^2}{2m^*w^2}, & |x| \le d \\ 0, & |x| > d \end{cases}. \tag{33}$$

Please notice that potential V_n depends on the eigen state, hence, the electron will encounter different heights of the potential barrier in different eigen states. Furthermore, it makes electron with different energy either transmitting or going through the barrier by tunneling. When electrons transmit the channel, the conductance will be changed and is expected to have the quantized value.

Let us express the transmission coefficient in dimensionless form as

$$T_n(\xi) = \left[1 + \frac{n^4 \sin^2\left(\pi\bar{d}\sqrt{\xi^2 - n^2}\right)}{4\xi^2(\xi^2 - n^2)}\right]^{-1}, \tag{34}$$

where $\xi = \sqrt{E}$, $\bar{d} = 2d/w$ is the aspect ratio of the channel. To display the quantization of the conductance, we conduct a combination consisting of all transmission coefficients which represents all electrons transmitting through all potential barriers. This combination is expressed in terms of the total transmission coefficients,

$$T_{Total}^{(N)}(\xi) = \sum_{n=1}^{N} T_n(\xi) = \sum_{n=1}^{N} \left[1 + \frac{n^4 \sin^2\left(\pi\bar{d}\sqrt{\xi^2 - n^2}\right)}{4\xi^2(\xi^2 - n^2)}\right]^{-1}. \tag{35}$$

Figure 9 illustrates the quantization of the total transmission coefficient. Take $N = 2$ as an example, $T_{Total}^{(N)}(\xi)$ is composed of $T_1(\xi)$ and$T_2(\xi)$:

$$T_{Total}^{(2)}(\xi) \approx \begin{cases} 0, & \xi < 1 \\ 1, & 1 \le \xi < 2, \\ 2, & \xi \ge 2 \end{cases} \tag{36}$$

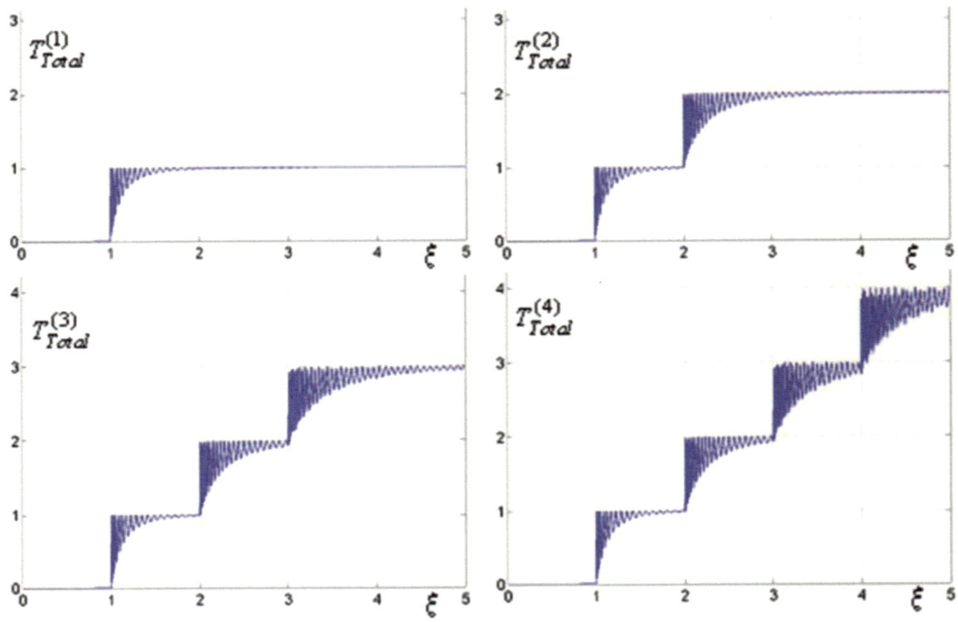

Figure 9.
The total transmission coefficients $T^{(N)}_{Total}(\xi)$ *display the step shape with the increasing of incident energy* ξ *for* $N = 1, 2, 3, 4$ *with* $\bar{d} = 10$. *[39].*

where we have ignored the rapid oscillations parts in the transmission coefficient (more detail refers to [39]). Eq. (36) shows the step structure illustrated in **Figure 9**, which has the same steps shape of the conductance shown in **Figure 8**. We have demonstrated that the total transmission coefficient is proportional to the total number of electrons passing the channel and it is relevant to the conductance in the channel.

5. Concluding remarks

Looking for the unifying theory of quantum and classical mechanics lasts for decades. Several approaches have been proposed, they share some viewpoints and contributions. We have learned that the quantum potential plays a switch role between the quantum and classical world. When the mass is getting larger and larger, the quantum potential will become smaller and smaller, and eventually becomes ignorable. Causality exists everywhere in the universe but hides itself in the microscopic world. What makes physicists miss the link that connects the two scale worlds is the statistical expression of the quantum world. It is impossible to extract the fundamental law from the probability interpretation. As the higher dimension is demanded, there are more evidences of causality emerging from the backbone of quantum mechanics. The complex weak measurement proposes the solid evidence of the complex space structure nature of the quantum world, and evokes the ontology return to the quantum kingdom. All quantum motions happen in complex space. All we can observe is a part of the whole appearance.

In Bohmian mechanics, the quantum potential is a product given by the transformation process which starts from the Schrödinger equation to the quantum HJ equation. In optimal guidance quantum motion formulation, the quantum potential naturally arises in the process of finding the minimum cost function. From the view point of the space geometry, the quantum potential exposits the geometric variation for the particle to lead its motion. This is what makes the quantum world quite different to the classical world as many quantum phenomena reveal. The quantum potential is so charming and plays the most important part that bridges the gap between the quantum and classical world.

Probability is a prescription to deal with the empirical data not to represent the essence of nature in such a small scale. We have demonstrated how to emerge the trajectory from the probability by expanding the dimensions to complex space. As meanwhile, we have pointed out how to reach the classical limit with increasing quantum numbers from the same ensemble of trajectories by adopting different statistical collection method. Take the advantage of the quantum potential, we are allowed to explain the force balanced condition in the hydrogen atom, moreover, we illustrate the formation of the shell structures which cohere with the shape of the electron clouds. The channels in 2D Nano-structure are shown to be related to the quantum potential and so does the conductance. We confirm that the quantized conductance is originated from the electron's transmission behavior. The ontology renders the reality of the identity to the quantum object. It cannot be done without the complex space structure. Complex space is essential for the quantum world and becomes the most crucial part of solving the quantum puzzle. It may proper to say that the causality returns to the quantum world and throughout the whole universe.

Author details

Ciann-Dong Yang[1] and Shiang-Yi Han[2*]

1 Department of Aeronautics and Astronautics, National Cheng Kung University, Tainan, Taiwan (R.O.C.), Republic of China

2 Department of Applied Physics, National University of Kaohsiung, Kaohsiung, Taiwan (R.O.C.), Republic of China

*Address all correspondence to: syhan.taiwan@gmail.com

References

[1] Einstein A. The motion of elements suspended in static liquids as claimed in the molecular kinetic theory of heat. Annals of Physics. 1905;**17**:549-560

[2] Holmes S. The Six Thatchers. Season 4 Episode 1. Available from: https://scatteredquotes.com/call-premonition-just-movement-web/

[3] Kocsis S et al. Observing the average trajectories of single photons in a two-slit interferometer. Science. 2011;**332**:1170-1173. DOI: 10.1126/science.1202218

[4] Murch KW, Weber SJ, Macklin C, Siddiqi I. Observing single quantum trajectories of a superconducting quantum bit. Nature. 2013;**502**:211-214

[5] Rossi M, Mason D, Chen J, Schliesser A. Observing and verifying the quantum trajectory of a mechanical resonator. Physical Review Letters. 2019;**123**:163601

[6] Zhou ZQ, Liu X, et al. Experimental observation of anomalous trajectories of single photons. Physical Review A. 2017;**95**:042121

[7] Aharonov Y, Albert DZ, Vaidman L. How the result of a measurement of a component of the spin of a spin-1/2 particle can turn out to be 100. Physical Review Letters. 1988;**60**:1351-1354

[8] Aharonov Y, Botero A. Quantum averages of weak values. Physical Review A. 2005;**72**:052111

[9] Mori T, Tsutsui I. Quantum trajectories based on the weak value. Progress of Theoretical and Experimental Physics. 2015;**2015**:043A01

[10] Shudo A, Ikeda KS. Complex classical trajectories and chaotic tunneling. Physical Review Letters. 1995;**74**:682-685

[11] Yang CD. Wave-particle duality in complex space. Annals of Physics. 2005;**319**:444-470

[12] Goldfarb Y, Degani I, Tannor DJ. Bohmian mechanics with complex action: A new trajectory-based formulation of quantum mechanics. The Journal of Chemical Physics. 2006;**125**:231103

[13] Chou CC, Wyatt RE. Quantum trajectories in complex space: One-dimensional stationary scattering problems. The Journal of Chemical Physics. 2008;**128**:154106

[14] John MV. Probability and complex quantum trajectories. Annals of Physics. 2008;**324**:220-231

[15] Sanz AS, Miret-Artes. Interplay of causticity and verticality within the complex quantum Hamilton-Jacobi formalism. Chemical Physics Letters. 2008;**458**:239-243

[16] Poirier B, Tannor D. An action principle for complex quantum trajectories. International Journal at the Interface Between Chemistry and Physics. 2012;**110**:897-908. DOI: 10.1080/00268976.2012.681811

[17] Dey S, Fring A. Bohm quantum trajectories from coherent states. Physical Review A. 2013;**88**:022116

[18] Yang CD, Su KC. Reconstructing interference fringes in slit experiments by complex quantum trajectories. International Journal of Quantum Chemistry. 2013;**113**:1253-1263

[19] Mahler DH, Rozema L, et al. Experimental nonlocal and surreal Bohmian trajectories. Science Advances. 2016;**2**:1501466

[20] Procopio LM, Rozema LA, et al. Single-photon test of hyper-complex

quantum theories using a metamaterial. Nature Communications. 2017;**8**:15044

[21] Davidson M. Bohmian trajectories for Kerr-Newman particles in complex space-time. Foundations of Physics. 2018;**11**:1590-1616

[22] Dressel J, Jordan AN. Significance of the imaginary part of the weak value. Physical Review A. 2012;**85**:012107

[23] de Aguiar MAM, Vitiello SA, Grigolo A. An initial value representation for the coherent state propagator with complex trajectories. Chemical Physics. 2010;**370**:42-50

[24] Kedem Y. Using technical noise to increase the signal-to-noise ratio of measurements via imaginary weak values. Physical Review A. 2012;**85**: 060102

[25] Petersen J, Kay KG. Wave packet propagation across barriers by semicalssical initial value methods. The Journal of Chemical Physics. 2015;**143**: 014107

[26] Rosenbrock HH. A stochastic variational treatment of quantum mechanics. Proceedings: Mathematical and Physical Sciences. 1995;**450**:417-437

[27] Wang MS. Stochastic interpretation of quantum mechanics in complex space. Physical Review Letters. 1997;**79**: 3319-3322

[28] Yang CD, Cheng LL. Optimal guidance law in quantum mechanics. Annals of Physics. 2013;**338**:167-185

[29] Bohm D. A suggested interpretation of quantum theory in terms of 'hidden' variables, I and II. Physical Review. 1952;**85**:166-193

[30] Leacock RA, Padgett MJ. Hamilton-Jacobi theory and the quantum action variable. Physical Review Letters. 1983; **50**:3-6

[31] de Castro AS, de Dutra AS. On the quantum Hamilton-Jacobi formalism. Foundations of Physics. 1991;**21**:649-663

[32] John MV. Modified de Broglie-Bohm approach to quantum mechanics. Foundation of Physics Letters. 2002;**15**: 329-343

[33] Yang CD. Quantum Hamilton mechanics: Hamilton equations of quantum motion, origin of quantum operators, and proof of quantization axiom. Annals of Physics. 2006;**321**: 2876-2926

[34] Gondran M. Numerical simulation of the double slit interference with ultracold atom. American Journal of Physics. 2005;**73**:507-515

[35] Floyd ER. Trajectory representation of a quantum Young's diffraction experiment. Foundations of Physics. 2007;**37**:1403-1420

[36] Yang CD, Han SY. Trajectory interpretation of correspondence principle: Solution of nodal issue. 2019. Available from: https://arxiv.org/abs/1911.04747

[37] Yang CD. Quantum dynamics of hydrogen atom in complex space. Annals of Physics. 2005;**319**:399-443

[38] Yang CD, Lee CB. Nonlinear quantum motions in 2D nano-channels part I: Complex potential and quantum trajectories. International Journal of Nonlinear Sciences and Numerical Simulation. 2010;**11**:297-318

[39] Yang CD, Lee CB. Nonlinear quantum motions in 2D nano-channels part II: Quantization and wave motion. International Journal of Nonlinear Sciences and Numerical Simulation. 2010;**11**:319-336

Chapter 6

Entropy in Quantum Mechanics and Applications to Nonequilibrium Thermodynamics

Paul Bracken

Abstract

Classical formulations of the entropy concept and its interpretation are introduced. This is to motivate the definition of the quantum von Neumann entropy. Some general properties of quantum entropy are developed, such as the quantum entropy which always increases. The current state of the area that includes thermodynamics and quantum mechanics is reviewed. This interaction shall be critical for the development of nonequilibrium thermodynamics. The Jarzynski inequality is developed in two separate but related ways. The nature of irreversibility and its role in physics are considered as well. Finally, a specific quantum spin model is defined and is studied in such a way as to illustrate many of the subjects that have appeared.

Keywords: classical, quantum, partition function, temperatures, entropy, irreversible

1. Introduction

The laws of thermodynamics are fundamental to the present understanding of nature [1, 2]. It is not surprising then to find they have a very wide range of applications beyond their original scope, such as to gravitation. The analogy between properties of black holes and thermodynamics could be extended to a complete correspondence, since a black hole in free space had been shown to radiate thermally with a temperature $T = \kappa/2\pi$, where κ is the surface gravity. One should be able to assign an entropy to a black hole given by $S_H = A_H/4$ where A_H is the surface area of the black hole [3]. In the nineteenth century, the problem of reconciling time asymmetric behavior with time symmetric microscopic dynamics became a central issue in this area of physics [4]. Lord Kelvin wrote about the subjection of physical phenomenon to microscopic dynamical law. If then the motion of every particle of matter in the universe were precisely reversed at any instant, the course of nature would be simply reversed for ever after [5]. Physical processes, on the other hand, are irreversible, such as conduction of heat and diffusion processes [6, 7]. It subsequently became apparent that not only is there no conflict between reversible microscopic laws and irreversible microscopic behavior, but there are extremely strong reasons to expect the latter from the former. There are many reasons; for example, there exists a great disparity between microscopic and macroscopic scales and the fact that the events we observe in the macroworld are determined not only by the microscopic dynamics but also by the initial conditions or state of the system.

In the twentieth century, it became clear that the microworld was described by a different kind of physics along with mathematical ideas that need not be taken into account in describing the macroworld. This is the subject of quantum mechanics. Even though the new quantum equations have similar symmetry properties as their classical counterparts, it also reveals numerous phenomena that can contribute at this level to the problems mentioned above. These physical phenomena which play various roles include the phenomenon of quantum entanglement, the effect of decoherence in general, and the theory of measurements as well.

The purpose of this is to study the subject of entropy as it applies to quantum mechanics [8, 9]. Its definition is to be relevant to very small systems at the atomic and molecular level. Its relationship to entropies known at other scales can be examined. It is also important to relate this information from this new area of physics to the older and more established theories of thermodynamics and statistical physics [10–15]. To summarize, many good reasons dictate that the arrow of time is specified by the direction of increase of the Boltzmann entropy, the von Neumann macroscopic entropy. To relate the quantum Boltzmann approach to irreversibility to measurement theory, the measuring apparatus must be included as a part of the closed quantum mechanical system.

2. Entropy and quantum mechanics

Boltzmann's great insight was to connect the second law of thermodynamics with phase space volume. This he did by making the observation that for a dilute gas, $\log|\Gamma_M|$ is proportional up to terms negligible compared to the system size, to the thermodynamic entropy of Clausius. He then extended his insight about the relation between thermodynamic entropy and $\log|\Gamma_M|$ to all macroscopic systems, no matter what their composition. This gave a macroscopic definition of the observationally measureable entropy of equilibrium macroscopic systems. With this connection established, he generalized it to define an entropy for systems not in equilibrium.

Clearly, the macrostate $M(\mathbf{x})$ is determined by $\mathbf{x}$, a point in phase space, and there are many such points, in fact a continuum, which correspond to the same M. Let Γ_M then be the region in Γ consisting of all microstates $\mathbf{x}$ corresponding to a given macrostate M. Boltzmann associated with each microstate $\mathbf{x}$ of a macroscopic system M a number S_B, which depends only on $M(\mathbf{x})$, such that up to multiplicative and additive constants is given by

$$S(\mathbf{x}) = S_B(M(\mathbf{x})) = k_B \log|\Gamma_M|. \tag{1}$$

This S is called the Boltzmann entropy of a classical system. The constant $k_B = 1.38 \cdot 10^{-16}$ erg/K is called Boltzmann's constant, and if temperature is measured in ergs instead of Kelvin, it may be set to one. Boltzmann argued that due to large differences in the sizes of Γ_M, $S_B(\mathbf{x}_t)$ will typically increase in a way which explains and describes the evolution of physical systems towards equilibrium.

The approach of Gibbs, which concentrates primarily on probability distributions or ensembles, is conceptually different from Boltzmann's. The entropy of Gibbs for a microstate $\mathbf{x}$ of a macroscopic system is defined for an ensemble density $\rho(\mathbf{x})$ to be

$$S_G(\rho) = -k_B \int_{\Gamma} \rho(\mathbf{x}) \log(\rho(\mathbf{x}))\, d\mathbf{x}. \tag{2}$$

In (2), $\rho(\mathbf{x})$ is the probability for the microscopic state of the system to be found in the phase space volume element $d\mathbf{x}$. Suppose $\rho(\mathbf{x})$ is taken to be the generalized microcanonical ensemble associated with a macrostate M

$$\rho_M(\mathbf{x}) = \begin{cases} |\Gamma_M|^{-1}, & \mathbf{x} \in \Gamma; \\ 0, & \text{otherwise}. \end{cases} \tag{3}$$

Then clearly

$$S_G(\rho_M) = k_B \log|\Gamma_M| = S_B(M). \tag{4}$$

The probability density for the system in the equilibrium macrostate $\rho_{M_{eq}}$ is the same as that for the microcanonical and equivalent to the canonical or grandcanonical ensemble when the system is of macroscopic size. The time development of S_B and S_G subsequent to some initial time when $\rho = \rho_M$ is very different unless $M = M_{eq}$ when there is no further systematic change in M or ρ. In fact, $S_G(\rho)$ never changes in time as long as $\mathbf{x}$ evolves according to Hamiltonian evolution, so ρ evolves according to the Liouville equation. Then S_G does not give any indication that the system is evolving towards equilibrium. Thus the relevant entropy for understanding the time evolution of macroscopic systems is S_B and not S_G.

From the standpoint of mathematics, these expressions for classical entropies can be unified under the heading of the Boltzmann-Shannon-Gibbs entropy [16]. A very general form of entropy which includes those mentioned can be defined in a mathematically rigorous way. To do so, let (Ω, A, μ) be a finite measure space, ν a probability measure that is absolutely continuous with respect to μ, and its Radon-Nikodym derivative $d\nu/d\mu$ exists. The generalized BSG entropy is defined to be

$$S_{BSG} = \int \frac{d\nu}{d\mu} \cdot \log\left(\frac{d\nu}{d\mu}\right) d\nu, \tag{5}$$

when the integrand is integrable.

This includes the classical Boltzmann-Gibbs entropy when $d\mu$ and $d\nu$ are given by

$$d\mu = \frac{d^{3N}p\, d^{3N}q}{\hbar^{3N}}, \qquad d\nu = \rho^{cl}\, d\mu. \tag{6}$$

It also includes the Shannon entropy appearing in information theory in which

$$\Omega = \{1, 2, \ldots\}, \quad \mu(\{1\}) = \mu(\{2\}) = \ldots = 1, \quad \nu(\{i\}) = 1. \tag{7}$$

In this case, (5) gives the entropy to be

$$S = -\sum_i \rho_i \log(\rho_i). \tag{8}$$

In attempting to translate these considerations to the quantum domain, it is immediately clear that a perfect analogy does not exist.

Although the situation is in many ways similar in quantum mechanics, it is not identical. The irreversible incompressible flow in phase space is replaced by the unitary evolution of wave functions in Hilbert space and velocity reversal of x by complex conjugation of the wave function. The analogue of the Gibbs entropy (2) of an ensemble is the von Neumann entropy of a density matrix ρ:

$$S_{vN}(\rho) = -k_B \operatorname{Tr}(\rho \log \rho). \tag{9}$$

This formula was given by von Neumann. It generalizes the classical expression of Boltzmann and Gibbs to the realm of quantum mechanics. The density matrix with maximal entropy is the Gibbs state. The range of S_{vN} is the whole of the extended real line $[0, \infty]$, so to every number ζ with $0 < \zeta \leq \infty$, there is a density matrix ρ such that $S_{vN}(\rho) = \zeta$. Like the classical $S_G(\rho)$, this does not alter in time for an isolated system evolving under Schrödinger evolution. It has value zero whenever ρ represents a pure stare. Similar to $S_G(\rho)$, it is not most appropriate for describing the time symmetric behavior of isolated macroscopic systems. The Szilard engine composed of an atom is an example in which the entropy of a quantum object is made use of. von Neumann discusses the macroscopic entropy of a system, so a macrostate is described by specifying values of a set of commuting macroscopic observable operators $\hat{A}$, such as particle number, energy, and so forth, to each of the cells that make up the system corresponding to the eigenvalues a_α, an orthogonal decomposition of the system's Hilbert space $\mathcal{H}$ into linear subspaces $\hat{\Gamma}_\alpha$ in which the observables $\hat{A}$ take the values a_α. Let Π_α the projection into $\hat{\Gamma}_\alpha$. von Neumann then defines the macroscopic entropy of a system with density matrix $\tilde{\rho}$ as

$$\tilde{S}_{mac}(\tilde{\rho}) = k_B \sum_{\alpha=1}^{N} p_\alpha(\tilde{\rho}) \log |\hat{\Gamma}_\alpha| - k_B \sum_{\alpha=1}^{N} p_\alpha(\tilde{\rho}) \log p_\alpha(\tilde{\rho}). \tag{10}$$

Here, $p_\alpha(\tilde{\rho})$ is the probability of finding the system with density matrix $\tilde{\rho}$ in the microstate M_α

$$p_\alpha(\tilde{\rho}) = Tr(\Pi_\alpha \tilde{\rho}), \tag{11}$$

and $|\hat{\Gamma}_\alpha|$ is the dimension of $\hat{\Gamma}_\alpha$. An analogous definition is made for a system which is represented by a wave function Ψ; simply replace $p_\alpha(\rho)$ by $p_\alpha(\Psi) = \langle \Psi, \Pi_\alpha \Psi \rangle$. In fact, $|\Psi\rangle\langle\Psi|$ just corresponds to a particular pure density matrix.

von Neumann justifies (10) by noting that

$$\tilde{S}_{mac}(\rho) = -k_B \operatorname{Tr}[\tilde{\rho} \log \tilde{\rho}] = S_{vN}(\tilde{\rho}), \tag{12}$$

for

$$\tilde{\rho} = \sum_\alpha \frac{p_\alpha}{|\hat{\Gamma}_\alpha|} \Pi_\alpha, \tag{13}$$

and $\tilde{\rho}$ is macroscopically indistinguishable from ρ.

A correspondence can be made between the partitioning of classical phase space Γ and the decomposition of Hilbert space $\mathcal{H}$ and to define the natural quantum analogues to Boltzmann's definition of $S_B(M)$ in (1) as

$$\hat{S}_B(M_\alpha) = k_B \log |\hat{\Gamma}_{M_\alpha}| \tag{14}$$

where $|\hat{\Gamma}_{M_\alpha}|$ is the dimension of $\hat{\Gamma}_{M_\alpha}$. With definition (14) the first term on the right of (10) is just what would be stated for the expected value of the entropy of a classical system of whose macrostate we were unsure. The second part of (10) will be negligible compared to the first term for a macroscopic system, classical or quantum, and going to zero when divided by the number of particles.

Note the difference that in the classical case, the state of the system is described by $\mathbf{x} \in \Gamma_\alpha$ for some α, so the system is always in one of the macrostates M_α. For a quantum system described by ρ or Ψ, this is not the case. There is no analogue of (1) for general ρ or Ψ. Even when the system is in a macrostate corresponding to a definite microstate at t_0, only the classical system will be in a unique macrostate at time t. The quantum system will in general evolve into a superposition of different macrostates, as is the case in the Schrödinger Cat paradox. In this wave function, Ψ corresponding to a particular macrostate evolves into a linear combination of wave functions associated with very different macrostates. The classical limit is obtained by a prescription in which the density matrix is identified with a probability distribution in phase space and the trace is replaced by integration over phase space. The superposition principle excludes partitions of the Hilbert space: an orthogonal decomposition is all that is relevant.

2.1 Properties of entropy functions

Entropy functions have a number of characteristic properties which should be briefly described in the quantum case. The set of observables will be the bounded, self-adjoint operators with discrete spectra in a Hilbert space. The set of normal states can be taken to be the density operators or positive operators of trace one.

The entropy functional satisfies the following inequalities. Let $\lambda_i > 0$ and $\sum_i \lambda_i = 1$. Then S has the concavity property:

$$S\left(\sum_i \lambda_i \rho_i\right) \geq \sum_i \lambda_i S(\rho_i), \tag{15}$$

with equality if all λ_i are equal.

Subadditivity holds with equality if and only if $\rho_i \rho_j = 0, i \neq j$

$$S\left(\sum_i \lambda_i \rho_i\right) \leq \sum_i \lambda_i S(\rho_i) - \sum_i \lambda_i \log \lambda_i. \tag{16}$$

and

$$S\left(\sum_i \lambda_i \rho_i\right) \leq S(T_B \rho) \leq S(\rho) - \sum_k p_k \log p_k \tag{17}$$

where the first equality holds iff $T_B \rho = \rho$ and the second iff $S(\rho_k) = S(\rho)$ for all k. The conditional entropy is defined to be

$$S(\rho_1 | \rho_2) = Tr(\rho_1 \log \rho_1 - \rho_1 \log \rho_2). \tag{18}$$

The formal expression will be interpreted as follows. If A, B are positive traceless operators with complete orthonormal sets of eigenstates $|a_i\rangle$ and $|b_i\rangle$, using a resolution of identity, $\sum_i \langle a_i | A \log A \, | a_i \rangle = \sum_{i,j} \langle a_i | A | b_j \rangle \langle b_j | \log A | a_i \rangle = \sum_{i,j} a_i \langle a_i | b_j \rangle \log a_i \langle b_j | a_i \rangle$ so that

$$\sum_j \langle b_j | A \log A - A \log B + B - A | b_j \rangle = \sum_j \langle a_i | A \log A - A \log B + B - A | a_i \rangle$$
$$= \sum_{i,j} |\langle a_i | b_i \rangle|^2 (a_i \log a_i - a_i \log b_j + b_j - a_i) = S(A|B). \tag{19}$$

Concavity of the function $x \log x$ ensures the terms of the final sum are nonnegative. In order that $S(\rho_1|\rho_2) < \infty$, it is necessary that $\pi_{\rho_1} \leq \pi_{\rho_2}$ where $\pi_W = \operatorname{supp} W$ is the support projection of W, so $\rho_1 < \rho_2$. From the definition, $S(\rho_1|\rho_2) \geq 0$ with equality if $\rho_1 = \rho_2$. If $\lambda\rho_1 \leq \rho_2$, for some $\lambda \in (0,1)$, $S(\rho_1|\rho_2) \leq -\log \lambda$ from operator monotony of $\log z$. If $\rho = \sum_i \lambda_i \rho_i$, then

$$S(\rho) = \sum_i \lambda_i S(\rho_i) + \sum_i \lambda_i S(\rho_i|\rho), \tag{20}$$

which gives (15) and (16). If T is a trace-preserving operator, then $\rho < T\rho$, and

$$S(T\rho) = S(\rho) + S(\rho|T\rho). \tag{21}$$

This is to say that T is entropy-increasing.

The concept of irreversibility is clearly going to be relevant to the subject at hand, so some thoughts related to it will be given periodically in what follows. A possible way to account for irreversibility in a closed system in nature is by the various types of course-graining. There are also strong reasons to suggest the arrow of time is provided by the direction of increase of the quantum form of the Boltzmann entropy. The measuring apparatus should be included as part of the closed quantum mechanical system in order to relate the quantum Boltzmann approach to irreversibility to the concept of a measurement. Let S_c be a composite system consisting of a macroscopic system S coupled to a measuring instrument $\mathcal{I}$, so $S_c = S + \mathcal{I}$, where $\mathcal{I}$ is a large but finite N-particle system. A set of course-grained mutually commuting extrinsic variables are provided whose eigenspaces correspond to the pointer positions of $\mathcal{I}$. von Neumann's picture of the measurement process is basic to the approach, but according to which, the coupling of S to $\mathcal{I}$ leads to the following effects. A pure state of S described by a linear combination $\sum_\alpha c_\alpha \psi_\alpha$ of its orthonormal energy eigenstates is converted into a statistical mixture of these states for which $|c_\alpha|^2$ is the probability of finding the system in state ψ_α. It also sends a certain set of classical or intercommuting, macroscopic variables $\mathcal{M}$ of $\mathcal{I}$ to values indicated by pointer readings that indicate which of the states is realized.

There is an amplification process of the $S - \mathcal{I}$ coupling where different microstates of S give rise to macroscopically different states of $\mathcal{I}$. If $\mathcal{I}$ is designed to have readings which are in one-to-one correspondence with the eigenstates of S, it may be assumed index α of its microstates goes from 1 to n. Denote the projection operator for subspace $\mathcal{K}$ by Π_α, then

$$\Pi_\alpha \Pi_\beta = \Pi_\alpha \delta_{\alpha\beta}, \qquad \sum_\alpha \Pi_\alpha = 1_{\mathcal{K}_\alpha}, \tag{22}$$

and each element of the abelian subalgebra of $\mathscr{B}$ takes the form with M_α scalars

$$M = \sum_\alpha M_\alpha \Pi_\alpha. \tag{23}$$

Define the projection operators:

$$\pi_\alpha = 1 \otimes \Pi_\alpha, \qquad \alpha = 1, \dots, n. \tag{24}$$

Suppose A is measured on system S, initially in a state of the composite system described by a density matrix ρ. The value p_α is obtained with probability $\tau_\alpha = \mathrm{Tr}\,(\rho\pi_\alpha)$. After the measurement, the state of the composite system is accounted for by the density matrix:

$$\rho_\alpha = \frac{1}{\tau_\alpha}\pi_\alpha\rho\pi_\alpha. \tag{25}$$

This is a mixture of states in each of which A has a definite value.

The transformation $\rho \to \tilde{\rho} = \sum_\alpha \pi_\alpha\rho\pi_\alpha$ may be viewed as a loss of information contained in non-diagonal terms $\psi_\alpha\rho\pi_{\alpha'}$ with $\alpha \neq \beta$ in $\sum_{\alpha\alpha'} \pi_\alpha\rho\pi_{\alpha'}$. When a sequence of measurements is carried out and a time evolution is permitted to occur between measurements leads one to assign to a sequence of events $\pi_{\alpha_1}(t_1)\pi_{\alpha_2}(t_2)\cdots\pi_{\alpha_n}(t_n)$ the probability distribution:

$$\mathcal{P}(\alpha) = \mathrm{Tr}\,\left(\pi_{\alpha_1}(t_n)\cdots\pi_{\alpha_1}(t_1)\rho\pi_{\alpha_1}(t_1)\cdots\pi_{\alpha_n}(t_n)\right), \tag{26}$$

where $\rho = \rho(0)$, over the set of histories, where the π_k satisfy (22) with Π replaced by the π. Let us define

$$D(\alpha', \alpha) = Tr\,\left(\pi_{\alpha_1}(t_1)\cdots\pi_{\alpha_n}(t_n)\rho\pi_{\alpha_n}(t_n)\cdots\pi_{\alpha_1}(t_1)\right). \tag{27}$$

The following definition can now be stated. A history is said to *decohere* if and only if

$$D(\alpha, \alpha') = \delta_{\alpha,\alpha'}\rho_\alpha. \tag{28}$$

A state is called decoherent with respect to the set of π_α if and only if

$$\pi_\alpha\rho(0)\pi_\beta = 0, \qquad \alpha \neq \beta. \tag{29}$$

This implies that $Tr\,(\pi_{\alpha'}\rho\pi_\alpha A) = 0$ for all $\alpha \neq \alpha'$, which is equivalent to $[\pi_\alpha, \rho] = 0$ for all α. In contrast to infinite systems where there is no need to refer to a choice of projections, decoherent mixed states over the macroscopic observables can be described by relations between the density matrix and the projectors. They would be of the form $\rho_m = |\Psi\rangle\langle\Psi|$ with $|\Psi\rangle = \sum_\alpha \lambda_\alpha\pi_\alpha\Phi_\alpha$ such that $\sum_\alpha |\lambda_\alpha|^2 = 1$ and $\Phi_\alpha \in \mathcal{H}$ and satisfy

$$\sum_{\alpha\neq\alpha'}(\pi_{\alpha'}\rho_m\pi_\alpha + \pi_\alpha\rho_m\pi_{\alpha'}) \neq 0. \tag{30}$$

The relative or conditional entropy between two states $S(\rho_1|\rho_2)$ was defined in (18), and it plays a crucial role. It is worth stating a few of its properties, as some are necessary for the theorem:

$$S(\rho_1|\rho_2) \geq 0. \tag{31}$$

$$S(\rho_1|\rho_2) = 0, \qquad \rho_1 = \rho_2. \tag{32}$$

$$S(\lambda\rho_1 + (1-\lambda)\rho_2|\lambda\sigma_1 + (1-\lambda)\sigma_2) \leq \lambda S(\rho_1|\sigma_1) + (1-\lambda)S(\rho_2|\sigma_2). \tag{33}$$

When γ is a completely positive map, or embedding

$$S(\rho_1 \cdot \gamma | \rho_2 \cdot \gamma) \leq S(\rho_1 | \rho_2). \tag{34}$$

The last two inequalities are known as joint concavity and monotonicity of the relative entropy. The following result may be thought of as a quantum version of the second law.

Theorem: Suppose the initial density matrix is decoherent at zero time (29) with respect to π_α and have finite entropy

$$\rho(0) = \sum_\alpha \pi_\alpha \rho(0) \pi_\alpha, \\ S(\rho(0)) = -k_B \mathrm{Tr}\,(\rho(0) \log(\rho(0))) < \infty, \tag{35}$$

and it is not an equilibrium state of the system. Let $\rho(t_f)$, for $t_f > 0$, be any subsequent state of the system, possibly an equilibrium state. Then for an automorphic, unitary time evolution of the system between $0 \leq t \leq t_f$

$$S(0) \leq S(t_f), \tag{36}$$

where $S(0) = S(t_f)$ if and only if **(e)** $\sum_{\alpha<\beta} \pi_\alpha \rho(t_f) \pi_\beta + \pi_\beta \rho(t_f) \pi_\alpha = 0$.

Proof: Set $\rho'(t_f) = \sum_\alpha \pi_\alpha \rho(t_f) \pi_\alpha = \rho(t_f) \cdot \gamma$, so ρ' is obtained from ρ by means of a completely positive map. It follows that

$$S(\rho'(t_f)|\rho'(0)) = -S(\rho'(t_f)) - k_B \sum_\alpha \mathrm{Tr}\,(\rho(t_f)\pi_\alpha \log(\rho(0))\pi_\alpha) \\ = -S(t_f) - k_B \mathrm{Tr}\,(\rho(t_f) \log(\rho(0)) \leq S(\rho(t_f)|\rho(0)) = -S(\rho(0)) - \mathrm{Tr}\,(\rho(t_f) \log(\rho(0))). \tag{37}$$

The first equality uses the cyclic property of the trace and the definition of ρ'. The second equality uses decoherence of $\rho(0)$, and the next inequality is a consequence of (34). The evolution is unitary and hence preserves entropy which is the last equality. This implies that $S(t) \geq S(0)$ and the equality condition **(e)** follows from (32). □

Of course, entropy growth as in the theorem is not necessarily monotonic in the time variable. For this reason, it is usual to refer to fixed initial and final states. For thermal systems, a natural choice of the final state is the equilibrium state of the system. It is the case in thermodynamics that irreversibility is manifested as a monotonic increase in the entropy. Thermodynamic entropy, it is thought, is related to the entropy of the states defined in both classical and quantum theory. Under an automorphic time evolution, the entropy is conserved. One application of an environment is to account for an increase. A type of course-graining becomes necessary together with the right conditions on the initial state to account for the arrow of time. In quantum mechanics, the course-graining seems to be necessary and may be thought of as a restriction of the algebra and can also be interpreted as leaving out unobservable quantum correlations. This may, for example, correspond to decoherence effects important in quantum measurements. Competing effects arise such as the fact that correlations becoming unobservable may lead to entropy increase. There is also the effect that a decrease in entropy might be due to nonautomorphic processes. Although both effects lead to irreversibility, they are not cooperative but rather contrary to one another. The observation that the second law does hold implies these nonautomorphic events must be rare in comparison with time scales relevant to thermodynamics.

3. Quantum mechanics and nonequilibrium thermodynamics

Some aspects of equilibrium thermodynamics are examined by considering an isothermal process. Since it is a quasistatic process, it may be decomposed into a sequence of infinitesimal processes. Assume initially the system has a Hamiltonian $H(\gamma)$ in thermal equilibrium at a temperature T. Boltzmann's constant is set to one. The state is given by the Gibbs density operator ρ. This expression can also be written in terms of the energy eigenvalues ε_n and eigenvectors $|n\rangle$ of H. The probability of finding the system in state $|n\rangle$ is

$$p_n = \langle n|\rho|n\rangle = \frac{e^{-\beta\varepsilon_n}}{Z}. \tag{38}$$

The average external energy U of the system is given as

$$U = \langle U\rangle = \mathrm{Tr}\,(H\rho) = \sum_n \varepsilon_n p_n. \tag{39}$$

When the parameter γ is changed to $\gamma + d\gamma$, both ε_n and p_n as well as U change to

$$dU = \sum_n \left[d\varepsilon_n p_n + \varepsilon_n dp_n\right]. \tag{40}$$

Each instantaneous infinitesimal process can be broken down into a part which is the work performed; the second is the heat transformed as the system relaxes to equilibrium. This breakup motivates us to define

$$\delta W = \sum_n (d\varepsilon_n)p_n, \qquad \delta Q = \sum_n \varepsilon_n dp_n, \tag{41}$$

so $dU = \delta Q + \delta W$, and δ is used to indicate that heat and work are not exact differentials. The free energy of the system is defined to be $F = -T\,\log Z$, so $dF = \sum_n (d\varepsilon_n)p_n$ which means

$$\delta W = dF. \tag{42}$$

By integrating over the infinitesimal segments, we find W is

$$W = \Delta F = \Delta U - Q. \tag{43}$$

Inverting Eq. (38) for p_n, we can solve for

$$\varepsilon_n = -T\,\log\,(Zp_n). \tag{44}$$

Substituting into the relation for δQ, we get two terms, one proportional to $\log(Z)$ and the other to $\log(p_n)$. The term with $\log(Z)$ when the p_k satisfy $\sum_k p_k = 1$ is

$$-T\sum_n \log(Z)dp_n = -T\log(Z)d\left(\sum_n p_n\right) = 0, \tag{45}$$

It remains to study

$$\delta Q = -T\sum_n dp_n \log(p_n). \tag{46}$$

By the chain rule

$$d\left(\sum_n p_n \log(p_n)\right) = \sum_n dp_n \log(p_n) + \sum_n dp_n = \sum_n dp_n \log(p_n). \tag{47}$$

So δQ is not a function of the state but is related to the variation of something that is. Define the entropy S as usual from (9), $S = -\sum_n p_n \log(p_n)$, and arrive at

$$\delta Q = T dS. \tag{48}$$

This relation only holds for infinitesimal processes. For finite and irreversible processes, there may be additional terms to the entropy change. This has been quite successful at describing many different types of physical system [17–19].

A deep insight has come recently into the properties of nonequilibrium thermodynamics which could be achieved by regarding work as a random variable. For example, consider a process in which a piston is used to compress a gas in a cylinder. Due to the nature of the gas and its chaotic motion, each time the piston is pressed, the gas molecules exert a back reaction with a different force. This means the work needed to achieve a given compression changes each time something is carried out.

Usually a knowledge of nonequilibrium processes is restricted to inequalities such as the Jarzynski inequality. He was able to show by interpreting work W as a random variable that an inequality can be obtained, even for a process performed arbitrarily far from equilibrium.

Suppose the system is always prepared in the same state initially. A process is carried out and the total work W performed is measured. Repeating this many times, a probability distribution for the work $\mathcal{P}(W)$ can be constructed. An average for W can be computed using $\mathcal{P}(W)$ as

$$\langle W \rangle = \int \mathcal{P}(W) dW. \tag{49}$$

Jarzynski showed that the statistical average of $e^{-\beta W}$ satisfies

$$\left\langle e^{-\beta W} \right\rangle = e^{-\beta \Delta F}, \tag{50}$$

where $\Delta F = F\left(T, \gamma_f\right) - F(T, \gamma_i)$. It holds for a process performed arbitrarily far from equilibrium. Now the inequality $W \geq \Delta F$ is contained in (50) and can be realized by applying Jensen's inequality, which states that $\left\langle e^{-\beta W} \right\rangle \geq e^{-\beta \langle W \rangle}$.

In macroscopic systems, individual measurements are usually very close to the average by the law of large numbers. For mictoscopic systems, this is usually not true. In fact, the individual realizations of W may be smaller than ΔF. These cases would be local violations of the second law but for large systems become extremely rare. If the function $\mathcal{P}(W)$ is known, the probability of a local violation of the second law is

$$\mathcal{P}(W < \Delta F) = \int_{-\infty}^{\Delta F} \mathcal{P}(W)\, dW. \tag{51}$$

To get (50) requires detailed knowledge of the system's dynamics, be it classical, quantum, unitary, or whatever.

Consider nonunitary quantum dynamics. Initially, the system has Hamiltonian $H_i = H(\gamma_i)$. The system was in thermal equilibrium with a bath at temperature T. The initial state of the system is the Gibbs thermal density matrix (38). Let ε_n^i and $|n\rangle$ denote the initial eigenvalues and eigenvectors of H_i as ε_n^i is obtained with probability $p_n = e^{-\beta \varepsilon_n^i}/Z$.

Immediately after this measurement, γ changes from $\gamma(0) = \gamma_i$ to $\gamma(\tau) = \gamma_f$ according to the rule $\gamma(t)$. If it is assumed the contact with the bath is very weak during this process, the state of the system evolves according to

$$|\psi(t)\rangle = U(t)\,|n\rangle, \tag{52}$$

where U is the unitary evolution operator which satisfies Schrödinger's equation, $i\partial_t U = H(t)U$, $U(0) = \mathbf{1}$.

The Hamiltonian is $H\left(\gamma_f\right)$ at the end and has energy levels ε_m^f, eigenvectors $|m\rangle$, so the probability ε_n^f measured is $|\langle m|\psi(\tau)\rangle|^2 = |\langle m|U(\tau)|n\rangle|^2$. This may be interpreted as the conditional probability a system in $|n\rangle$ will be in $|m\rangle$ after time τ.

No heat has been exchanged with the environment, so any change in the environment has to be attributed to the work performed by the external agent and is

$$W = \varepsilon_m^f - \varepsilon_n^i, \tag{53}$$

where both ε_n^i and ε_m^f are fluctuating and change during each realization of the experiment. The first ε_n^i is random due to thermal fluctuations and ε_m^f is random due to quantum fluctuations in W as a random variable by (53).

To get an expression for $\mathcal{P}(W)$ obtained by repeating the process several times, this is a two-step measurement process. From probability theory, if A, B are two events, the total probability $p(A|B)$ that both events have occurred is

$$p(A,B) = p(A|B)p(B), \tag{54}$$

where $p(B)$ is the probability B which occurs and $p(A|B)$ is the conditional probability B that has occurred. The probability of both events that have occurred is $|\langle m|U(\tau)|n\rangle|^2 p_n$. Since we are interested in the work performed, we write

$$\mathcal{P}(W) = \sum_{n,m} |\langle m|U(\tau)|n\rangle|^2 p_n\, \delta\left(W - \left(\varepsilon_m^f - \varepsilon_n^i\right)\right). \tag{55}$$

And some over all allowed events, weighted by their probabilities, and arrange the terms according to the values $\varepsilon_m^f - \varepsilon_n^i$. In most systems, there are present a rather large number of allowed levels, and even more allowed differences $\varepsilon_m^f - \varepsilon_n^i$. It is more efficient to use the Fourier transform

$$\mathcal{G}(y) = \left\langle e^{iyW}\right\rangle = \int_{-\infty}^{\infty} \mathcal{P}(W) e^{iyW}\, dW. \tag{56}$$

This has the inverse Fourier transform

$$\mathcal{P}(W) = \frac{1}{2\pi}\int_{-\infty}^{\infty} dy\, \mathcal{G}(y) e^{-iyW}. \tag{57}$$

Using (55), we obtain that

$$\mathcal{G}(y) = \sum_{n,m} |\langle m|U|n\rangle|^2 p_n e^{iy(\varepsilon^f_m - \varepsilon^i_n)} = \sum_{n,m} \left\langle n|U^\dagger e^{iy\varepsilon^f_m}|m\right\rangle \left\langle m|Ue^{-iy\varepsilon^i_n} p_n|n\right\rangle$$
$$= \sum_{n,m} \langle n|U^\dagger e^{iyH_f}|m\rangle\langle m|Ue^{-iyH_i}\rho|n\rangle = \mathrm{Tr}\left(U^\dagger(\tau)e^{iyH_f}U(\tau)e^{-iyH_i}\rho\right). \tag{58}$$

Hence, it may be concluded that

$$\mathcal{G} = Tr\left(U^\dagger(\tau)e^{iyH_f}U(\tau)e^{-iyH_i}\rho\right). \tag{59}$$

This turns out to be somewhat easier to work with than $\mathcal{P}(W)$, and (59) plays a similar role as Z in equilibrium statistical mechanics. From $G(y)$, the statistical moments of W can be found by expanding

$$\mathcal{G}(y) = \left\langle e^{iyW}\right\rangle = 1 + iy\langle W\rangle - \frac{y^2}{2}\left\langle W^2\right\rangle - \frac{y^3}{6}\left\langle W^3\right\rangle + \cdots. \tag{60}$$

A formula for the quantum mechanical formula for the moments can be found as well. The average work is $\langle W\rangle = \langle H_f\rangle - \langle H_i\rangle$, where for any operator A, we have $\langle A\rangle_t = Tr\left(U^\dagger(t)AU(t)\rho\right)$ as the expectation value of A at time t. This follows from the fact that the state of the system at t is $\rho(t) = U(t)\rho U(t)^\dagger$. From the definition of $\mathcal{G}$, it ought to be the case that $G(y = i\beta) = \langle e^{-\beta W}\rangle$. However, ρ in (38) and (59) yields

$$\mathcal{G}(i\beta) = \frac{1}{Z_i}\mathrm{Tr}\left(U^\dagger e^{-\beta H_f}U\right) = \frac{1}{Z_i}\mathrm{Tr}\left(e^{\beta H_f}\right) = \frac{Z_f}{Z_i}. \tag{61}$$

Using $Z = e^{-\beta F}$, (61) yields (50)

$$\mathcal{G}(iy) = \left\langle e^{-\beta W}\right\rangle = e^{-\beta\Delta F}. \tag{62}$$

Nothing has been assumed about the speed of this process. Thus inequality (50) must hold for a process arbitrarily far from equilibrium.

4. Heat flow from environment approach

There is another somewhat different way in which the Jarzynski inequality can be generalized to quantum dynamics. In a classical system, the energy of the system can be continuously measured as well as the flow of heat and work. Continuous measurement is not possible in quantum mechanics without disrupting the dynamics of the system [20].

A more satisfactory approach is to realize that although work cannot be continuously measured, the heat flow from the environment can be measured. To this end, the system of interest is divided into a system of interest and a thermal bath. The ambient environment is large, and it rapidly decoheres and remains at thermal equilibrium, uncorrelated and unentangled with the system. Consequently, we can measure the change in energy of the bath $(-Q)$ without disturbing the dynamics of the system. The open-system Jarzynski identity is expressed as

$$\left\langle e^{-\beta W}\right\rangle = \left\langle e^{-\beta E_f}e^{\beta Q}e^{\beta E_i}\right\rangle = e^{-\beta\Delta F}. \tag{63}$$

For a system that has equilibrated with Hamiltonian H interacting with a thermal bath at temperature T, the equilibrium density matrix is $\rho = e^{\beta H}/Z = e^{-\beta F - \beta H}$, where $\beta = 1/K_BT$. The dynamics of an open quantum system is described by a quantum operator $\tilde{\rho} = S\rho$, a linear trace-preserving, complete positive map of operators. Any such complete positive superoperator has an operator-sum representation

$$S\rho = \sum_{\alpha} A_\alpha \rho A_\alpha^\dagger. \tag{64}$$

Conversely, any operator-sum represents a complete positive superoperator. The set of operators $\{A_\alpha\}$ is often called Krauss operators. The superoperator is trace-preserving and conserves probability if $\sum_\alpha A_\alpha^\dagger A_\alpha = \mathbf{I}$. In the simplest case, the dynamics of an isolated quantum system is described by a single unitary operator $U^\dagger = U^{-1}$.

The interest here is in the dynamics of a quantum system governed by a time-dependent Hamiltonian weakly coupled to an extended, thermal environment. Let the total Hamiltonian be

$$H = H^S(t) \otimes I^B + \mathbf{I}^S \otimes H^B + \varepsilon H^{int}, \tag{65}$$

where $\mathbf{I}^S$ and $\mathbf{I}^B$ are system and bath identity operators, $H^S(t)$ the system Hamiltonian, H^B the bath Hamiltonian, and H^{int} the bath-system interaction with ε a small parameter. Assume initially the system and environment are uncorrelated such that the initial combined state is $\rho^S \otimes \rho^B_{eq}$, where ρ^B_{eq} is the thermal density equilibrium matrix of the bath.

By following the unitary dynamics of the combined total system for a finite time and measuring the final state of the environment, a quantum operator description of the system dynamics can also be obtained:

$$\begin{aligned} S(s,t)\rho^S &= \mathrm{Tr}_B U\left(\rho^S \otimes \rho^B_{eq}\right)U^\dagger = \sum_{i,f} \langle b_f|U(\rho^S \otimes (\sum_i \frac{e^{-\beta\varepsilon_i^B}}{Z^B}|b_i\rangle\langle b_i|))U^\dagger|b_f\rangle \\ &= \frac{1}{Z_B}\sum_{i,f} e^{-\beta\varepsilon_i^B}\langle b_f|U|b_i\rangle\rho^S\langle b_i|U^\dagger|b_f\rangle. \end{aligned} \tag{66}$$

Here U is the unitary evolution operator of the total system

$$U = \exp\left(\frac{i}{\hbar}\int_s^t H(\tau)d\tau\right), \tag{67}$$

and Tr_B is the partial trace over the bath degrees of freedom, $\{\varepsilon_i^B\}$ are the energy eigenvalues, $\{|b\rangle\}$ is the orthonormal energy eigenvectors of the bath, and Z^B is the bath partition function. Assume the bath energy states are nondegenerate. Then (66) implies the Krauss operators for this dynamics are

$$A_{i,f} = \frac{1}{\sqrt{Z_B}} e^{-\beta\varepsilon_i^B/2}\langle b_f|U|b_i\rangle. \tag{68}$$

Suppose the environment is large, with a characteristic relaxation time short compared with the bath-system interactions, and the system-bath coupling ε is small. The environment remains near thermal equilibrium, unentangled and uncorrelated with the system. The system dynamics of each consecutive time interval can be described by a superoperator derived as in (66) which can then be chained together to form a quantum Markov chain:

$$\rho(t) = S(t-1,t)\cdots S(s+1,s+2)S(s,s+1)\rho. \tag{69}$$

The Hermitian operator of a von Neumann-type measurement can be broken up into a set of eigenvalues λ_σ and orthonormal projection operators π_σ such that $H = \sum_\sigma \lambda_\sigma \pi_\sigma$. In a more general sense, the measured operator of a positive operator-valued measurement need not be projectors or orthonormal. The probability of observing the a-th outcome is

$$p_a = \mathrm{Tr}\left(A_a \rho A_a^\dagger\right). \tag{70}$$

The state of the system after this interaction is

$$\tilde{\rho}_a = \frac{A_a \rho A_a^\dagger}{\mathrm{Tr}\left(A_a \rho A_a^\dagger\right)}. \tag{71}$$

The result of the measurement can be represented by using a Hermitian map superoperator $\mathcal{A}$:

$$\mathcal{A} = \sum_\alpha a_\alpha A_\alpha \rho A_\alpha^\dagger. \tag{72}$$

An operator-value sum maps Hermitian operators into Hermitian operators:

$$[\mathcal{A}H]^\dagger = \left[a_\alpha A_\alpha H A_\alpha^\dagger\right]^\dagger = \sum_\alpha a_\alpha \left(A^\dagger\right)^\dagger H^\dagger A_\alpha^\dagger = \mathcal{A}H. \tag{73}$$

In the other direction, any Hermitian map has an operator-value-mean representation. Hermitian maps provide a particularly concise and convenient representation of sequential measurements and correlation functions. For example, suppose Hermitian map $\mathcal{A}$ represents a measurement at time 0, $\mathcal{C}$ is a different measurement at time t, and the quantum operation S_t represents the system evolution between the measurements. The expectation value of a single measurement is

$$\langle a \rangle = \mathrm{Tr}(\mathcal{A}\rho) = \sum_\alpha a_\alpha \mathrm{Tr} A_\alpha \rho A_\alpha^\dagger = \sum_\alpha p_\alpha a_\alpha. \tag{74}$$

The correlation function $\langle b(t)a(0)\rangle$ can be expressed as

$$\langle b(t)a(0)\rangle = \mathrm{Tr}\left(\mathcal{B}S_t\mathcal{A}\rho(0)\right) = \sum_{\alpha,\beta} a_\alpha b_\beta \mathrm{Tr}\, B_\alpha\left(S_t\left(A_\alpha \rho(0) A_\alpha^\dagger\right)\right)B_\beta^\dagger. \tag{75}$$

It may be shown that just as every Hermitian operator represents some measurement on the Hilbert space of pure states, every Hermitian map can be associated with some measurement on the Liouville space of mixed states.

A Hermitian map representation of heat flow can now be constructed under assumptions that the bath and system Hamiltonian are constant during the measurement and the bath-system coupling is very small. A measurement on the total system is constructed, and thus the bath degrees of freedom are projected out. This leaves a Hermitian map superoperator that acts on the system density matrix alone. Let us describe the measurement process and mathematical formulation together.

Begin with a composite system which consists of the bath, initially in thermal equilibrium weakly coupled to the system:

$$\rho^S \otimes \rho_{eq}^B. \tag{76}$$

Measure the initial energy eigenstate of the bath so based on (76):

$$\left(I^S \otimes |b_i\rangle\langle b_i|\right)\left(\rho^S \otimes \rho_{eq}^B\right)\left(I^S \otimes |b_j\rangle\langle b_j|\right). \tag{77}$$

Now allow the system to evolve together with the bath for some time:

$$U\left(I^S \otimes |b_i\rangle\langle b_i|\right)\left(\rho^S \otimes \rho_{eq}^B\right)\left(I^S \otimes |b_j\rangle\langle b_j|\right)U^\dagger. \tag{78}$$

Finally, measure the final energy eigenstate of the bath:

$$\left(I^S \otimes |b_i\rangle\langle b_f|\right)U\left(I^S \otimes |b_i\rangle\langle b_i|\right)\left(\rho^S \otimes \rho_{eq}^B\right)\left(I^S \otimes |b_j\rangle\langle b_j|\right)U^\dagger\left(I^S \otimes |b_f\rangle\langle b_f|\right). \tag{79}$$

Taking the trace over the bath degrees of freedom produces the final normalized system density matrix where trace over S gives the probability of observing the given initial and final bath eigenstates. Multiply by the Boltzmann weighted heat, and sum over the initial and final bath states to obtain the desired average Boltzmann weighted heat flow:

$$\begin{aligned}\left\langle e^{\beta Q}\right\rangle = \sum_{i,f} e^{-\beta\left(\varepsilon_f^B-\varepsilon_i^B\right)} Tr_S Tr_B\left(I^S \otimes |b_f\rangle\langle b_f|\right)U\left(I^S \otimes |b_i\rangle\langle b_i|\right)\\ \left(\rho^S \otimes \rho_{eq}^B\right)\left(I^S \otimes |b_j\rangle\langle b_j|\right)U^\dagger\left(I^S \otimes |b_j\rangle\langle b_j|\right).\end{aligned} \tag{80}$$

Replace the heat bath Hamiltonian by $I^S \otimes H^B = H - H^S(t) \otimes I^B - \varepsilon H^{int}$. The total Hamiltonian commutes with the unitary dynamics and cancels. The interaction Hamiltonian can be omitted in the small coupling limit giving

$$\left\langle e^{\beta Q}\right\rangle = \mathrm{Tr}_S \mathrm{Tr}_B\left(e^{\beta H^S/2} \otimes I^S\right)U\left(e^{-\beta/2H^S} \otimes I^B\right)\left(\rho^S \otimes \rho_{eq}^B\right)\left(e^{-\beta H^S/2} \otimes I^B\right)U^\dagger\left(e^{\beta H^S/2} \otimes I^B\right) \tag{81}$$

Collecting the terms acting on the bath and system separately and replacing the Krauss operators describing the reduced dynamics of the system, the result is

$$\begin{aligned}\left\langle e^{\beta Q}\right\rangle &= \mathrm{Tr}_S e^{\beta H^S/2}\left(Tr_B\left(Ue^{-\beta H^S/2}\rho^S e^{-\beta H^S/2}\right) \otimes \rho_{eq}^B U^\dagger\right)e^{\beta H^S/2}\\ &= \mathrm{Tr}_S \sum_\alpha e^{\beta H^S/2} A_\alpha e^{-\beta H^S/2} \rho^S e^{\beta H^S/2} A_\alpha^\dagger e^{-\beta H^S/2}.\end{aligned} \tag{82}$$

To summarize, it has been found that the average Boltzmann weighted heat flow is represented by

$$\left\langle e^{\beta Q}\right\rangle = \mathrm{Tr}\left(\mathcal{R}^{-1} \mathcal{S} \mathcal{R} \rho^S\right). \tag{83}$$

where S represents the reduced dynamics of the system. The Hermitian map superoperator $\mathcal{R}_t$ is given by

$$\mathcal{R}_t \rho = e^{-\beta H_t/2} \rho e^{\beta H_t/2}. \tag{84}$$

The paired Hermitian map superoperators act at the start and end of a time interval. They give a measure of the change in the energy of the system over that interval. This procedure does not disturb the system beyond that already incurred by coupling the system to the environment. The Jarzynski inequality now follows by applying this Hermitian map and quantum formalism. Discretize the experimental time into a series of discrete intervals labeled by an integer t.

The system Hamiltonian is fixed within each interval. It changes only in discrete jumps at the boundaries. The heat flow can be measured by wrapping the superoperator time evolution of each time interval S_t along with the corresponding Hermitian map measurements $\mathcal{R}_t^{-1} S \mathcal{R}_t$. In a similar fashion, the measurement of the Boltzmann weighted energy change of the system can be measured with $\langle e^{-\beta \Delta E} \rangle = Tr\, \mathcal{R}_\tau S \mathcal{R}_\tau^{-1}$. The average Boltzmann weighted work of a driven, dissipative quantum system can be expressed as

$$\langle e^{-\beta W} \rangle = \mathrm{Tr} \left(\mathcal{R}_\tau \prod_t \left(\mathcal{R}_t^{-1} S_t \mathcal{R}_t \right) \mathcal{R}_\tau^{-1} \rho_0^{eq} \right), \tag{85}$$

In (85), ρ_{eq}^t is the system equilibrium density matrix when the system Hamiltonian is H_t^S.

This product actually telescopes due to the structure of the energy change Hermitian map (84) and the equilibrium density matrix (65). This leaves only the free energy difference between the initial and final equilibrium ensembles, as can be seen by writing out the first few terms

$$\begin{aligned}
\langle e^{-\beta W} \rangle &= Tr\left[\mathcal{R}_\tau \left(\mathcal{R}_\tau^{-1} S_\tau \mathcal{R}_\tau \right) \cdots \left(\mathcal{R}_2^{-1} S_2 \mathcal{R}_2 \right) \left(\mathcal{R}_1^{-1} S_1 \mathcal{R}_1 \right) \mathcal{R}_0^{-1} \rho_{eq}^0 \right] \\
&= \mathrm{Tr}\left[{}_\tau \left(\mathcal{R}_\tau^{-1} S_\tau \mathcal{R}_\tau \right) \cdots \left(\mathcal{R}_2^{-1} S_2 \mathcal{R}_2 \right) \left(\mathcal{R}_1^{-1} S_1 \mathcal{R}_1 \right) \frac{I}{Z(0)} \right] \\
&= \mathrm{Tr}[\mathcal{R}_\tau \left(\mathcal{R}_\tau^{-1} S_\tau \mathcal{R}_\tau \right) \cdots \left(\mathcal{R}_2^{-1} S_2 \mathcal{R}_2 \right) \left(\mathcal{R}_1^{-1} S_1 \rho_{eq}^1 \frac{Z(1)}{Z(0)} \right] \\
\cdots &= \frac{Z(\tau)}{Z(0)} = e^{-\beta \Delta F} = e^{-\beta \Delta F}.
\end{aligned} \tag{86}$$

In the limit in which the time intervals are reduced to zero, the inequality can be expressed in the continuous Lindblad form:

$$\langle e^{-\beta W} \rangle = \mathrm{Tr}\, \mathcal{R}(t) \exp \left[\int_0^t \mathcal{R}(\xi)^{-1} S(\xi) \mathcal{R}(\xi) d\xi \right] \mathcal{R}(0)^{-1} \rho_0^{eq} = e^{-\beta \Delta F}. \tag{87}$$

5. A model quantum spin system

A magnetic resonance experiment can be used to illustrate how these ideas can be applied in practice. A sample of noninteracting spin-1/2 particles are placed in a strong magnetic field B_0 which is directed along the z direction. Denote by $\sigma_j, j = x, y, z$ the usual Pauli matrices and $\mathbf{1}$ the 2×2 identity matrix. It is assumed the motion of the system is unitary. Then the spin is governed by the Hamiltonian:

$$H_0 = -\frac{1}{2} B_0 \sigma_z. \tag{88}$$

In units where $\hbar$ is one, B_0 represents the characteristic precession frequency of the spin. Since H_o is diagonal in the $|\pm\rangle$ basis that diagonalizes σ_z, the matrix exponential and partition function are given by

$$e^{-H/T} = \begin{pmatrix} e^{B_0/2T} & 0 \\ 0 & e^{-B_0/2T} \end{pmatrix}, \qquad Z = \mathrm{Tr}\left(e^{-H/T}\right) = 2\cosh\left(\frac{B_0}{2T}\right), \tag{89}$$

If we set $\tilde{\sigma}$ to be the equilibrium magnetization of the system, $\tilde{\sigma} = \langle \sigma_x \rangle_{th}$, the thermal density matrix is

$$\rho = \rho_{th} = \frac{1}{2}\begin{pmatrix} 1+\tilde{\sigma} & 0 \\ 0 & 1-\tilde{\sigma} \end{pmatrix}, \qquad \tilde{\sigma} = \tanh\left(\frac{B_0}{T}\right). \tag{90}$$

and $\tilde{\sigma}$ corresponds to the parametric response of a spin-1/2 particle.

The work segment is implemented by introducing a very small field of amplitude B rotating in the xy plane with frequency ω. The work parameter is governed by the field

$$\mathbf{B} = B(\sin(\omega t), \cos(\omega t), 0). \tag{91}$$

Typically, $B_0 \approx \omega T$ and $B \approx 0.01T$, so we may approximate $B << B_0$. The total Hamiltonian is the combination

$$H(t) = -\frac{B_0}{2}\sigma_z - \frac{B}{2}\left(\sigma_z \sin(\omega t) + \sigma_y \cos(\omega t)\right). \tag{92}$$

The oscillating field plays the role of a perturbation which although weak may initiate transitions between the up and down spin states and will be most frequent at the resonance condition $\omega = B_0$, so the driving frequency matches the natural oscillation frequency.

The time evolution operator $U(t)$ is calculated now. To do this, define a new operator $V(t)$ by means of the equation

$$U(t) = e^{i\omega t\sigma_z/2} V(t). \tag{93}$$

Substituting (43) into the evolution equation for $U(t)$, $i\partial_t U = H(t)U$, $U(0) = 1$. It is found that $V(t)$ obeys the Schrödinger equation:

$$i\frac{\partial V}{\partial t} = \tilde{H}(t)V, \qquad V(0) = 1, \tag{94}$$

It is found that $V(t)$ satisfies

$$i\frac{\partial V}{\partial t} = \frac{1}{2}\left(\omega\sigma_z - B_0\sigma_z - Be^{-i\omega t\sigma_z/2}(\sigma_x \sin(\omega t) + \sigma_y \cos(\omega t))e^{i\omega\sigma_z/2}\right)V(t). \tag{95}$$

Using the commutation relations of the Pauli matrices and the fact that

$$e^{-i\omega\sigma_z} = \mathbf{1}\cos\left(\frac{\omega t}{2}\right) - i\sigma_z \sin\left(\frac{\omega t}{2}\right), \tag{96}$$

it is found that the terms in the evolution equation can be simplified

$$\begin{aligned} e^{-i\alpha\sigma_z}\sigma_x e^{i\alpha\sigma_z} &= (\mathbf{1}\cos\alpha - i\sigma_x \sin\alpha)\sigma_x(\mathbf{1}\cos\alpha + i\sigma_z \sin\alpha) \\ &= \sigma_x + 2\sin\alpha\cos\alpha\sigma_y - 2i\sigma_z\sigma_y \sin^2\alpha = \sigma_x + 2\sin\alpha\cos\alpha\sigma_y \sin^2\alpha, \end{aligned} \tag{97}$$

$$e^{-i\alpha\sigma_z}\sigma_y e^{i\alpha\sigma_z} = (\sigma_y \cos\alpha - i\sigma_x\sigma_y \sin\alpha(\mathbf{1}\cos\alpha + i\sigma_z \sin\alpha) = \sigma_y - 2\sin\alpha\cos\alpha\sigma_x + 2i\sigma_z\sigma\sin^2\alpha. \tag{98}$$

By means of these results, it remains to simplify

$$e^{-i\omega t\sigma_z/2}(\sigma_z \sin(\omega t) + \sigma_y \cos(\omega t))e^{i\omega t\sigma_z/2}$$

$$= \sigma_z(\sin\omega t - \sin\omega t + \cos\omega t \sin\omega t - \cos\omega t \sin\omega t) + \sigma_y(\sin^2\omega t + \cos\omega t - \cos\omega t + \cos^2\omega t) = \sigma_y. \tag{98a}$$

Taking these results to (95), we arrive at

$$i\frac{\partial V}{\partial t} = H_1 V, \quad H_1 = -\frac{1}{2}(B_0 - \omega)\sigma_z - \frac{1}{2}B\sigma_y. \tag{99}$$

This means $V(t)$ evolves according to a time-dependent Hamiltonian, so the solution can be written as

$$V(t) = e^{-iH_1 t}, \tag{100}$$

and the full-time evolution operator is given by

$$U(t) = e^{i\omega t\sigma_z/2}e^{-iH_1 t}. \tag{101}$$

Since the operators σ_y and σ_z do not commute, the exponentials in (101) cannot be using the usual addition rule.

To express (100) otherwise, suppose $\mathbf{M}$ is an arbitrary matrix such that $\mathbf{M}^2 = \mathbf{1}$. When α is an arbitrary parameter, power series expansion of $e^{-i\alpha\mathbf{M}}$ yields

$$e^{-i\alpha\mathbf{M}} = \mathbf{1}\cos(\alpha) - i\mathbf{M}\sin(\alpha). \tag{102}$$

Now H_1 can be put in equivalent form

$$H_1 = \frac{\Omega}{2}(\sigma_z \cos\vartheta + \sigma_y \sin\vartheta),$$
$$\Omega = \sqrt{(B_0 - \omega)^2 + B^2}, \qquad \tan\vartheta = \frac{B}{B_0 - \omega}, \tag{103}$$

Since $\sigma_i^2 = \mathbf{1}$, it follows that

$$(\sigma_z \cos\vartheta + \sigma_y \sin\vartheta)^2 = \mathbf{1}. \tag{104}$$

Consequently, (100) can be used to prove that $V(t)$ is given by

$$e^{-iH_1 t} = \mathbf{1}\cos\left(\frac{\Omega}{2}t\right) + i(\sigma_z\cos\vartheta + \sigma_y\sin\vartheta)\sin\left(\frac{\Omega}{2}t\right)$$
$$= \begin{pmatrix} \cos\left(\frac{\Omega}{2}t\right) + i\cos\vartheta\sin\left(\frac{\Omega}{2}t\right) & \sin\vartheta\sin\left(\frac{\Omega}{2}t\right) \\ -\sin\vartheta\sin\left(\frac{\Omega}{2}t\right) & \cos\left(\frac{\Omega}{2}t\right) - i\cos\vartheta\sin\left(\frac{\Omega}{2}t\right) \end{pmatrix} \tag{105}$$

Since

$$e^{i\omega\sigma_z t/2} = \begin{pmatrix} e^{i\omega t/2} & 0 \\ 0 & e^{-i\omega t/2} \end{pmatrix} \tag{106}$$

the evolution operator is then given by

$$U(t) = \begin{pmatrix} u(t) & v(t) \\ -v^*(t) & u^*(t) \end{pmatrix}. \tag{107}$$

The functions $u(t)$ and $v(t)$ in (107) are given as

$$u(t) = e^{i\omega t/2}\left(\cos\left(\frac{\Omega}{2}t\right) + i\sin\vartheta \sin\left(\frac{\Omega}{2}t\right)\right), \qquad v(t) = e^{i\omega t/2}\cdot \sin\vartheta \cdot \sin\left(\frac{\Omega}{2}t\right). \tag{108}$$

Apart from a phase factor, the final result depends only on Ω and ϑ, and these in turn depend on B_0, B, and ω through (108). To understand the physics of $U(t)$ a bit better, suppose the system is initially in the pure state $|+\rangle$. The probability will be found in state $|-\rangle$ after time t is

$$|\langle -|U(t)|+\rangle|^2 = |v|^2. \tag{109}$$

This expression represents the transition probability per unit time a transition will occur. Since the unitarity condition $U^\dagger U = \mathbf{1}$ implies that $|u|^2 + |v|^2 = 1$, we conclude $|u|^2$ is the probability when no transition occurs. Note v is proportional to $\sin\vartheta$, which gives a physical meaning to ϑ. It represents the transition probability and reaches a maximum when $\omega = B_0$ at resonance where $\Omega = B$, so u and v simplify to

$$u(t) = e^{i\omega t/2}\cos\left(\frac{B}{2}t\right), \qquad v(t) = e^{i\omega t/2}\sin\left(\frac{B}{2}t\right). \tag{110}$$

Now that $U(t)$ is known, the evolution of any observable A can be calculated

$$\langle A\rangle_t = Tr\left(U^\dagger(t) A\, U(t)\rho\right). \tag{111}$$

If A is replaced by σ_z in (111), we obtain

$$\begin{aligned}\langle\sigma_z\rangle_t &= \mathrm{Tr}\begin{pmatrix} u^*(t) & -v(t) \\ v^*(t) & u(t) \end{pmatrix}\sigma_z\begin{pmatrix} u(t) & v(t) \\ -v^*(t) & u^*(t) \end{pmatrix}\frac{1}{2}\begin{pmatrix} 1+\tilde{\sigma} & 0 \\ 0 & 1-\tilde{\sigma} \end{pmatrix} \\ &= \tilde{\sigma}\left(|u|^2 - |v|^2\right) = \tilde{\sigma}\left(1 - 2|v|^2\right).\end{aligned} \tag{112}$$

Substituting $|v|^2$, this takes the form

$$\langle\sigma_z\rangle_t = \tilde{\sigma}\left(\cos^2\vartheta + \sin^2\vartheta\cos(\Omega t)\right) = \tanh\left(\frac{B_0}{2T}\right)\left(\cos^2\vartheta + \sin^2\vartheta\cos(\Omega t)\right). \tag{113}$$

Consider the average work. Suppose $B << B_0$, so the unperturbed Hamiltonian H_0 can be used instead of the full Hamiltonian $H(t)$ when expectation values of quantities are calculated which are related to the energy. Let us determine the energy of the system at any t by taking operator A to be H_0:

$$\langle H_0\rangle_t = -\frac{B_0}{2}\langle\sigma_z\rangle_t = -\frac{1}{2}B_0\,Tr\,U^\dagger(t)\sigma_z U(t)\rho)$$
$$= -\frac{1}{2}B_0\,\mathrm{Tr}\left(\begin{pmatrix} u^* & -v \\ v^* & u \end{pmatrix}\begin{pmatrix} 1 & 0 \\ 0 & -1 \end{pmatrix}\begin{pmatrix} u & v \\ -v^* & u^* \end{pmatrix}\frac{1}{2}\begin{pmatrix} 1+\tilde{\sigma} & 0 \\ 0 & 1-\tilde{\sigma} \end{pmatrix}\right)$$
$$= -\frac{B_0}{4}\,\mathrm{Tr}\left(\begin{pmatrix} u^* & -v \\ v^* & u \end{pmatrix}\begin{pmatrix} u & v \\ v^* & -u^* \end{pmatrix}\begin{pmatrix} 1+\tilde{\sigma} & 0 \\ 0 & 1-\tilde{\sigma} \end{pmatrix}\right) = -\frac{1}{4}B_0\tilde{\sigma}\left(1-2|v|^2\right).$$
(114)

The average work at time t is simply the difference between the energy st time t_1 and $t=0$. Since $v(0)=0$, this difference is

$$\begin{aligned}\langle W\rangle_t &= -\frac{B_0}{2}\tilde{\sigma}\left(1-2|v|^2\right) + \frac{B_0}{2}\tilde{\sigma} = \tilde{\sigma}B_0|v|^2 \\ &= \tilde{\sigma}B_0\sin^2\vartheta\sin^2\left(\frac{\Omega}{2}t\right) = \tilde{\sigma}B_0\frac{B}{\Omega^2}\sin^2\left(\frac{\Omega}{2}t\right),\end{aligned} \tag{115}$$

since $\sin^2\vartheta = 1-\cos^2\vartheta = B^2/\Omega^2$. The average work oscillates indefinitely with frequency $\Omega/2$. This is a consequence of the fact the time evolution is unitary. The amplitude multiplying the average work is proportional to the initial magnetization $\tilde{\sigma}$ and B^2/Ω^2, so the ratio is a Lorentzian function.

The equilibrium free energy is $F=-T\,\log Z$ where $Z=2\cosh(B_0/2T)$. The free energy of the initial state at $t=0$ and final state at any arbitrary time is the same yielding

$$\Delta F = 0. \tag{116}$$

This is a consequence of the fact that $B<<B_0$. According to $\langle W\rangle \geq F$, it should be expected that

$$\langle W\rangle_t \geq \Delta F = 0. \tag{117}$$

Given the matrices for $U(t)$ and ρ that have been determined so far, the function $\mathcal{G}$ can be computed:

$$\mathcal{G}(y) = Tr\left(U^\dagger(y)e^{ixH_f}U(y)e^{-iyH_i}\rho\right)$$
$$= \mathrm{Tr}\begin{pmatrix} u^* & -v \\ v^* & u \end{pmatrix}\begin{pmatrix} e^{-iyB_0/2} & 0 \\ 0 & e^{iyB_0/2} \end{pmatrix}\begin{pmatrix} u & v \\ -v^* & u^* \end{pmatrix}\begin{pmatrix} e^{iyB_0/2} & 0 \\ 0 & e^{-iyB_0/2} \end{pmatrix}\begin{pmatrix} \frac{1}{2}(1+\tilde{\sigma}) & 0 \\ 0 & \frac{1}{2}(1-\tilde{\sigma}) \end{pmatrix}$$
$$= |u|^2 + \frac{1}{2}\left((1+\tilde{\sigma})e^{iyB_0} + (1-\tilde{\sigma}e^{-iyB_0}\right)|v|^2.$$
(118)

Set $x=i\beta$ and recall use definition (42) for $\tilde{\sigma}$ in the second term of (118) to give

$$\begin{aligned}&\left(1+\tanh\left(\frac{\beta}{2}B_0\right)\right)e^{-\beta B_0} + \left(1-\tanh\left(\frac{\beta}{2}B_0\right)\right)e^{\beta B_0} \\ &= 2\cosh(\beta B_0) - 2\tanh\left(\frac{\beta}{2}B_0\right)\sinh(\beta B_0) = 1.\end{aligned} \tag{119}$$

Substituting (119) into (118), we can conclude

$$\langle e^{-\beta W} \rangle = \mathcal{G}(i\beta) = |u|^2 + |v|^2 = 1. \tag{120}$$

This is the Jarzynski inequality, since it is the case that $\Delta F = 0$ here. The statistical moments of the work can be obtained by writing an expression for $\mathcal{G}$ into a power series

$$\mathcal{G}(y) = |u|^2 + |v|^2 \left(1 + i\tilde{\sigma} B_0 y - \frac{1}{2} B_0^2 y^2 + \cdots \right). \tag{121}$$

From (121), the first and second moments can be obtained; for example

$$\langle W \rangle = \tilde{\sigma} B_0 |v|^2, \qquad \langle W^2 \rangle = B_0^2 |v|^2. \tag{122}$$

As a consequence, the variance of the work can be determined

$$\text{var}(W) = \langle W^2 \rangle - \langle W \rangle^2 = B_0^2 |v|^2 - \tilde{\sigma}^2 B_0^2 |v|^4 - B_0^2 |v|^2 \left(1 - \tilde{\sigma}^2 \, v|^2 \right). \tag{123}$$

A final calculation that may be considered is the full distribution of work $\mathcal{P}(W)$. Now $\mathcal{P}(W)$ is the inverse Fourier transform of $\mathcal{G}(y)$:

$$\mathcal{P}(W) = \frac{1}{2\pi} \int_{-\infty}^{\infty} dy \, \mathcal{G}(y) e^{-iyW}. \tag{124}$$

Using the Fourier integral form of the delta function, (124) can be written as

$$\mathcal{P}(W) = \frac{1}{2\pi} \int_{-\infty}^{\infty} dy \, \mathcal{G}(y) e^{-iyW}$$

$$= \frac{1}{2\pi} \int_{-\infty}^{\infty} dy \left(|u(t)|^2 + |v(t)|^2 \left[\frac{1}{2} (1 + \tilde{\sigma}) e^{iB_0 y} + \frac{1}{2} (1 - \tilde{\sigma}) e^{-iB_0 y} \right] \right) e^{-iyW}. \tag{125}$$

$$= |u|^2 \delta(W) + \frac{1}{2} |v(t)|^2 \delta(W - B_0) + \frac{1}{2} |v(t)|^2 (1 + \tilde{\sigma}) \delta(W + B_0).$$

Work taken as a random variable can take three values $W = 0, +B_0, -B_0$ where B_0 is the energy spacing between the up and down states. The event $W = B_0$ corresponds to the case where the spin was originally up and then reversed, so an up-down transition. The energy change is $(B_0/2) - (-B_0/2) = B_0$. Similarly, $W = -B_0$ is the opposite flip from this one, and $W = 0$ is the case with no spin flip.

The second law would have us think that $W > 0$, but a down-up flip should have $W = -B_0$, so $\mathcal{P}(W = -B_0)$ is the probability of observing a local violation of the second law. However, since $\mathcal{P}(W = \pm B_0)$ is proportional to $1 \pm \tilde{\sigma}$, up-down flips are always more likely than down-up. This ensures that $\langle W \rangle \geq 0$, so violations of the second law are always exceptions to the rule and never dominate.

The work performed by an external magnetic field on a single spin-1/2 particle has been studied so far. The energy differences mentioned correspond to the work. For noninteracting particles, energy is additive. Hence the total work $\langle \mathcal{W} \rangle$ which is performed during a certain process is the sum of works performed on each individual particle $\mathcal{W} = W_1 + \cdots + W_N$. Since the spins are all independent and energy is an extensive variable, it follows that $\langle \mathcal{W} \rangle = N \langle W \rangle$. where $\langle W \rangle$ is the average work from (115).

6. Conclusions

We have tried to give an introduction to this frontier area that lies in between that of thermodynamics and quantum mechanics in such a way as to be comprehensible. There are many other areas of investigation presently which have had interesting repercussions for this area as well. There is a growing awareness that entanglement facilitates reaching equilibrium [21–23]. It is then worth mentioning that the ideas of einselection and entanglement with the environment can lead to a time-independent equilibrium in an individual quantum system and statistical mechanics can be done without ensembles. However, there is really a lot of work yet to be done in these blossoming areas and will be left for possible future expositions.

Author details

Paul Bracken
Department of Mathematics, University of Texas, Edinburg, TX, USA

*Address all correspondence to: paul.bracken@utrgv.edu

References

[1] Boltzmann L. Vorlesungen über Gastheorie. Leipzig: Barth; 1872

[2] Landau L, Lifschitz E. Statistical Mechanics. Oxford Press, Oxford; 1978

[3] Gibbons E, Hawking S. Euclidean Quantum Gravity. Singapore: World Scientific; 1993

[4] Lebowitz J. From time-symmetric microscopic dynamics to time-asymmetric macroscopic behavior an overview. In: Gallavotti G, Reuter W, Yngyason J, editors. Boltzmann's Legacy. Zürich: European Mathematical Society; 2008

[5] Bracken P. A quantum version of the classical Szilard engine. Central European Journal of Physics. 2014;**12**:1-8

[6] Bracken P. Quantum dynamics, entropy and quantum versions of Maxwell's demon. In: Bracken P, editor. Recent Advances in Quantum Dynamics. Croatia: IntechOpen; 2016. pp. 241-263

[7] Fermi E. Thermodynamics. Mineola, NY: Dover; 1956

[8] Allahverdyan A, Balian R, Nieuwenhuisen T. Understanding quantum measurement from the solution of dynamical models. Physics Reports. 2013;**525**:1-166

[9] Narnhofer H, Wreszinski WF. On reduction of the wave packet, decoherence, irreversibility and the second law of thermodynamics. Physics Reports. 2014;**541**:249-273

[10] Linblad G. Entropy, information and quantum mechanics. Communications in Mathematical Physics. 1973;**33**:305-322

[11] Lindblad G. Expectations and entropy inequalities for finite quantum systems. Communications in Mathematical Physics. 1974;**39**:111-119

[12] Lindblad G. Completely positive maps and entropy inequalities. Communications in Mathematical Physics. 1973;**40**:147-151

[13] Lieb E. The stability of matter. Reviews of Modern Physics. 1976;**48**: 553-569

[14] Fano U. Description of states in quantum mechanics and density matrix techniques. Reviews of Modern Physics. 1957;**29**:74-93

[15] Ribeiro W, Landi GT, Semião F. Quantum thermodynamics and work fluctuations with applications to magnetic resonance. American Journal of Physics. 2016;**84**:948-957

[16] Wehrl A. General properties of entropy. Reviews of Modern Physics. 1978;**50**:221-260

[17] Bracken P. A quantum Carnot engine in three dimensions. Advanced Studies in Theoretical Physics. 2014;**8**: 627-633

[18] Lieb E. The classical limit of quantum spin systems. Communications in Mathematical Physics. 1973;**31**:327-340

[19] Lieb E, Liebowitz J. The constitution of matter: Existence of thermodynamics for systems composed of electrons and nuclei. Advances in Mathematics. 1972; **9**:316-398

[20] Crooks J. On the Jarzynski relation for dissipative systems. Journal of Statistical Mechanics: Theory and Experiment. 2008;**10023**:1-9

[21] Vedral V. The role of relative entropy in quantum information theory.

Reviews of Modern Physics. 2002;**74**: 197-234

[22] Zurek WH. Decoherence, einselection, and the quantum origins of the classical. Reviews of Modern Physics. 2000;**9**:715-775

[23] Zurek WH. Eliminating ensembles from equilibrium statistical physics, Maxwell's demon, Szilard's engine and thermodynamics via entanglement. Physics Reports. 2018;**755**:1-21

Chapter 7

Equations of Relativistic and Quantum Mechanics (without Spin)

Vahram Mekhitarian

Abstract

A relativistically invariant representation of the generalized momentum of a particle in an external field is proposed. In this representation, the dependence of the potentials of the interaction of the particle with the field on the particle velocity is taken into account. The exact correspondence of the expressions of energy and potential energy for the classical Hamiltonian is established, which makes identical the solutions to the problems of mechanics with relativistic and nonrelativistic approaches. The invariance of the proposed representation of the generalized momentum makes it possible to equivalently describe a physical system in geometrically conjugate spaces of kinematic and dynamic variables. Relativistic invariant equations are proposed for the action function and the wave function based on the invariance of the representation of the generalized momentum. The equations have solutions for any values of the constant interaction of the particle with the field, for example, in the problem of a hydrogen-like atom, when the atomic number of the nucleus is $Z > 137$. Based on the parametric representation of the action, the expression for the canonical Lagrangian, the equations of motion, and the expression for the force acting on the charge are derived when moving in an external electromagnetic field. The Dirac equation with the correct inclusion of the interaction for a particle in an external field is presented. In this form, the solutions of the equations are not limited by the value of the interaction constant. The solutions of the problem of charge motion in a constant electric field, the problems for a particle in a potential well and the passage of a particle through a potential barrier, the problems of motion in an exponential field (Morse), and also the problems of a hydrogen atom are given.

Keywords: quantum mechanics, relativistic invariant equations

1. Introduction

–To doubt everything or to believe everything are two equally convenient solutions; both dispense with the necessity of reflection.

Henri Poincaré (1854-1912)

–I know, I know, but suppose – just suppose! – the purity of the circle has blinded us from seeing anything beyond it!
I must begin all over with new eyes, I must rethink everything!

Hypathia (~360-415 AD)

In 1913, Bohr, based on the Balmer empirical formulas, constructed a model of atom based on the quantization of the orbital momentum [1], which was subsequently supplemented by the more general Sommerfeld quantization rules. In those years, naturally, the presence of a spin or an intrinsic magnetic moment of the particle or, especially, spin-orbit interaction, or interaction with the nuclear spin, was not supposed.

In 1916, Sommerfeld, within the framework of relativistic approaches, derived a formula for the energy levels of a hydrogen-like atom, without taking into account the spin [2]. Sommerfeld proceeded from the model of the Bohr atom and used the relativistic relation between the momentum **p** and the energy E of a free particle with the mass m.

$$E^2-(\mathbf{p}c)^2=\left(mc^2\right)^2, \tag{1}$$

where c is the speed of light.

In an external field with a four-dimensional potential $(\varphi,\mathbf{A})$, it was supposed that for a particle with the charge q this relation can also be used if we subtract the components of the four-dimensional momentum of the field $(q\varphi,q\mathbf{A})$ from the expression for the generalized particle momentum:

$$(E-q\varphi)^2-(\mathbf{p}c-q\mathbf{A})^2=\left(mc^2\right)^2. \tag{2}$$

In the case of the Coulomb potential $\varphi=Z|e|/r$, where e is the charge of electron, r is the distance from the nucleus, and Z is an atomic number, we obtain in spherical coordinates

$$p_{\mathrm{r}}{}^2+r^2p_{\varphi}{}^2=p_{\mathrm{r}}{}^2+\frac{L^2}{r^2}=\frac{\left(E+Ze^2/r\right)^2-\left(mc^2\right)^2}{c^2} \tag{3}$$

where L is the angular momentum. The Bohr-Sommerfeld quantization conditions take the form

$$\begin{aligned}&\oint p_{\varphi}d\varphi=\hbar n_{\varphi},\\&\oint p_{\mathrm{r}}dr=\oint\sqrt{\frac{\left(E+Ze^2/r\right)^2-\left(mc^2\right)^2}{c^2}-\frac{L^2}{r^2}}dr=\hbar n_{\mathrm{r}},\end{aligned} \tag{4}$$

where n_{φ} and n_r are the orbital and radial quantum numbers, respectively. For the energy levels, Sommerfeld obtained the formula

$$E_{n,l}=\frac{mc^2}{\sqrt{1+\frac{(Z\alpha)^2}{\left(n-\frac{(Z\alpha)^2}{l+1/2+\sqrt{(l+1/2)^2-(Z\alpha)^2}}\right)^2}}}, \tag{5}$$

where the principal quantum number $n=n_{\mathrm{r}}+l+1=1,\ 2,\ 3,\ \ldots,l=0,\ 1\ ,\ 2,\ 3,\ \ldots,\ n-1$, and $\alpha=1/137.036$ is the fine structure constant. However, in a paper published in 1916 [3], Sommerfeld '*made a fortunate mistake*' [4] and the derived formula was presented in the following form

$$E_{n,l} = \frac{mc^2}{\sqrt{1 + \frac{(Z\alpha)^2}{\left(n - \frac{(Z\alpha)^2}{l+1+\sqrt{(l+1)^2-(Z\alpha)^2}}\right)^2}}}. \tag{6}$$

The formula (6) perfectly described all the peculiarities of the structure of the spectrum of hydrogen and other similar atoms with the limiting for those years accuracy of measurements, and there was no doubt about the correctness of the formula itself. Therefore, the Sommerfeld formula was perceived as empirical, and instead of the quantum number l, a '*mysterious*' internal quantum number with half-integer values $j = 1/2,\ 3/2,\ 5/2,\ \ldots,\ n + 1/2$ was introduced, and formula (6) was used in the representation

$$E_{n,j} = \frac{mc^2}{\sqrt{1 + \frac{Z\alpha^2}{\left(n - \frac{Z^2\alpha^2}{j+1/2+\sqrt{(j+1/2)^2-Z^2\alpha^2}}\right)^2}}}, \tag{7}$$

where $n = n_r + j + 1/2 = 1,\ 2,\ 3,\ \ldots,\ j = 1/2,\ 3/2,\ 5/2,\ \ldots,\ n + 1/2$, and l possess the values $l = 0$ at $j = 1/2$ and $l = j \pm 1/2$ for others. This formula coincides with the result of an exact solution of the relativistic Dirac equations in 1928 [5] for a particle with the spin 1/2 with the classical expression for the potential energy of an immobile charge in the Coulomb field of a nucleus with an atomic number Z in the form $U(r) = Ze^2/r$.

Formula (7) also indicated a strange limitation of value the charge of a nucleus with the atomic number $Z < 137$, above which the formula is losing its meaning. It was also evident that within the framework of the approaches outlined, the strong and gravitational interactions, the motions of the planets are not described. The problem $Z < 137$ or $\alpha > 1$ remains the unresolved problem of relativistic quantum mechanics. Expanding the formula (7) over the order of powers $Z\alpha^2$ in the Taylor series, with an accuracy of expansion up to the terms by the powers $Z\alpha^6$, we obtain

$$E_{n,j} = mc^2 - \frac{(Z\alpha)^2}{2n^2} - \frac{(Z\alpha)^4}{2n^3}\left(\frac{1}{j+1/2} - \frac{3}{4n}\right) + \ldots \tag{8}$$

In 1925–1926, Schrödinger worked on the derivation of the equation for the wave function of a particle describing the De Broglie waves [6]. The derivation of the equation also was based on the relativistic relation (1) between the momentum $\mathbf{p}$ and the energy E of the particle, which he presented with the help of the operators of squares of energy and momentum in the form of an equation for the wave function

$$\left(i\hbar\frac{\partial}{\partial t}\right)^2\Psi - c^2\left(-i\hbar\frac{\partial}{\partial \mathbf{r}}\right)^2\Psi = \left(mc^2\right)^2\Psi \tag{9}$$

Like Sommerfeld, Schrödinger used the following representation for a particle in an external field

$$\left(i\hbar\frac{\partial}{\partial t} - q\varphi\right)^2\Psi - c^2\left(-i\hbar\frac{\partial}{\partial \mathbf{r}} - \frac{q}{c}\mathbf{A}\right)^2\Psi = \left(mc^2\right)^2\Psi \tag{10}$$

In the case of stationary states of a charged particle in the field of the Coulomb potential for a hydrogen atom it was necessary to solve the equation

$$\frac{d^2\psi}{d\mathbf{r}^2} + \frac{2m}{\hbar^2}\left(\frac{E^2 - m^2c^4}{2mc^2} - \frac{E}{mc^2}q\varphi(\mathbf{r}) + \frac{q^2}{2mc^2}\varphi^2(\mathbf{r})\right)\psi = 0 \tag{11}$$

As can be seen, the quadratic expression of potential energy $q^2\varphi^2(\mathbf{r})/2mc^2$ is present in the equation with a positive sign and in the case of attracting fields, the solutions lead to certain difficulties. When approaching the singularity point, due to the negative sign, the attractive forces increase and the presence of the singularity leads to known limitations on the magnitude of the interactions (**Figure 1**).

Next, the wave vector k is represented as

$$k_1 = \frac{1}{\hbar c}\sqrt{E^2 - (mc^2)^2}, \qquad k_2 = \frac{1}{\hbar c}\sqrt{(E-U)^2 - (mc^2)^2} \tag{12}$$

and when considering the problem of the passage of a particle with energy E through a potential barrier $U = q\varphi(\mathbf{r})$ (**Figure 2**), the height of which is greater than the doubled resting energy of the particle $U > 2mc^2$, the transmission coefficient becomes unity, regardless of the height of the barrier (Klein paradox) [7].

Another difficulty is that, as the solution of the particle problem in a potential well shows, at a sufficient depth, a particle with a wavelength $\lambda\!\!\!^{-} = \hbar/mc$ can have bound states (can be localized) in a well width narrower than the wavelength of the particle $d < \lambda\!\!\!^{-}/2$ (**Figure 3**), which contradicts the fundamental principle of quantum mechanics—the Heisenberg's uncertainty principle.

Also, the solution of the problem of a hydrogen-like atom is limited by the value of the ordinal number of the atomic nucleus $Z \le 68$ (for the Dirac equation, the restriction of the atomic number is $Z \le 137$). The same in relativistic mechanics—when considering strong interactions, the solution of the Hamilton-Jacoby relativistic equation indicates the so-called "*particle fall on the center*" [8].

In order to get rid of the quadratic term or reverse its sign, in recent years it has been proposed to represent potential energy in the Klein-Gordon and Dirac equations as the difference of squares from the expressions of scalar and vector potentials

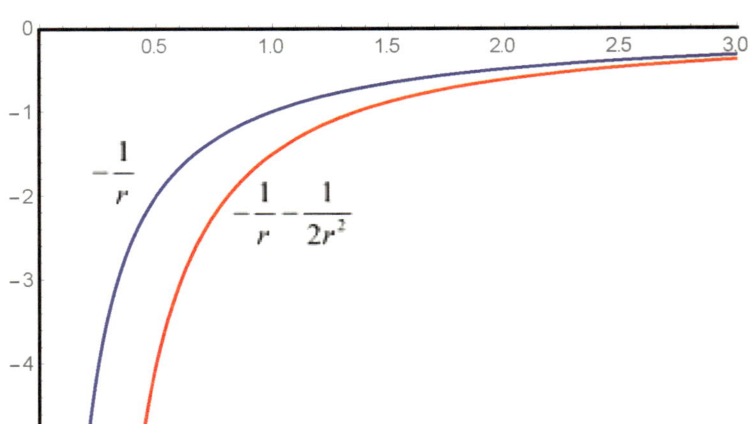

Figure 1.
The sample dependency of the attractive field potential $-1/r$ and potential interaction energy $-1/r - 1/2r^2$ in the Klein-Gordon equations.

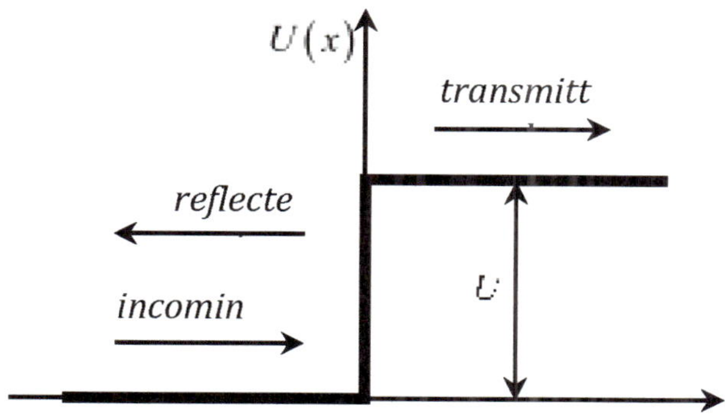

Figure 2.
Passage of a particle through a potential barrier U.

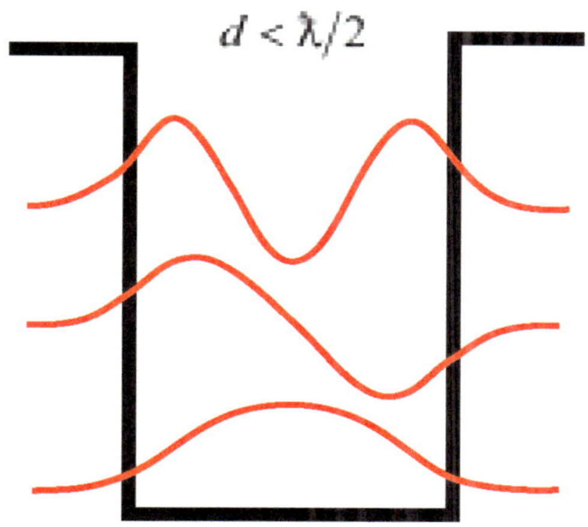

Figure 3.
A particle with a wavelength ƛ can be localized in a well width d < ƛ/2.

(S-wave equation) [9–11]. Such a mathematical formalism corrects the situation, but from a physical point of view such representations are in no way justified, and the fields corresponding to such pseudo-potentials do not exist in nature.

Things are even worse with the presence of a quadratic term of the vector field, because of the sign of which we obtain non-existent states in nature and solutions that contradict experience.

$$\frac{d^2\psi}{d\mathbf{r}^2} - 2i\frac{q}{\hbar c}\mathbf{A}(\mathbf{r})\cdot\frac{d\psi}{d\mathbf{r}} + \frac{2m}{\hbar^2}\left(\frac{E^2 - m^2c^4}{2mc^2} - \frac{q^2}{2mc^2}\mathbf{A}^2(\mathbf{r})\right)\psi = 0 \tag{13}$$

According to the solutions of the equations of quantum mechanics and Hamilton-Jacoby, it turns out that a charged particle in a magnetic field, in addition to rotating in a circle, also has radial vibrations—Landau levels [12] (even in the case of zero orbital momentum).

$$\frac{\hbar^2}{2M}\left(R'' + \frac{1}{\rho}R' - \frac{m^2}{\rho^2}\right) + \left(E - \frac{p_z^2}{2M} - \frac{M\omega_H^2}{8}\rho^2 - \frac{\hbar\omega_H m}{2}\right)R = 0. \tag{14}$$

$$E = \hbar\omega_H\left(n_\rho + \frac{|m| + m + 1}{2}\right) + \frac{p_z^2}{2M}$$

Over these 90 years, especially in very accurate cyclotron resonance experiments, none has detected the electron radial vibrations and the Landau levels.

Solving this equation, Schrödinger, like Sommerfeld, received the formula (5), which described the structure of the hydrogen spectrum not exactly. Moreover, from the solution of the problem for a particle in a potential well, it turns out that a particle with a wavelength $\lambda\!\!\!^{-} = \hbar/mc$ has bound states (is placed) in a well of arbitrary size and, in particular, much smaller than $\lambda\!\!\!^{-}/2$. This fact contradicts the fundamental principle of the quantum (wave) theory, the principle of uncertainty.

In 1925 Schrödinger sent this work to the editors of 'Annalen der Physik' [13], but then took the manuscript, refused the relativistic approaches and in 1926 built a wave equation based on the classical Hamiltonian expression, the Schrödinger equation [14].

$$\mathrm{H} = \frac{\mathbf{p}^2}{2m} + U; \quad \rightarrow \quad i\hbar\frac{\partial}{\partial t}\Psi = \left(\frac{1}{2m}\left(-i\hbar\frac{\partial}{\partial \mathbf{r}}\right)^2 + U\right)\Psi \tag{15}$$

Equation described the spectrum of the hydrogen atom only qualitatively, however, it did not have any unreasonable restrictions or singular solutions in the form of the Sommerfeld-Dirac formula. Klein [15], Fock [16] and Gordon [17] published the relativistic equation based on the wave equation for a particle without spin in 1926; it is called the Klein-Fock-Gordon equation.

With the discovery of the spin, the situation changed drastically, and in 1926 Heisenberg and Jordan [18] showed that, within the Pauli description of the spin of an electron, half the energy of the spin-orbit interaction is equal to a term with a power of α^4 in the Taylor series expansion of the Sommerfeld formula *equation reference goes here*.

Why exactly the half, Thomas tried to explain this in 1927 by the presence of a relativistic precession of an electron in the reference frame of motion along the orbit [19]. The energy of the Thomas precession is exactly equal to half the value of the energy of the spin-orbit interaction with the inverse (positive) sign, which should be added to the energy of the spin-orbit interaction. However, the incorrect assumption that the Thomas precession frequency is identical in both frames of reference and the absence of a common and correct derivation for non-inertial (rotating) frames of reference raised doubts about the correctness of such approaches. The reason for the appearance of half the energy of the spin-orbit interaction in the Sommerfeld formula is still under investigation and is one of the unresolved problems in modern physics.

On the other hand, both in the derivation of the Sommerfeld formula and at the solution of the Klein-Fock-Gordon equation for the hydrogen atom problem [20], neither the spin nor the spin-orbit interaction energy was taken into account initially. Therefore, the obtained fine splitting can in no way be owing to the spin-orbit interaction. This is a relativistic but purely mechanical effect, when the mass (inertia) of a particle is already depends on the velocity of motion along the orbit (of the angular momentum), because of which the radial motion of the electron changes, and vice versa. Just this dependence, which results in the splitting of the energy levels of the electron, and to the impossibility of introducing only one, the principal quantum number. Nevertheless, even with this assumption, the order of splitting of the levels according to formula (8) contradicts to the logic; it turns out to be that the greater the orbital angular momentum, the lesser the energy of the split level.

The matrix representation of the second-order wave Eq. (9) by a system of equations of the first order is the Dirac construction of the relativistic electron equation [21] (the Dirac matrices are the particular representation of the

Clifford-Lipschitz numbers [22]). In the standard representation the Dirac equation for a free particle has the form [23].

$$\begin{aligned} \hat{\varepsilon}\,\phi - \boldsymbol{\sigma}\cdot\hat{\mathbf{p}}\,\chi &= mc\,\phi, \\ -\hat{\varepsilon}\,\chi + \boldsymbol{\sigma}\cdot\hat{\mathbf{p}}\,\varphi &= mc\,\chi, \end{aligned} \tag{16}$$

where

$$\mathbf{1} = \begin{pmatrix} 1 & 0 \\ 0 & 1 \end{pmatrix}, \quad \sigma_x = \begin{pmatrix} 0 & 1 \\ 1 & 0 \end{pmatrix}, \quad \sigma_y = \begin{pmatrix} 0 & -i \\ i & 0 \end{pmatrix}, \quad \sigma_z = \begin{pmatrix} 1 & 0 \\ 0 & -1 \end{pmatrix} \tag{17}$$

are the Pauli matrices (the unit matrix in the formulas is omitted).

For a particle in an external field, Eq. (16) is usually written in the form

$$\begin{aligned} \left(\hat{\varepsilon} - \frac{q}{c}\varphi\right)\phi - \boldsymbol{\sigma}\cdot\left(\hat{\mathbf{p}} - \frac{q}{c}\mathbf{A}\right)\chi &= mc\,\phi, \\ -\left(\hat{\varepsilon} - \frac{q}{c}\varphi\right)\chi + \boldsymbol{\sigma}\cdot\left(\hat{\mathbf{p}} - \frac{q}{c}\mathbf{A}\right)\phi &= mc\,\chi, \end{aligned} \tag{18}$$

where for an invariant representation in the case of a free particle, the equations are composed for the difference between the generalized momentum and the momentum of the field.

In the case of the potential energy of an immobile charge in a Coulomb field, we obtain the Sommerfeld-Dirac formula as a result of an exact solution of this particular equation. There, again, although for a system with spin 1/2 the energy of the spin-orbit interaction is not taken into account initially, but the half is obtained from the exact solution of the hydrogen atom problem.

More accurate measurements of Lamb in 1947 and subsequent improvements in the spectrum of the hydrogen atom revealed that, in addition to the lines with the maximum j, all the others are also split and somewhat displaced (the Lamb shift). To harmonize the results of the theory with more accurate experimental data on the spectrum of the hydrogen atom, one had to propose other solutions and approaches than were laid down by the derivation of the Dirac equation.

The new theoretical approaches had yield nothing and only supplemented the theory with the illogical and non-physical proposals to overcome the emerging singularity of solutions: the renormalization, the finite difference of infinities with the desired value of the difference, and so on. The accounting for the size of the nucleus corrected only the Z value into the bigger value, but did not solve the $Z > 137$ problem. An incredible result was also obtained for the hydrogen atom problem that the electron is located, most likely, at the center of the atom, that is, in the nucleus.

The results of solution of the problem for a particle in a potential well both in the case of the Klein-Fock-Gordon equation and of the Dirac equation contradict to the basic principle of quantum mechanics, to the uncertainty principle. From the solutions, it turns out to be that a particle can be in a bound state in a well with any dimensions, in particular, with the size much smaller than the wavelength of the particle itself, $A = \hbar/mc$ [23].

Despite Dirac himself proposed a system of linear first-degree relativistic equations in the matrix representation that described the system with spin 1/2, the contradictions did not disappear, and he himself remained unhappy with the results of his theory. As Dirac wrote in 1956 [24], the development of relativistic electron theory can now be considered as an example of how incorrect arguments sometimes lead to a valuable result. In the 70s, it became clear that the relativistic theory of quantum mechanics does not exist, and new, fundamental approaches and equations should be sought for constructing a consistent theory of relativistic quantum mechanics. And in the 80s, Dirac already spoke about the insuperable difficulties of the existing quantum theory and the need to create a new one [25].

The reason for the failure of these theories is quite simple—it is in the ignoring of the dependence of the interaction energy with the field on the velocity of the particle. The generalized momentum of the system, the particle plus the external field, is the sum of the relativistic expression for the mechanical momentum of the particle and the field momentum in the case of interaction with the immobile particle

$$\mathbf{P} = (\varepsilon,\ \mathbf{p}) = \frac{1}{c}\left(\frac{mc^2}{\sqrt{1-\beta^2}} + q\varphi,\quad \frac{mc^2}{\sqrt{1-\beta^2}}\boldsymbol{\beta} + q\mathbf{A},\quad \mathbf{P}^2 \neq inv\right), \tag{19}$$

which is not an invariant representation of the particle velocity. To construct some invariant from such a representation, an 'invariant' relation was used in all cases in the form of a difference between the generalized momentum of the system and the field momentum in the case of interaction with the immobile particle

$$(\varepsilon - q\varphi,\ \mathbf{p} - q\mathbf{A}) = \frac{1}{c}\left(\frac{mc^2}{\sqrt{1-\beta^2}},\quad \frac{mc^2}{\sqrt{1-\beta^2}}\boldsymbol{\beta}\right),\quad (\varepsilon - q\varphi)^2 - (\mathbf{p} - q\mathbf{A})^2 = (mc)^2 \tag{20}$$

Obviously, the permutation of the components of the generalized momentum for the construction of the invariant does not solve the posed problem. The statement that the expression (20) is the mechanical momentum of a particle and therefore is an invariant is unproven and it is necessary to apprehend the formula (20) as an empirical. Therefore, at high velocities or strong interactions, an unaccounted dependence of the energy of particle interaction with the field on the velocity of the particle motion, which results to the erroneous results or the impossibility of calculations.

In [26], an invariant representation of the generalized momentum of the system was suggested, where the dependence of the interaction energy of the particle with the field on the velocity was taken into account:

$$\mathbf{P} = (\varepsilon,\ \mathbf{p}) = \frac{1}{c}\left(\frac{mc^2 + q\varphi + q\boldsymbol{\beta}\cdot\mathbf{A}}{\sqrt{1-\beta^2}},\quad \frac{(mc^2 + q\varphi)\boldsymbol{\beta} + q\mathbf{A}_{\parallel}}{\sqrt{1-\beta^2}} + q\mathbf{A}_{\perp}\right) \tag{21}$$

$$\mathbf{P}^2 = \varepsilon^2 - \mathbf{p}^2 = \frac{(mc^2 + q\varphi)^2 - (q\mathbf{A})^2}{c^2}, \tag{22}$$

which is the four-dimensional representation of the generalized momentum of the system based on the expression for the generalized momentum of an immobile particle in a state of rest

$$\mathbf{P}_0 = (\varepsilon_0,\ \mathbf{p}_0) = \frac{1}{c}(mc^2 + q\varphi,\ q\mathbf{A}) \tag{23}$$

whose invariant is always equal to the expression (19) regardless of the state of the system.

The application of variational principles to construct the relativistic and quantum theory was based on the principles of construction the mechanics with the help of the Lagrangian of the system [27], which originally was not intended for relativistic approaches. The Lagrangian construction is parametric with the one time variable $\tau = ct$, singled out from the variables of the four-dimensional space (the rest are represented by the dependence on this variable τ) and contains the total differential with respect to this variable, the velocity of the particle. Such a construction is unacceptable because of the impossibility to apply the principle of invariance of the representation of variables and the covariant representation of the action of the system.

In [28], to construct the relativistic theory on the basis of variational principles, the canonical (non- parametric) solutions of the variational problem for canonically defined integral functionals have been considered and the canonical solutions of the variational problems of mechanics in the Minkowski spaces are written. Because of unifying the variational principles of least action, flow, and hyperflow, the canonically invariant equations for the generalized momentum are obtained. From these equations, the expressions for the action function and the wave function are obtained as the general solution of the unified variational problem of mechanics.

Below, we present the generalized invariance principle and the corresponding representation of the generalized momentum of the system, the equations of relativistic and quantum mechanics [29], give the solutions of the problems of charge motion in a constant electric field, the problems for a particle in a potential well and the passage of a particle through a potential barrier, the problems of motion in an exponential field (Morse), the problems of charged particle in a magnetic field, and also the problems of a hydrogen atom are given.

2. Principle of invariance

2.1 Generalization of the principle of invariance

The principle of invariance of the representation of a generalized pulse is applicable also in the case of motion of a particle with the velocity $\mathbf{v}$ and in the case of a transition to a reference frame moving with the velocity $\mathbf{V}$.

The four-dimensional momentum of a particle $\mathbf{P}$ with the rest mass m moving with the velocity $\boldsymbol{\beta} = \mathbf{v}/c$ is represented in the form.

$$\mathbf{P} = (\varepsilon,\ \mathbf{p}) = \left(\frac{mc}{\sqrt{1-\beta^2}},\ \ \frac{mc}{\sqrt{1-\beta^2}}\boldsymbol{\beta}\right), \mathbf{P}^2 = \varepsilon^2 - \mathbf{p}^2 = (mc)^2 \tag{24}$$

This is the property of invariance of the representation of the four-dimensional momentum $\mathbf{P}$ in terms of the velocity of the particle $\boldsymbol{\beta} = \mathbf{v}/c$.

If to consider the representation of the four-dimensional momentum of an immobile particle with a mass m by transition into the reference frame moving with the velocity $\boldsymbol{\beta}' = \mathbf{V}/c$, for the four-dimensional particle momentum $\mathbf{P}$ we have.

$$\mathbf{P} = (\varepsilon,\ \mathbf{p}) = \left(\frac{mc}{\sqrt{1-\beta^2}},\ \ \frac{mc}{\sqrt{1-\beta^2}}\boldsymbol{\beta}'\right), \mathbf{P}^2 = \varepsilon^2 - \mathbf{p}^2 = (mc)^2. \tag{25}$$

This is a property of invariance of the representation of the four-dimensional momentum $\mathbf{P}$ through the velocity of the reference system $\boldsymbol{\beta}' = \mathbf{V}/c$.

For an invariant of the system I, we have

$$I^2 = \mathbf{P}^2 = \varepsilon^2 - \mathbf{p}^2 = (\varepsilon_0)^2 = (mc)^2. \tag{26}$$

At $\boldsymbol{\beta} = \boldsymbol{\beta}' = 0$, we obtain

$$\mathbf{P} = (\varepsilon, \ \mathbf{p})\big|_{\boldsymbol{\beta}=\boldsymbol{\beta}'=0} = \varepsilon_0(1, \ 0) = mc(1, \ 0). \tag{27}$$

Thus, the generalized momentum of a particle has an invariant representation on the particle velocity $\mathbf{v}$ and the velocity of the reference system $\mathbf{V}$. This property should be considered because of the general principle of the relativity of motion. Accordingly, the generalized momentum of the particle $\mathbf{P}$ is an invariant regardless of the state of the system.

If a charged particle is in an external electromagnetic field with potentials $(\varphi, \ \mathbf{A})$, then the stationary charge sees the field exactly with such potentials. If the charge has a nonzero velocity $\mathbf{v}$, then it will interact with the field differently. To determine the interaction for a charge moving with the velocity $\mathbf{v}$, one can start from the principle of the relativity of motion. The effective values of the force or interaction with the field of the charge moving with the velocity v are the same as in the case when the charge is immobile, and the field moves with the velocity $-\mathbf{v}$ (in the laboratory frame of reference).

The fact that the interaction of a charged particle with a field depends on the speed of motion is evidently represented in the formula for the Liénard-Wiechert potential [8].

More clearly, this can be demonstrated by an example of the Doppler effect for two atoms in the field of a resonant radiation, when one of the atoms is at rest and the other moves with the velocity $\mathbf{v}$ (**Figure 4**).

The atom, which is at rest, absorbs a photon, and the moving one does not absorb or interacts weakly with the field, because of the dependence of the interaction on the velocity of the atom. It is also known that the acting field for an atom moving with the velocity v corresponds to the interaction with the field moving with the velocity $-\mathbf{v}$.

2.2 Invariant representation of the generalized momentum

Thus, for a moving charge, the effective values of the potentials $(\varphi', \ \mathbf{A}')$ (in the laboratory frame of reference) can be written in the form [8]

$$(\varphi', \mathbf{A}') = \left(\frac{\varphi + \boldsymbol{\beta} \cdot \mathbf{A}}{\sqrt{1-\beta^2}}, \quad \mathbf{A}_\perp + \frac{\mathbf{A}_\parallel + \varphi \boldsymbol{\beta}}{\sqrt{1-\beta^2}} \right). \tag{28}$$

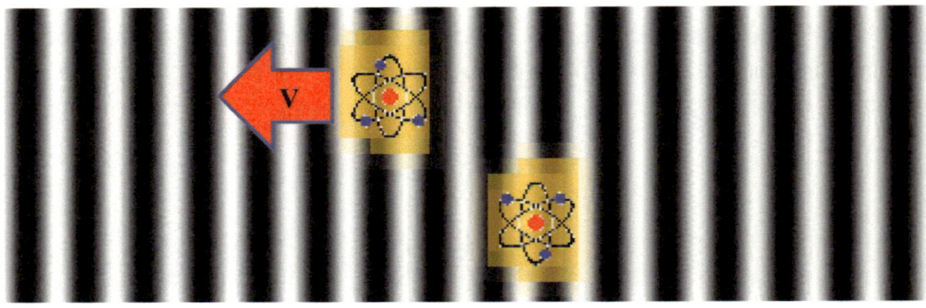

Figure 4.
Two atoms in the field of a resonant radiation.

If one represents the generalized momentum of the particle in the form

$$\mathbf{P} = \frac{1}{c}\left(\frac{mc^2}{\sqrt{1-\beta^2}} + q\varphi', \quad \frac{mc^2}{\sqrt{1-\beta^2}}\boldsymbol{\beta} + q\mathbf{A}'\right), \tag{29}$$

where φ' and $\mathbf{A}'$ already effective values of the interaction potentials of the particle moving with velocity $\mathbf{v}$ in a field with the potentials φ and $\mathbf{A}$, we obtain

$$\mathbf{P} = \frac{1}{c}\left(\frac{mc^2 + q\varphi + q\boldsymbol{\beta}\cdot\mathbf{A}}{\sqrt{1-\beta^2}}, \quad \frac{(mc^2 + q\varphi)\boldsymbol{\beta} + q\mathbf{A}_\parallel}{\sqrt{1-\beta^2}} + q\mathbf{A}_\perp\right). \tag{30}$$

The expression (30) can be represented in the form

$$\mathbf{P} = \left(\frac{mc^2 + q\varphi + q\boldsymbol{\beta}\cdot\mathbf{A}}{c\sqrt{1-\beta^2}}, \quad \frac{mc^2 + q\varphi + q\boldsymbol{\beta}\cdot\mathbf{A}}{c\sqrt{1-\beta^2}}\boldsymbol{\beta} + \frac{q}{c}\mathbf{A} - \frac{q}{c}\frac{1}{1+\sqrt{1-\beta^2}}(\mathbf{A}\cdot\boldsymbol{\beta})\boldsymbol{\beta}\right) \tag{31}$$

or

$$\mathbf{P} = \left(\varepsilon, \quad \varepsilon\boldsymbol{\beta} + \frac{q}{c}\mathbf{A} - \frac{q}{c}\frac{1}{1+\sqrt{1-\beta^2}}(\mathbf{A}\cdot\boldsymbol{\beta})\boldsymbol{\beta}\right). \tag{32}$$

This transformation can be presents in matrices form

$$\{\varepsilon', \ \mathbf{p}'\} = \{\varepsilon, \ \mathbf{p}\} + \hat{\mathrm{T}}\{\varepsilon, \ \mathbf{p}\} \tag{33}$$

where a Lorentz transformation have a form

$$\hat{1} + \hat{\mathrm{T}} = \begin{Vmatrix} 1 & 0 & 0 & 0 \\ 0 & 1 & 0 & 0 \\ 0 & 0 & 1 & 0 \\ 0 & 0 & 0 & 1 \end{Vmatrix} + (\gamma - 1)\begin{Vmatrix} 1 & 0 & 0 & 0 \\ 0 & 0 & 0 & 0 \\ 0 & 0 & 0 & 0 \\ 0 & 0 & 0 & 0 \end{Vmatrix} + \gamma\begin{Vmatrix} 0 & \beta_1 & \beta_2 & \beta_3 \\ \beta_1 & 0 & 0 & 0 \\ \beta_2 & 0 & 0 & 0 \\ \beta_3 & 0 & 0 & 0 \end{Vmatrix} + (\gamma - 1)\begin{Vmatrix} 0 & 0 & 0 & 0 \\ 0 & \frac{\beta_1\beta_1}{\beta^2} & \frac{\beta_1\beta_2}{\beta^2} & \frac{\beta_1\beta_3}{\beta^2} \\ 0 & \frac{\beta_2\beta_1}{\beta^2} & \frac{\beta_2\beta_2}{\beta^2} & \frac{\beta_2\beta_3}{\beta^2} \\ 0 & \frac{\beta_3\beta_1}{\beta^2} & \frac{\beta_3\beta_2}{\beta^2} & \frac{\beta_3\beta_3}{\beta^2} \end{Vmatrix} \tag{34}$$

The matrices of the invariant representation of a four-dimensional vector, which preserve the vector module in four-dimensional space, form the Poincare group (inhomogeneous Lorentz group). In addition to displacements and rotations, the group contains space-time reflection representations $\hat{\mathrm{P}}$, $\hat{\mathrm{T}}$ and inversion $\hat{\mathrm{P}}\hat{\mathrm{T}} = \hat{\mathrm{I}}$.

For the module I of the four-dimensional vector of the generalized momentum $\mathbf{P}$, we have

$$I^2 = \mathbf{P}^2 = \varepsilon^2 - \mathbf{p}^2 = \frac{(mc^2 + q\varphi)^2 - (q\mathbf{A})^2}{c^2}, \tag{35}$$

which is the four-dimensional representation of the generalized momentum of the system on the basis of the expression of the generalized momentum of a particle in the state of rest

$$\mathbf{P}_0 = (\varepsilon_0,\ \mathbf{p}_0) = \frac{1}{c}(mc^2 + q\varphi,\ q\mathbf{A}), \tag{36}$$

whose invariant is defined by the expression (30).

Thus, the generalized momentum of the particle in an external field is not only invariant relative to the transformations at the transition from one reference system to another but also has an invariant representation in terms of the velocity of motion of the particle (30); at each point of space, the value of the invariant I is determined by the expression (35). This property has not only the representation of the proper momentum of the particle (the mechanical part), but also the generalized momentum of the particle in general.

Let us generalize this result to the case of representation of the generalized momentum of any systems and interactions, arguing that, regardless of the state (the motion) of the system, the generalized four-dimensional momentum always has an invariant representation

$$\mathbf{P} = (\varepsilon,\ \mathbf{p}) \ \Rightarrow\ \mathbf{P}^2 = \varepsilon^2 - \mathbf{p}^2 = \varepsilon_0{}^2 - \mathbf{p}_0{}^2 = \pm I^2 = \text{inv}, \tag{37}$$

where ε и p are the energy and momentum of the system, respectively, and the invariant is determined by the modulus of sum of the components of the generalized momentum of the system ε_0 and $\mathbf{p}_0$ at rest. If the particles interact with the field in the form $\varepsilon_0 + \alpha\varphi$, the invariants of the generalized momentum of the system are represented by the expressions [25].

$$\begin{aligned}
\mathbf{P}_+{}^2 &= (\varepsilon_0 + \alpha\varphi)^2 - (\alpha\mathbf{A})^2 = \varepsilon_0{}^2 + 2\varepsilon_0\alpha\varphi + (\alpha\varphi)^2 - (\alpha\mathbf{A})^2,\\
\mathbf{P}_-{}^2 &= (\alpha\varphi)^2 - (\varepsilon_0\mathbf{n} + \alpha\mathbf{A})^2 = -\varepsilon_0{}^2 - 2\varepsilon_0\alpha\mathbf{n}\cdot\mathbf{A} + (\alpha\varphi)^2 - (\alpha\mathbf{A})^2,\\
\mathbf{P}_0{}^2 &= (\varepsilon_0 + \alpha\varphi)^2 - (\varepsilon_0\mathbf{n} + \alpha\mathbf{A})^2 = 2\varepsilon_0\alpha(\varphi - \mathbf{n}\cdot\mathbf{A}) + (\alpha\varphi)^2 - (\alpha\mathbf{A})^2.
\end{aligned} \tag{38}$$

Let us represent the expression for the invariant $\varepsilon^2 - p^2$ (35) in the following form

$$\varepsilon^2 = \frac{E^2}{c^2} = \mathbf{p}^2 + \frac{(mc^2 + q\varphi)^2 - (q\mathbf{A})^2}{c^2} = \mathbf{p}^2 + m^2c^2 + 2mq\varphi + \frac{q^2}{c^2}(\varphi^2 - \mathbf{A}^2) \tag{39}$$

and divide it by $2m$. Grouping, we obtain the Hamiltonian H of the system in the form

$$\mathrm{H} = \frac{\varepsilon^2 - m^2c^2}{2m} = \frac{E^2 - m^2c^4}{2mc^2} = \frac{\mathbf{p}^2}{2m} + q\varphi + \frac{q^2}{2mc^2}(\varphi^2 - \mathbf{A}^2), \tag{40}$$

that is, we obtain the formula for the correspondence between the energy of the system E and the energy of the system in the classical meaning H. The correspondence in the form $\mathrm{H} = \mathbf{p}^2/2m + U(\tau, r)$ [26] will be complete and accurate if we determine the potential energy of interaction U and the energy of system in the classical meaning as

$$U = q\varphi + \frac{q^2}{2mc^2}\left(\varphi^2 - \mathbf{A}^2\right), \quad \mathrm{H} = \frac{E^2 - m^2c^4}{2mc^2} \quad \Rightarrow \quad E = \pm mc^2\sqrt{1 + \frac{2\mathrm{H}}{mc^2}}. \tag{41}$$

For example, the potential energy U of the electron in the field of the Coulomb potential $\varphi = Ze/r$ and in a homogeneous magnetic field $\mathbf{B}$ with the vector potential $\mathbf{A} = [\mathbf{r} \times \mathbf{B}]/2$ is

$$U = -e\varphi + \frac{e^2}{2mc^2}\left(\varphi^2 - \mathbf{A}^2\right) = -\frac{Ze^2}{r} + \frac{1}{2mc^2}\frac{Z^2e^4}{r^2} - \frac{e^2B^2}{8mc^2}r^2\sin^2\theta. \tag{42}$$

Note, whatever is the dependence of the potential φ, the possible minimum potential energy $U_{\min} = -mc^2/2$, and the potential energy as a function of the vector potential is always negative. The hard constraint of the classical potential energy value $U_{\min} = -mc^2/2$, which does not depend on the nature of the interactions, results in the fundamental changes in the description of interactions and the revision of the results of classical mechanics. At short distances, the origination of repulsion for attraction forces caused by the uncertainty principle is clearly reflected in the expression for the potential energy of the particle.

Many well-known expressions of the potential energy of interaction with attractive fields have a repulsive component in the form of half the square of these attractive potentials—Kratzer [30], Lennard-Jones [31], Morse [32], Rosen [33] and others. Expression (41) justifies this approach, which until now is phenomenological or the result of an appropriate selection for agreement with experimental data.

The Hamiltonian H can be called the energy and its value remains constant in the case of conservation of energy E, but the value of Hand its changes differ from the true values of the energy E and changes of its quantity. Thus, the classical approaches are permissible only in the case of low velocities, when $\mathrm{H} \ll mc^2$ and the energy expression can be represented in the form

$$E = mc^2\sqrt{1 + \frac{2\mathrm{H}}{mc^2}} \approx mc^2 + \mathrm{H}. \tag{43}$$

3. Equations of relativistic mechanics

3.1 Canonical Lagrangian and Hamilton-Jacoby equation

Let us use the parametric representation of the Hamilton action in the form [28].

$$S = -\int_{t_1,\ \mathbf{r}_1}^{t_2,\ \mathbf{r}_2} (\varepsilon dt - \mathbf{p}\cdot d\mathbf{r}) = -\int_{\mathbf{R}_1}^{\mathbf{R}_2} \mathbf{P}\cdot d\mathbf{R} = -\int_{\mathbf{R}_1}^{\mathbf{R}_2} \mathbf{P}\cdot\frac{d\mathbf{R}}{ds}ds = -\int_{\mathbf{R}_1}^{\mathbf{R}_2} (\mathbf{P}\cdot\mathbf{V})ds \rightarrow \min, \tag{44}$$

where ds is the four-dimensional interval and $\mathbf{V}$ is the four-dimensional generalized velocity.

The functional that takes into account the condition of the invariant representation of the generalized momentum $\mathbf{P}^2 = \varepsilon^2 - \mathbf{p}^2 = I^2 = inv$, can be composed by the method of indefinite Lagrange coefficients in the form

$$S = \int_{s_1}^{s_1}\left(-\mathbf{P}\cdot\mathbf{V} + \frac{\mathbf{P}^2 - I^2}{2\lambda}\right)ds = \int_{s_1}^{s_1}\left(\frac{(\mathbf{P} - \lambda\mathbf{V})^2 + \lambda^2 - I^2}{2\lambda} - \frac{}{2\lambda}\right)ds \rightarrow \min, \tag{45}$$

where $\lambda = \lambda(s)$ is the given parameter, determined by the condition of invariance of the representation. Because λ and I are given and they do not depend on the velocity, we have an explicit solution in the form

$$\mathbf{P} - \lambda \mathbf{V} = 0, \qquad \lambda = \pm I(\tau, \mathbf{r}), \tag{46}$$

where the four-dimensional momentum is represented in the form

$$\mathbf{P} = I\mathbf{V} = \sqrt{\varepsilon^2 - \mathbf{p}^2}\left(\frac{1}{\sqrt{1-\eta^2}}, \quad \frac{\boldsymbol{\eta}}{\sqrt{1-\eta^2}}\right). \tag{47}$$

Thus, the action is represented in the form

$$S = \int_{s_1}^{s_2} I ds = \int_{s_1}^{s_2} \sqrt{\varepsilon^2 - \mathbf{p}^2} ds = \int_{\tau_1}^{\tau_2} \sqrt{\varepsilon^2 - \mathbf{p}^2}\sqrt{1-\eta^2} d\tau \tag{48}$$

and the canonical Lagrangian of the system is given by

$$\mathrm{L} = I\sqrt{1-\eta^2} = \sqrt{\varepsilon^2 - \mathbf{p}^2}\sqrt{1-\eta^2}. \tag{49}$$

The correctness of the presented parametrization is confirmed by the obtained expressions for the generalized momentum and energy from the Lagrangian of the system in the form

$$\begin{aligned} \varepsilon &= \boldsymbol{\eta}\frac{\partial \mathrm{L}}{\partial \boldsymbol{\eta}} - \mathrm{L} = \frac{I}{\sqrt{1-\eta^2}} = \frac{\sqrt{\varepsilon^2 - \mathbf{p}^2}}{\sqrt{1-\eta^2}}, \\ \mathbf{p} &= \frac{\partial \mathrm{L}}{\partial \boldsymbol{\eta}} = \frac{I}{\sqrt{1-\eta^2}}\boldsymbol{\eta} = \varepsilon \boldsymbol{\eta}, \end{aligned} \tag{50}$$

which coincide with the initial representations of the generalized momentum and energy. Accordingly, the Lagrange equation of motion takes the form

$$\frac{d\mathbf{p}}{d\tau} = -\frac{I}{\varepsilon}\frac{\partial I}{\partial \mathbf{r}}. \tag{51}$$

If we multiply Eq. (50) by $\mathbf{p} = \varepsilon\boldsymbol{\eta}$ scalarly, after reduction to the total time differential, we obtain,

$$\frac{d\varepsilon^2}{d\tau} = \frac{\partial I^2}{\partial \tau}. \tag{52}$$

If the invariant is clearly independent of time, then the energy ε is conserved and the equation of motion is represented in the form of the Newtonian equation

$$\frac{d\boldsymbol{\eta}}{d\tau} = -\frac{I}{\varepsilon^2}\frac{\partial I}{\partial \mathbf{r}}. \tag{53}$$

For a particle in an external field we have

$$\mathrm{L} = -\frac{1}{c}\sqrt{(mc^2 + q\varphi)^2 - (q\mathbf{A})^2}\sqrt{1-\eta^2/c^2}. \tag{54}$$

Using the explicit form of the generalized momentum (32) with the accuracy of the expansion to the power of β^2, we obtain the equation of motion in the form

$$\frac{d}{d\tau}\left(\varepsilon - \frac{q}{2c}\mathbf{A}\cdot\boldsymbol{\beta}\right)\boldsymbol{\beta} = q\mathbf{E} + q[\boldsymbol{\beta}\times\mathbf{B}] - \frac{\partial}{\partial\mathbf{r}}\left(\frac{q}{c}\mathbf{A}\cdot\boldsymbol{\beta} + \frac{q^2}{2mc^2}\left(\varphi^2 - \mathbf{A}^2\right)\right) \tag{55}$$

where the velocity-dependent components of the force are present. In particular, the velocity-dependent force is present in the Faraday law of electromagnetic induction [34], which is absent in the traditional expression for the Lorentz force.

The Hamilton-Jacobi equation is represented in the form

$$\left(\frac{\partial S}{\partial\tau}\right)^2 - \left(\frac{\partial S}{\partial\mathbf{r}}\right)^2 = \frac{\left(mc^2 + q\varphi\right)^2 - \left(q\mathbf{A}\right)^2}{c^2} \tag{56}$$

and it reflects the invariance of the representation of the generalized momentum. The well-known representations of the Hamilton-Jacobi Eq. (8) also contain the differential forms of potentials—the components of the electric and the magnetic fields.

3.2 Motion of a charged particle in a constant electric field

Let us consider the motion of a charged particle with the mass m and charge $-q$ in the constant electric field between the plane electrodes with the potential difference U and the distance l between them. For one-dimensional motion, taking the cathode location as the origin and anode at the point $x = l$, from (56) we have

$$\left(\frac{\partial S}{\partial\tau}\right)^2 - \left(\frac{\partial S}{\partial x}\right)^2 = \frac{\left(mc^2 + qU(1 - x/l)\right)^2}{c^2}. \tag{57}$$

Let us represent the action S in the form

$$S = -Et + f(x), \tag{58}$$

where $E = mc^2 + qU$ is an the electron energy at the origin on the surface of the cathode under voltage $-U$; as a result, from (57) we obtain

$$S = -Et + \frac{1}{c}\int\sqrt{E^2 - \left(mc^2 + qU - qU\frac{x}{l}\right)^2}\,dx. \tag{59}$$

We find the solution from the condition $\partial S/\partial E = \text{const}$. As a result of integration, we obtain

$$t = \frac{l}{c}\left(1 + \frac{1}{\alpha}\right)\arccos\left(1 - \frac{\alpha}{1+\alpha}\frac{x}{l}\right), \qquad \alpha = qU/mc^2 \tag{60}$$

or

$$x = l\frac{1+\alpha}{\alpha}\left(1 - \cos\left(\frac{\alpha}{1+\alpha}\frac{ct}{l}\right)\right), \qquad t \le \frac{l}{c}\left(\frac{1+\alpha}{\alpha}\right)\arccos\left(\frac{1}{1+\alpha}\right). \tag{61}$$

The well-known solution in the framework of the traditional theory [8] is the following:

$$t = \frac{l}{\alpha c}\sqrt{\left(1+\alpha\frac{x}{l}\right)^2 - 1} \text{ or } x = l\frac{(\alpha ct/l)^2}{1+\sqrt{1+(\alpha ct/l)^2}}, \qquad t \le \frac{l}{c}\sqrt{1+\frac{2}{\alpha}}. \tag{62}$$

In the ultrarelativistic limit $qU \gg mc^2$, the ratio of the flight time of the gap between the electrodes $(x = l)$ is equal to $\pi/2$ according to formulas (60) and (62) (**Figure 5**).

The electron velocity $v = dx/dt$ when reaching the anode is

$$v = c\sqrt{1 - 1/(1+\alpha)^2}. \tag{63}$$

3.3 Problem of the hydrogen-like atom

Let us consider the motion of an electron with the mass m and charge $-e$ in the field of an immobile nucleus with the charge Ze. Then the problem reduces to an investigation of the motion of the electron in the centrally symmetric electric field with the potential $-Ze^2/r$.

Choosing the polar coordinates (r, φ) in the plane of motion, we obtain the Hamilton-Jacobi equation in the form

$$\left(\frac{\partial S}{\partial \tau}\right)^2 - \left(\frac{\partial S}{\partial r}\right)^2 - \frac{1}{r^2}\left(\frac{\partial S}{\partial \varphi}\right)^2 - \frac{\left(mc^2 - Ze^2/r\right)^2}{c^2} = 0. \tag{64}$$

Let us represent the action S in the form

$$S = -Et + M\varphi + f(r), \tag{65}$$

where E and M are the constant energy and angular momentum of the moving particle, respectively. As a result, we obtain

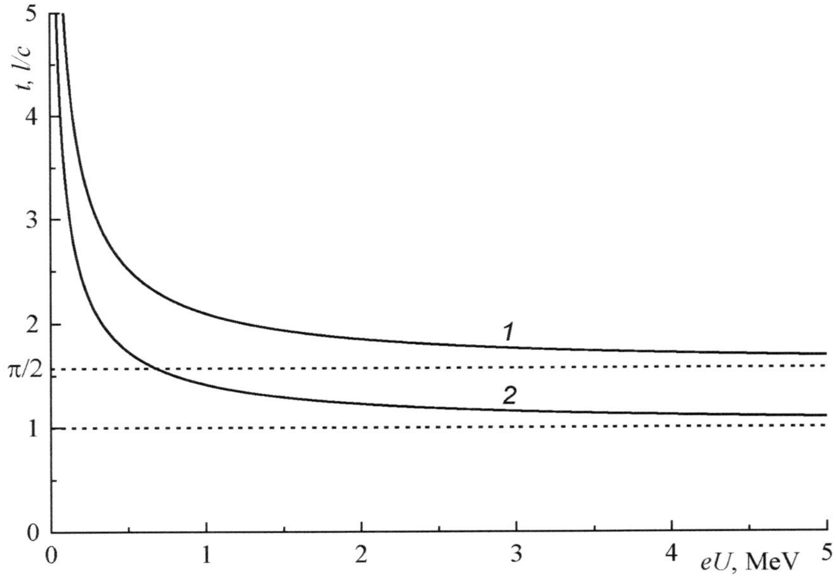

Figure 5.
Dependence of the flight time of the gap between the electrodes on the applied voltage according to the formula (60) and (curve 1) and (62) (curve 2) in l/c units.

$$S = -Et + M\varphi + \frac{1}{c}\int\sqrt{E^2 - (mc^2)^2 + 2mc^2\frac{Ze^2}{r} - \frac{M^2c^2 + (Ze^2)^2}{r^2}}dr. \tag{66}$$

We find trajectories from the condition $\partial S/\partial M =$ const, with use of which we obtain,

$$\varphi = \int\frac{Mc}{\sqrt{E^2 - (mc^2)^2 + 2mc^2\frac{Ze^2}{r} - \frac{M^2c^2 + Ze^2}{r^2}}}d\frac{1}{r}, \tag{67}$$

which results in the solution

$$r = \frac{(Mc)^2 + (Ze^2)^2}{mc^2Ze^2}\frac{1}{1 + \sqrt{\left(\frac{E}{mc^2}\right)^2\left(1 + \left(\frac{Mc}{Ze^2}\right)^2\right) - \left(\frac{Mc}{Ze^2}\right)^2}\cos\left(\varphi\sqrt{1 + \left(\frac{Ze^2}{Mc}\right)^2}\right)}. \tag{68}$$

The coefficient of the repulsive effective potential is essentially positive, that is, $M^2c^2 + (Ze^2)^2 > 0$ therefore, any fall of the particle onto the center is impossible. The minimum radius $r_{\min} = r_0(Z+1)$, where $r_0 = e^2/mc^2$ is the classical radius of an electron.

The secular precession is found from the condition

$$\varphi\sqrt{1 + (Ze^2/Mc)^2} = 2\pi, \tag{69}$$

whence, we obtain

$$\Delta\varphi = 2\pi - \frac{2\pi}{\sqrt{1 + (Ze^2/Mc)^2}} \approx \pi\left(\frac{Ze^2}{Mc}\right)^2, \tag{70}$$

that has the opposite sign as compared with the solution in [8]. The reason for the discrepancy of the sign is the unaccounted interaction of the self-momentum with the rotating field, that is, the spin-orbit interaction.

4. Equations of the relativistic quantum mechanics

Using the principle of the invariant representation of the generalized momentum

$$\mathbf{P}^2 = \varepsilon^2 - \mathbf{p}^2 = I^2 = \text{inv}, \tag{71}$$

it is possible to compose the corresponding equation of the relativistic quantum mechanics by representing the energy and momentum variables by the corresponding operators $\hat{\varepsilon} = i\hbar\partial/\partial\tau$ and $\hat{\mathbf{p}} = -i\hbar\partial/\partial\mathbf{r}$:

$$\begin{gathered}(\hat{\varepsilon})^2\Psi - (\hat{\mathbf{p}})^2\Psi = \left(i\hbar\frac{\partial}{\partial\tau}\right)^2\Psi - \left(-i\hbar\frac{\partial}{\partial\mathbf{r}}\right)^2\Psi = \\ (\varepsilon^2 - \mathbf{p}^2)\Psi + i\hbar\left(\frac{\partial\varepsilon}{\partial\tau} + \text{div}\mathbf{p}\right) = I^2\Psi + i\hbar\left(\frac{\partial\varepsilon}{\partial\tau} + \text{div}\mathbf{p}\right),\end{gathered} \tag{72}$$

and

$$(\hat{\varepsilon}\Psi)^2 - (\hat{\mathbf{p}}\Psi)^2 = \left(i\hbar\frac{\partial\Psi}{\partial\tau}\right)^2 - \left(-i\hbar\frac{\partial\Psi}{\partial\mathbf{r}}\right)^2 = (\varepsilon^2 - \mathbf{p}^2)\Psi^2 = I^2\Psi^2. \tag{73}$$

The case of conservative systems, when any energy losses or sources in space are absent, corresponds to the relation $\partial\varepsilon/\partial\tau + \mathrm{div}\mathbf{p} = 0$. In this way,

$$\begin{cases} \dfrac{\partial^2\Psi}{\partial\tau^2} - \dfrac{\partial^2\Psi}{\partial\mathbf{r}^2} = -\dfrac{I^2}{\hbar^2}\Psi \\ \left(\frac{\partial\Psi}{\partial\tau}\right)^2 - \left(\frac{\partial\Psi}{\partial\mathbf{r}}\right)^2 = -\dfrac{I^2}{\hbar^2}\Psi^2. \end{cases} \tag{74}$$

For the charged particle in an external field with an invariant in the form of (30), the equations will take the form

$$\begin{cases} \dfrac{\partial^2\Psi}{\partial\tau^2} - \dfrac{\partial^2\Psi}{\partial\mathbf{r}^2} = -\dfrac{(mc^2+q\varphi)^2 - (q\mathbf{A})^2}{\hbar^2c^2}\Psi \\ \left(\frac{\partial\Psi}{\partial\tau}\right)^2 - \left(\frac{\partial\Psi}{\partial\mathbf{r}}\right)^2 = -\dfrac{(mc^2+q\varphi)^2 - (q\mathbf{A})^2}{\hbar^2c^2}\Psi^2. \end{cases} \tag{75}$$

For stationary states we obtain

$$\begin{cases} \dfrac{\partial^2\Psi}{\partial\mathbf{r}^2} + \dfrac{E^2 - (mc^2+q\varphi)^2 + (q\mathbf{A})^2}{\hbar^2c^2}\Psi = 0 \\ \left(\frac{\partial\Psi}{\partial\mathbf{r}}\right)^2 + \dfrac{E^2 - (mc^2+q\varphi)^2 + (q\mathbf{A})^2}{\hbar^2c^2}\Psi^2 = 0. \end{cases} \tag{76}$$

Rewriting the equations taking into account the formulas of the classical correspondence (40), we will obtain the equations for the wave function in the traditional representation

$$\begin{gathered} \Delta\Psi + \frac{2m}{\hbar^2}(\mathrm{H} - U)\Psi = 0, \\ \left(\frac{\partial\Psi}{\partial\mathbf{r}}\right)^2 + \frac{2m}{\hbar^2}(\mathrm{H} - U)\Psi^2 = 0, \end{gathered} \tag{77}$$

the first of which formally coincides with the Schrödinger equation for the wave function of stationary states.

For the action function S associated with the wave function by the representation $\Psi = A\ exp\,(-iS/\hbar)$ or $S = i\hbar\ln\Psi + i\hbar\ln A$, we will obtain

$$\begin{cases} \dfrac{\partial^2 S}{\partial\mathbf{r}^2} = 0 \\ \left(\frac{\partial S}{\partial\mathbf{r}}\right)^2 - \dfrac{E^2 - (mc^2+q\varphi)^2 + (q\mathbf{A})^2}{c^2} = 0 \end{cases} \Rightarrow \begin{cases} \dfrac{\partial^2 S}{\partial\mathbf{r}^2} = 0 \\ \left(\frac{\partial S}{\partial\mathbf{r}}\right)^2 - 2m(\mathrm{H} - U) = 0, \end{cases} \tag{78}$$

which represents the exact classical correspondence instead of the quasiclassical approximation [12]. Note, the equations similar to (78) also follow from the Eq. (46) in [12] if we demand for an exact correspondence and equate to zero the real and imaginary parts.

4.1 Particle in the one-dimensional potential well

Let us consider the particle of mass m in a one-dimensional rectangular potential well of the form

$$V(x) = \begin{cases} 0, & 0 \ge x \ge a \\ -V_0, & 0 \le x \le a. \end{cases} \tag{79}$$

From the first equation of system (70) we have

$$\frac{d^2\Psi}{dx^2} + \frac{E^2 - (mc^2 + V(x))^2}{\hbar^2 c^2}\Psi = 0. \tag{80}$$

Then, $U_0 = -V_0 + {V_0}^2/(2mc^2)$ corresponds to the potential energy of the particle in the well in the classical meaning. In the latter case, it is known [12] that the bound state with the energy $H = 0 (E = mc^2)$ arises under the conditions

$$U_0 = -\frac{\pi^2\hbar^2}{2ma^2}n^2 = -V_0 + \frac{{V_0}^2}{2mc^2} \ge -\frac{mc^2}{2}, \quad a \ge \frac{\pi\hbar}{mc}n = \frac{\lambda}{2}n, \tag{81}$$

$$\begin{aligned} E_n &= mc^2\left(1 - \sqrt{1 - \frac{\pi^2\hbar^2}{m^2c^2a^2}n^2}\right) = mc^2\left(1 - \sqrt{1 - \left(\frac{\lambda}{2a}n\right)^2}\right) \\ &= mc^2\frac{\left(\frac{\lambda}{2a}n\right)^2}{1 + \sqrt{1 - \left(\frac{\lambda}{2a}n\right)^2}}, \end{aligned} \tag{82}$$

where $\lambda = 2\pi\lambda\!\!\!^{-} = 2\pi\hbar/mc = h/mc$. Maximum depth of the classic well is equal to $U_0 = -mc^2/2$ at $V_0 = mc^2$. The condition for the existence of the bound state with an energy $H = 0$ $(E = mc^2)$ in a potential well of size a is expressed by the relation

$$a = \lambda n/2, \qquad n = 1,\ 2,\ 3\ldots \tag{83}$$

In the three-dimensional case, the bound state with the energy $H = 0$ $(E = mc^2)$ arises under the same conditions [23] for a spherical well with a diameter d and depth V_0 with the $d = \lambda n/2$, $n = 1, 2, 3\ldots$.

The solution of this simple example is fundamental and accurately represents the uncertainty principle $\Delta x \Delta p \ge \hbar/2$. It clearly represents the wave property of the particle, clearly showing that the standing wave exists only at the condition $a \ge \lambda/2$ when the geometric dimensions of the well are greater than half the wavelength of the particle.

4.2 Penetration of a particle through a potential barrier

Let us consider the problem of penetration of a particle through the rectangular potential barrier [23] with the height V_0 and width a. Then, $U_0 = V_0 + {V_0}^2/(2mc^2)$ corresponds to the potential energy of the particle in the well in the classical meaning, and $H = (E^2 - m^2c^4)/2mc^2$ corresponds to the energy. Substituting these expressions into the solution of the Schrödinger equation for the rectangular potential barrier, we obtain for the transmission coefficient D of the particle penetrating through the potential barrier at $E > |V_0 + mc^2|$

$$D=\left[1+\frac{\left(\left(\frac{V_0}{mc^2}+1\right)^2-1\right)^2}{4\left(\left(\frac{E}{mc^2}\right)^2-1\right)\left(\left(\frac{E}{mc^2}\right)^2-\left(\frac{V_0}{mc^2}+1\right)^2\right)}\sin^2\left(\frac{a}{\lambdabar}\sqrt{\left(\frac{E}{mc^2}\right)^2-\left(\frac{V_0}{mc^2}+1\right)^2}\right)\right]^{-1} \tag{84}$$

and at $E<\left|V_0+mc^2\right|$

$$D=\left[1+\frac{\left(\left(\frac{V_0}{mc^2}+1\right)^2+1-2\left(\frac{E}{mc^2}\right)^2\right)^2}{4\left(\left(\frac{E}{mc^2}\right)^2-1\right)\left(\left(\frac{V_0}{mc^2}+1\right)^2-\left(\frac{E}{mc^2}\right)^2\right)}\sinh^2\left(\frac{a}{\lambdabar}\sqrt{\left(\frac{V_0}{mc^2}+1\right)^2-\left(\frac{E}{mc^2}\right)^2}\right)\right]^{-1} \tag{85}$$

where $\lambdabar=\hbar/mc$ is the de Broglie wavelength of the particle. As can be seen, the barrier is formed only in the energy range $-2mc^2>V_0>mc^2$.

For the problem of the passage of a particle with energy E through a potential barrier U (**Figure 2**) the wave vector k is represented as

$$k_1=\frac{1}{\hbar c}\sqrt{E^2-(mc^2)^2},\qquad k_2=\frac{1}{\hbar c}\sqrt{E^2-(mc^2+U)^2} \tag{86}$$

and if the particle energy does not exceed the potential barrier, then the transmission coefficient is zero, regardless of the height of the barrier and not have. In this case, there is no contradiction similar to the Klein paradox.

4.3 Charged particle in a magnetic field

The vector potential of a uniform magnetic field **A** along the **z** axis direction in the cylindrical coordinate system (ρ,φ,z) has components $A_\varphi=H\rho/2$, $A_\rho=A_z=0$ and Eq. (76) takes the form

$$\frac{\hbar^2}{2M}\left(R''+\frac{1}{\rho}R'\right)+\left(E-\frac{\hbar^2m^2}{2M}\frac{1}{\rho^2}+\frac{M\omega_H^2}{8}\rho^2-\frac{p_z^2}{2M}\right)R=0, \tag{87}$$

where m – angular quantum number, M – mass of electron, H– magnetic field value, $\omega_H=eH/Mc$. In this case, the equation below differs from the known [12] one by the absence of the field linear term $\hbar\omega_H m/2$ and the sign of a quadratic term $M\omega_H^2\rho^2/8$.

In this form, the Eq. (87) does not have a finite solution depending on the variable ρ and, provided R = const, we have

$$\begin{gathered}R''+\frac{1}{\rho}R'=0,\\ \left(E-\frac{\hbar^2m^2}{2M\rho^2}+\frac{M\omega_H^2}{8}\rho^2-\frac{p_z^2}{2M}\right)R=0.\end{gathered} \tag{88}$$

Or

$$E-\frac{\hbar^2m^2}{2M}\frac{1}{\rho^2}+\frac{M\omega_H^2}{8}\rho^2-\frac{p_z^2}{2M}\equiv 0. \tag{89}$$

From (89) we have for the energy levels

$$W = Mc^2\sqrt{1 + m^2\left(\frac{\lambda\!\!\!^{-}}{\rho}\right)^2 - \left(\frac{\rho}{2\rho_H}\right)^2 + \left(\frac{p_z}{Mc}\right)^2}. \tag{90}$$

where $\rho_H = c/\omega_H$ (magnetic event horizon), and ρ as a constant parameter.

If an electron is excited by a magnetic field from a state of rest, then $W = Mc^2$ and from (88) we obtain

$$-\frac{\hbar^2 m^2}{2M\rho^2} + \frac{M\omega_H^2}{8}\rho^2 = 0, \tag{91}$$

or

$$m\hbar\omega_H = M(\rho\omega_H)^2/2 = \frac{Mc^2}{2}\left(\frac{\rho}{\lambda\!\!\!^{-}_H}\right)^2 \tag{92}$$

From (92) for a magnetic flux quantum we have

$$\frac{e}{hc}H\pi\rho^2 = \frac{e}{hc}\Phi = m, \qquad \Delta\Phi = \frac{hc}{e}. \tag{93}$$

We get the same results when solving the Hamilton-Jacobi equation.

4.4 Particle in the field with Morse potential energy

We determine the energy levels for a particle moving in a field with a potential $\varphi(x) = -\varphi_0 e^{-x/d}$.

According to (41), for the potential energy of interaction $V(x)$ with the field $\varphi(x)$ we obtain the expression of the potential Morse energy (**Figure 6**)

$$V(x) = -q\varphi_0 e^{-x/d} + \frac{1}{2mc^2}\left(q\varphi_0 e^{-x/d}\right)^2 = mc^2\left(-\frac{q\varphi_0 e^{-x/d}}{mc^2} + \frac{1}{2}\left(\frac{q\varphi_0}{mc^2}e^{-x/d}\right)^2\right). \tag{94}$$

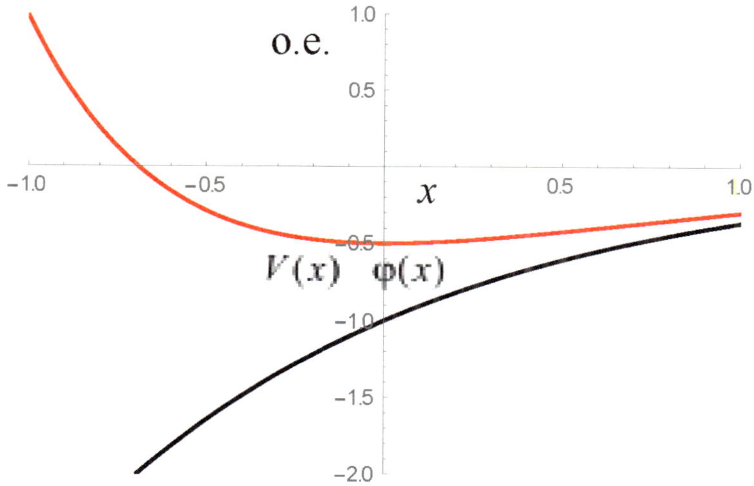

Figure 6.
The exponential potential of the field $\varphi(x)$ and Morse potential energy of interaction $V(x)$.

Schrödinger equation takes the form

$$\frac{d^2\psi}{dx^2}+\frac{2m}{\hbar^2}\left(E-mc^2\left(-\frac{q\varphi_0}{mc^2}e^{-x/d}+\frac{1}{2}\left(\frac{q\varphi_0}{mc^2}e^{-x/d}\right)^2\right)\right)\psi=0. \tag{95}$$

Following the procedure for solving Eq. (95) in [12], introducing a variable (taking values in the interval [0,∞]) and the notation

$$\xi=2d\frac{q\varphi_0}{mc^2}\frac{\lambda\!\!\!^{-}}{d}e^{-x/d},\qquad s=\frac{d}{\lambda\!\!\!^{-}}\sqrt{-\frac{2E}{mc^2}},\qquad n=\frac{d}{\lambda\!\!\!^{-}}-\left(s+\frac{1}{2}\right), \tag{96}$$

We get

$$\frac{d^2\psi}{d\xi^2}+\frac{1}{\xi}\frac{d\psi}{d\xi}+\left(-\frac{1}{4}+\frac{n+s+1/2}{\xi}-\frac{s^2}{\xi^2}\right)\psi=0. \tag{97}$$

Given the asymptotic behavior of function ψ for $\xi\to\infty$ and $\xi\to 0$, after substituting $\psi=e^{-\xi/2}\xi^s w(\xi)$ we obtain

$$\xi w''+(2s+1-\xi)w'+nw=0 \tag{98}$$

equation of degenerate hypergeometric function (Kummer function).

$$w={}_1F_1(-n,2s+1,\xi) \tag{99}$$

A solution satisfying the finiteness condition for $\xi=0$ and when $\xi\to\infty$ the w turns to infinity no faster than a finite degree ξ is obtained for a generally positive n. Moreover, the Kummer function ${}_1F_1$ reduces to a polynomial.

In accordance with (96) and (99), we obtain values for energy levels W (**Figure 7**)

$$W=mc^2\sqrt{1-\left(1-\frac{\lambda\!\!\!^{-}}{2d}(2n+1)\right)^2}. \tag{100}$$

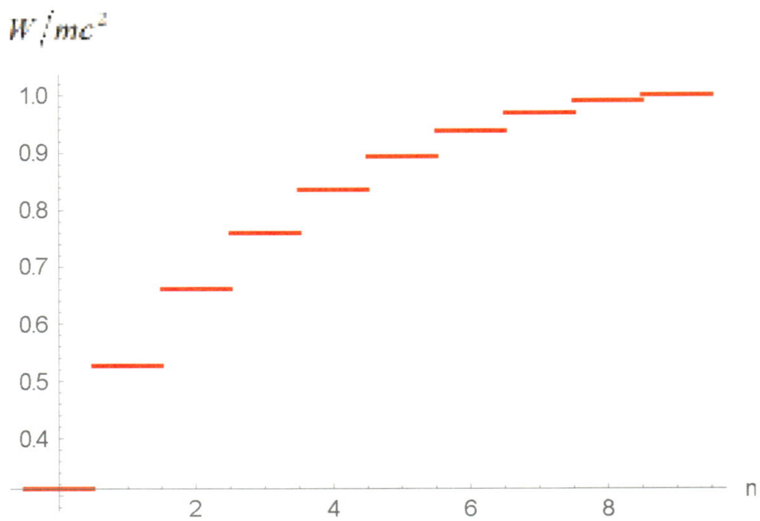

Figure 7.
The dependence of the energy of particle W on the quantum number n(100) at $d=10\lambda\!\!\!^{-}$ in units of mc^2.

For the binding energy in the ground state W_0 for $n = 0$ of (100) we have (**Figure 8**).

$$W_0 = mc^2 - mc^2\sqrt{1 - \left(1 - \frac{\lambda}{2d}\right)^2} \tag{101}$$

Because parameter s is determined to be positive (96) $s = d/\lambda - n - 1/2 \geq 0$ and $n \leq d/\lambda - 1/2$, then at $n = 0$ the minimum value is $d = \lambda/2$, which reflects the Heisenberg uncertainty principle. The maximum binding energy of a particle $mc^2 - W$ is limited from above by a value mc^2 regardless of the nature and magnitude of the interaction (**Figure 9**).

The interaction constant $q\varphi_0/mc^2$ (97) does not have any limitation on the value and is not included in the expression for energy levels (100) and only determines the spatial properties of the wave function (99) through variable ξ (**Figures 9** and **10**).

We emphasize that despite the fact that the potential energy for a stationary particle $V(x)$ has a depth of $mc^2/2$, the maximum binding energy for a moving

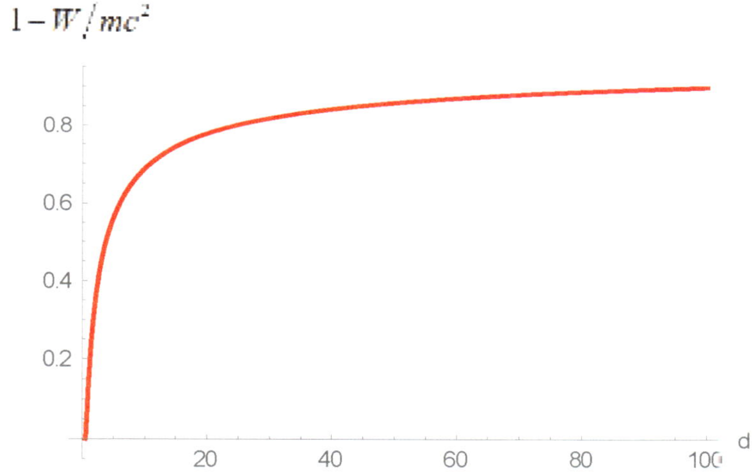

Figure 8.
The dependence of the binding energy of the ground state $mc^2 - W_0$ (101) on the size $d \geq \lambda/2$ in units of mc^2.

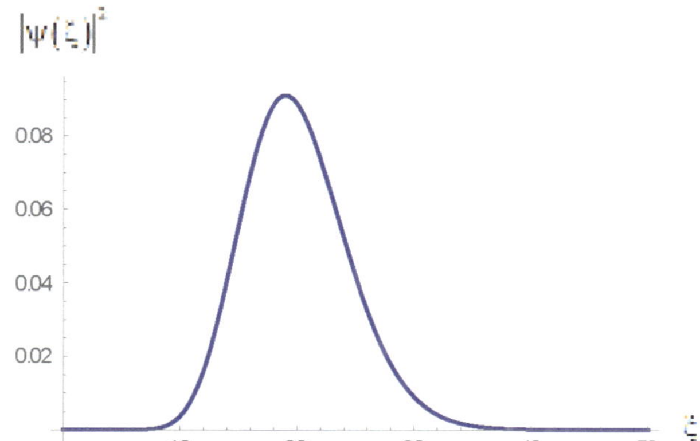

Figure 9.
Dependency of function $|\psi(\xi)|^2$ at $d = 10\lambda$ and $n = 0$.

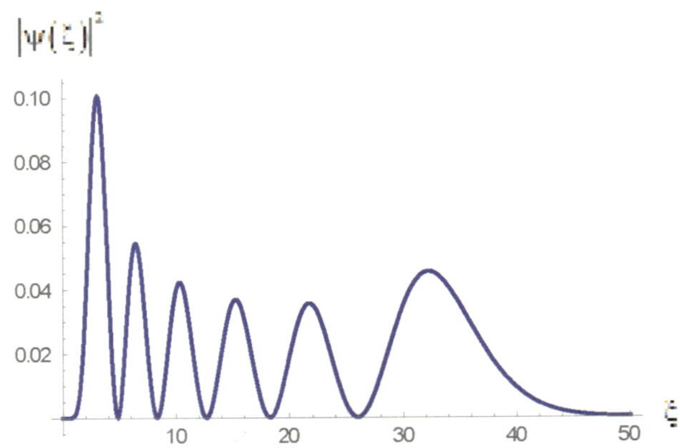

Figure 10.
Dependency of function $|\psi(\xi)|^2$ *at* $d = 10\lambda$ *and* $n = 5$.

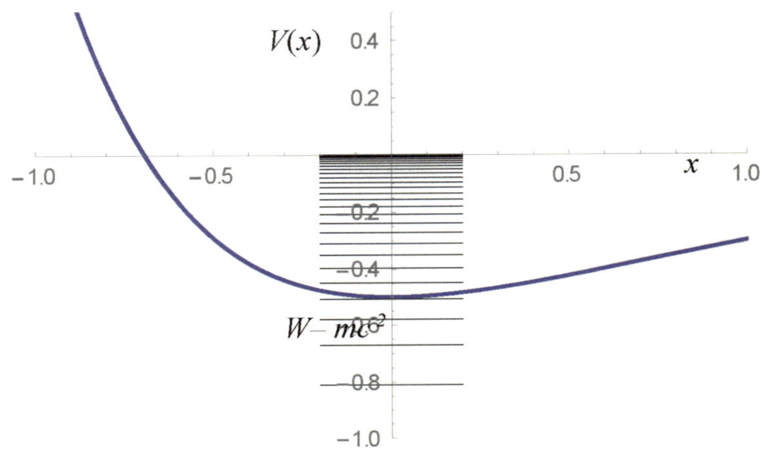

Figure 11.
The dependency of the potential energy of the interaction of Morse $V(x)$ *and energy levels of the particle* $W - mc^2$ *at* $d = 27\lambda$ *in units of* mc^2.

particle in the ground state is equal to mc^2 (**Figure 11**), which is a relativistic effect of the particle's motion in the ground state - in the ground state, the particle not at rest.

4.5 Problem of the hydrogen-like atom

The motion of a charged particle in the Coulomb field can be described as a motion in the field of an atomic nucleus (without the spin and magnetic moment) with the potential energy $-Ze^2/r$.

In spherical coordinates, Eq. (70) for the wave function takes the form

$$\frac{1}{r^2}\frac{\partial}{\partial r}\left(r^2\frac{\partial\Psi}{\partial r}\right)+\frac{1}{r^2\sin\theta}\frac{\partial}{\partial\theta}\left(\sin\theta\frac{\partial\Psi}{\partial\theta}\right)+ \\ \frac{1}{r^2\sin^2\theta}\frac{\partial^2\Psi}{\partial\varphi^2}+\frac{1}{\hbar^2c^2}\left(E^2-\left(mc^2-\frac{Ze^2}{r}\right)^2\right)\Psi=0. \tag{102}$$

Separating the variables

$$\Psi = \Phi_m(\varphi)Y_{l,\ m}(\theta)R_{\ n_R,\ l}(r) \tag{103}$$

and introducing the notations [12]

$$\begin{aligned} &\alpha = \frac{e^2}{\hbar c}, \qquad \rho = \frac{mZe^2}{\hbar^2}\frac{2r}{N} = Z\alpha\frac{mc}{\hbar}\frac{2r}{N}, \qquad M^2 = \hbar^2 l(l+1), \\ &\mathrm{H}_n = \frac{E_n{}^2 - m^2c^4}{2mc^2} = -\frac{mZ^2e^4}{\hbar^2}\frac{1}{2N^2} = -mc^2Z^2\alpha^2\frac{1}{2N^2}, \\ &s(s+1) = l(l+1) + Z^2\alpha^2 \ \Rightarrow \ s = -1/2 + \sqrt{(l+1/2)^2 + Z^2\alpha^2} \end{aligned} \tag{104}$$

(only the positive root is taken for s), for stationary states we have

$$\begin{aligned} &\frac{d^2\Phi}{d\varphi^2} = -m^2\Phi, \\ &\frac{1}{\sin\theta}\frac{d}{d\theta}\left(\sin\theta\frac{dY}{d\theta}\right) - \frac{m^2}{\sin^2\theta}Y = -l(l+1)Y, \\ &\frac{d^2R}{d\rho^2} + \frac{2}{\rho}\frac{dR}{d\rho} - \frac{s(s+1)}{\rho^2}R = -\left(\frac{n_r}{\rho} - \frac{1}{4}\right)R, \end{aligned} \tag{105}$$

where $m = \pm 0,\ \pm 1,\ \pm 2,\ \dots, l = 0,\ 1,\ 2,\ 3,\ \dots,\ |m| < l$ and $s = -1/2 + \sqrt{(l+1/2)^2 + Z^2\alpha^2}$.

The solution of Eq. (88) formally coincides with the well-known Fuse solution for the molecular Kratzer potential in the form $U = \frac{A}{r^2} - \frac{B}{r} = \frac{Z^2e^4}{2mc^2}\frac{1}{r^2} - Ze^2\frac{1}{r}$ at the condition, that $n - s - 1 = n_r$ must be a positive integer or zero. According to (87), we obtain the energy levels

$$\begin{aligned} &\mathrm{H}_{n,j} = -mc^2\frac{Z^2\alpha^2}{2\left(n_r + 1/2 + \sqrt{(l+1/2)^2 + Z^2\alpha^2}\right)^2}, \\ &E_{n,j} = mc^2\sqrt{1 - \frac{Z^2\alpha^2}{\left(n_r + 1/2 + \sqrt{(l+1/2)^2 + Z^2\alpha^2}\right)^2}}, \end{aligned} \tag{106}$$

where the radial quantum number $n_r = 0, 1, 2, \dots$. Introducing the principal quantum number $n = n_r + l + 1/2$, $l < n$ $(n = 1, 2, 3, \dots)$, we finally obtain

$$E_{n,j} = mc^2\sqrt{1 - \frac{Z^2\alpha^2}{\left(n + \frac{Z\alpha^2}{l+1/2+\sqrt{(l+1/2)^2+Z^2\alpha^2}}\right)^2}}. \tag{107}$$

For the ground state with the $l = 0$ and $n = 1$, we have

$$E_0 = \frac{mc^2}{\sqrt{1/2 + \sqrt{1/4 + Z^2\alpha^2}}}, \quad s = \frac{Z^2\alpha^2}{1/2 + \sqrt{1/4 + Z^2\alpha^2}} \tag{108}$$

without any restrictions for the value of Z. In this case, $1–s > 0$ and there is no fall of the particle on the center [8], and the probability of finding the particle at the center (in the nucleus) is always equal to zero.

In this case, the obtained fine splitting is in no way connected with the spin-orbit interaction and is due to the relativistic dependence of the mass on the orbital and radial velocity of motion, which results to the splitting of the levels.

4.6 Dirac equations

In the standard representation, the Dirac equations in compact notation for a particle have the form [21].

$$\begin{aligned} \hat{\varepsilon}\phi - \boldsymbol{\sigma}\cdot\hat{\mathbf{p}}\,\chi &= mc\phi, \\ \hat{\varepsilon}\chi + \boldsymbol{\sigma}\cdot\hat{\mathbf{p}}\,\phi &= mc\chi. \end{aligned} \tag{109}$$

In addition, for the particle in an external field they can be represented in theorm

$$\begin{aligned} \hat{\varepsilon}\phi - \boldsymbol{\sigma}\cdot\hat{\mathbf{p}}\,\chi &= \left(mc + \frac{q}{c}\varphi\right)\phi + \frac{q}{c}\boldsymbol{\sigma}\cdot\mathbf{A}\chi, \\ \hat{\varepsilon}\chi + \boldsymbol{\sigma}\cdot\hat{\mathbf{p}}\,\phi &= \left(mc + \frac{q}{c}\varphi\right)\chi - \frac{q}{c}\boldsymbol{\sigma}\cdot\mathbf{A}\phi. \end{aligned} \tag{110}$$

By writing the wave equations for the wave functions, we obtain

$$\begin{aligned} \left(\frac{\partial^2}{\partial\tau^2} - \frac{\partial^2}{\partial\mathbf{r}^2}\right)\phi &= -\frac{(mc^2+q\varphi)^2 - (q\mathbf{A})^2}{\hbar^2c^2}\phi - \frac{q}{\hbar c}\boldsymbol{\sigma}\cdot(\mathbf{B} - i\mathbf{E})\chi, \\ \left(\frac{\partial^2}{\partial\tau^2} - \frac{\partial^2}{\partial\mathbf{r}^2}\right)\chi &= -\frac{(mc^2+q\varphi)^2 - (q\mathbf{A})^2}{\hbar^2c^2}\chi + \frac{q}{\hbar c}\boldsymbol{\sigma}\cdot(\mathbf{B} - i\mathbf{E})\phi, \end{aligned} \tag{111}$$

where we used the properties of the Pauli matrices. It is easy to verify that the functions ϕ and χ differ only in the constant phase $\phi = \chi e^{\pm i\pi} = -\chi$ and the equations can be completely separated and only one equation can be used, bearing in mind that (111) can be of a variable sign

$$\left(\frac{\partial^2}{\partial\tau^2} - \frac{\partial^2}{\partial\mathbf{r}^2}\right)\Psi = -\frac{(mc^2+q\varphi)^2 - (q\mathbf{A})^2}{\hbar^2c^2}\Psi \pm \frac{q}{\hbar c}\boldsymbol{\sigma}\cdot(\mathbf{B} - i\mathbf{E})\Psi. \tag{112}$$

In the case of a stationary state, the standard representation of the wave Eq. (110) has the form

$$\begin{aligned} \left(\varepsilon - mc - \frac{q}{c}\varphi\right)\,\upphi &= \boldsymbol{\sigma}\cdot\left(\mathbf{p} + \frac{q}{c}\mathbf{A}\right)\,\chi, \\ \left(\varepsilon + mc + \frac{q}{c}\varphi\right)\,\chi &= \boldsymbol{\sigma}\cdot\left(\mathbf{p} - \frac{q}{c}\mathbf{A}\right)\,\upphi. \end{aligned} \tag{113}$$

4.7 Dirac equations solution for a hydrogen-like atom

For a charge in a potential field with the central symmetry [23], we have

$$\begin{pmatrix} \varphi \\ \chi \end{pmatrix} = \begin{pmatrix} \dfrac{f(r)}{r}\Omega_{jlm} \\ (-1)^{1+l-l'}\dfrac{g(r)}{r}\Omega_{jl'm} \end{pmatrix}. \tag{114}$$

After substituting (96) into (95), we obtain

$$\begin{cases} f' + \frac{\chi}{r}f - \left(\varepsilon + mc - \frac{Ze^2}{c}\frac{1}{r}\right)g = 0 \\ g' - \frac{\chi}{r}g + \left(\varepsilon - mc + \frac{Ze^2}{c}\frac{1}{r}\right)f = 0, \end{cases} \qquad \begin{cases} j = |l \pm 1/2|, \quad j_{\max} = l_{\max} + 1/2 \\ \chi = -1, \quad l = 0 \\ \chi = \pm(j + 1/2). \end{cases} \tag{115}$$

Let us represent the functions f and g in the form

$$\begin{aligned} f &= \sqrt{mc + \varepsilon}e^{-\rho/2}\rho^{\gamma}(Q_1 + Q_2), \\ g &= \sqrt{mc - \varepsilon}e^{-\rho/2}\rho^{\gamma}(Q_1 - Q_2), \end{aligned} \tag{116}$$

where

$$\rho = 2\lambda r/\hbar, \qquad \lambda = \sqrt{(mc)^2 - \varepsilon^2}, \qquad \gamma = \sqrt{\chi^2 + Z^2\alpha^2}, \qquad \alpha = \frac{e^2}{\hbar c}. \tag{117}$$

Substituting (116) into the Eq. (117), for the sum and difference of the equations we have

$$\begin{aligned} &\rho Q_1' + (\gamma - Z\alpha mc/\lambda)Q_1 + (\chi - Z\alpha\varepsilon/\lambda)Q_2 = 0, \\ &\rho Q_2' + (\gamma + Z\alpha mc/\lambda - \rho)Q_2 + (\chi + Z\alpha\varepsilon/\lambda)Q_1 = 0. \end{aligned} \tag{118}$$

Close to $\rho = 0$, the system of equations always has a solution, because

$$\gamma^2 - (Z\alpha mc/\lambda)^2 = \chi^2 - (Z\alpha\varepsilon/\lambda)^2. \tag{119}$$

Then

$$Q_2 = -\frac{\gamma - Z\alpha mc/\lambda}{\chi - Z\alpha\varepsilon/\lambda}Q_1 = -\frac{\chi + Z\alpha\varepsilon/\lambda}{\gamma + Z\alpha mc/\lambda}Q_1. \tag{120}$$

Forming equations of the second order and solving with respect to Q_1 and Q_2, we obtain

$$\begin{aligned} &\rho Q_1'' + (2\gamma + 1 - \rho)Q_1' - (\gamma - Z\alpha mc/\lambda)Q_1 = 0, \\ &\rho Q_2'' + (2\gamma + 1 - \rho)Q_2' - (\gamma + 1 - Z\alpha mc/\lambda)Q_2 = 0. \end{aligned} \tag{121}$$

With allowance for (121), the solution of these equations is

$$\begin{aligned} &Q_1 = AF(\gamma - Z\alpha mc/\lambda,\ 2\gamma + 1,\ \rho), \\ &Q_2 = -A\frac{\gamma - Z\alpha mc/\lambda}{\chi - Z\alpha\varepsilon/\lambda}F(\gamma + 1 - Z\alpha mc/\lambda,\ 2\gamma + 1,\ \rho), \end{aligned} \tag{122}$$

where $F(\alpha, \beta, z)$ is the degenerate hypergeometric function and A is the normalization constant of the wave function. The function $F(\alpha, \beta, z)$ reduces to a polynomial, if the parameter α is equal to an integer negative number or zero. Therefore, finite solutions for the functions f and g are

$$\gamma - \frac{Z\alpha mc}{\lambda} = -n_r. \tag{123}$$

From expressions (117), we obtain

$$f=A\sqrt{mc+\varepsilon}e^{-\rho/2}\rho^{\gamma-1}\left(F(-n_{\mathrm{r}},\quad 2\gamma+1,\quad \rho)+\frac{n_r}{\chi-Z\alpha\varepsilon/\lambda}F(1-n_{\mathrm{r}},\quad 2\gamma+1,\quad \rho)\right),$$
$$g=A\sqrt{mc-\varepsilon}e^{-\rho/2}\rho^{\gamma-1}\left(F(-n_{\mathrm{r}},\quad 2\gamma+1,\quad \rho)-\frac{n_r}{\chi-Z\alpha\varepsilon/\lambda}F(1-n_{\mathrm{r}},\quad 2\gamma+1,\quad \rho)\right),$$
$$(124)$$

where $n_r=0,1,2,\ \ldots$ is the radial quantum number. For the energy levels, we obtain from the condition (117)

$$\frac{\varepsilon_{p,\chi}}{mc}=\sqrt{1-\frac{Z^2\alpha^2}{\left(n_{\mathrm{r}}+\sqrt{\chi^2+Z^2\alpha^2}\right)^2}} \tag{125}$$

and taking into account the obtained values of χ, we finally have

$$E_{n,j}=mc^2\sqrt{1-\frac{Z^2\alpha^2}{\left(n_{\mathrm{r}}+\sqrt{(j+1/2)^2+Z^2\alpha^2}\right)^2}}=$$
$$mc^2\sqrt{1-\frac{Z^2\alpha^2}{\left(n+\frac{Z\alpha^2}{j+1/2+\sqrt{(j+1/2)^2+Z^2\alpha^2}}\right)^2}}, \tag{126}$$

where the principal quantum number $n=n_r+j+1/2$. Besides $j=n-1/2$, all other levels with $j<n-1/2$ are degenerated twice in the orbital angular momentum $l=|j\pm 1/2|$. The ground state energy for $n=1$ and $j=1/2$ is

$$E_0=\frac{mc^2}{\sqrt{1+Z^2\alpha^2}} \tag{127}$$

without any limitations for the value of Z. In this case $\gamma-1=-1+\sqrt{1+Z^2\alpha^2}>0$, and no falling of particle on the center is observed, and the probability to find the particle in the center (in the nucleus) is always equal to zero.

In the resulting formula (126), the order of sequence of the fine splitting levels is inverse relative to the order of sequence in the well-known Sommerfeld-Dirac formula. If to compare the expansions in a series in the degree of the fine-structure constant of two formulas

$$\frac{E_n}{mc^2}=\frac{1}{2}\left(1-\frac{1}{n^2}\right)\alpha^2+\left(\frac{1}{8}+\frac{3}{8n^4}-\frac{1}{2n^3(j+1/2)}\right)\alpha^4,\ \frac{\Delta E_{3/2,1/2}}{mc^2}=\frac{\alpha^4}{32}, \tag{128}$$

$$\frac{E_n}{mc^2}=\frac{1}{2}\left(1-\frac{1}{n^2}\right)\alpha^2+\left(-\frac{3}{8}-\frac{1}{8n^4}+\frac{1}{2n^3(j+1/2)}\right)\alpha^4,\ \frac{\Delta E_{3/2,1/2}}{mc^2}=-\frac{\alpha^4}{32}, \tag{129}$$

then the difference will be equal to

$$\frac{\Delta E_n}{mc^2}=\frac{\alpha^4}{2}-\frac{\alpha^4}{2n^4}-\frac{\alpha^4}{n^3(j+1/2)}, \tag{130}$$

where the last term is the expression for the spin-orbit interaction energy. Thus, to obtain the true value of the energy levels of the hydrogen atom, it is necessary to add the energy of the spin-orbit interaction in formula (126) in the form (130). This is completely justified, because such an interaction was not initially included in Eq. (115) and was not reflected in the final result.

5. Conclusion

The principle of invariance is generalized and the corresponding representation of the generalized momentum of the system is proposed; the equations of relativistic and quantum mechanics are proposed, which are devoid of the above-mentioned shortcomings and contradictions. The equations have solutions for any values of the interaction constant of the particle with the field, for example, in the problem of a hydrogen-like atom, when the atomic number of the nucleus $Z > 137$. The equations are applicable for different types of particles and interactions.

Based on the parametric representation of the action and the canonical equations, the corresponding relativistic mechanics based on the canonical Lagrangian is constructed and the equations of motion and expression are derived for the force acting on the charge moving in an external electromagnetic field.

The matrix representation of equations of the characteristics for the action function and the wave function results in the Dirac equation with the correct enabling of the interaction. In this form, the solutions of the Dirac equations are not restricted by the value of the interaction constant and have a spinor representation by scalar solutions of the equations for the action function and the wave function.

The analysis of the solutions shows the full compliance with the principles of the relativistic and quantum mechanics, and the solutions are devoid of any restrictions on the nature and magnitude of the interactions.

The theory of spin fields and equations for spin systems will be described in subsequent works.

Author details

Vahram Mekhitarian
Institute for Physical Research, Armenian National Academy, Ashtarak, Armenia

*Address all correspondence to: vm@ipr.sci.am

References

[1] Bohr N. On the constitution of atoms and molecules. Philosophical Magazine. 1913;**26**:1-25. DOI: 10.1080/14786441308634955

[2] Uhlenbeck GE, Goudsmit S. Ersetzung der Hypothese vom unmechanischen Zwang durch eine Forderung bezüglich des inneren Verhaltens jedes einzelnen Elektrons. Naturwissenschaften. 1925;**13**:953-954. DOI: 10.1007/BF01558878

[3] Sommerfeld A. Zur Quantentheorie der Spektrallinien. Ann. Phys. 1916;**356**: 1-94. DOI: 10.1002/andp.19163561702

[4] Granovskii YI. Sommerfeld formula and Dirac's theory. Physics-Uspekhi. 2004;**47**:523-524. DOI: 10.1070/PU2004v047n05ABEH001885

[5] Dirac PAM. The quantum theory of the electron. Proceedings of the Royal Society. 1928;**117**:610-624. DOI: 10.1098/rspa.1928.0023

[6] De Broglie ML. Ondes et quanta. Comptes Rendus. 1923;**177**:507-510

[7] Klein O. Die reflexion von elektronen an einem potentialsprung nach der relativistischen dynamik von Dirac. Zeitschrift für Physik. 1929;**53**:157-165. DOI: 10.1007/BF01339716

[8] Landau LD, Lifshitz EM. The Classical Theory of Fields. Vol. 2. 4th ed. Oxford: Butterworth-Heinemann; 1980. p. 444

[9] Talukar B, Yunus A, Amin MR. Continuum states of the Klein-Gordon equation for vector and scalar interactions. Physics Letters A. 1989; **141**:326-330. DOI: 10.1016/0375-9601(89)90058-3

[10] Chetouani L, Guechi L, Lecheheb A, Hammann TF, Messouber A. Path integral for Klein-Gordon particle in vector plus scalar Hulthén-type potentials. Physica A. 1996;**234**:529-544. DOI: 10.1016/S0378-4371(96)00288-9

[11] Dominguez-Adame F. Bound states of the Klein-Gordon equation with vector and scalar Hulthén-type potentials. Physics Letters A. 1989;**136**:175-177. DOI: 10.1016/0375-9601(89)90555-0

[12] Landau LD, Lifshitz EM. Quantum Mechanics: Non-Relativistic Theory. 3rd ed. Vol. 3. Oxford: Butterworth-Heinemann; 1981. p. 689

[13] Hoffman D. Erwin Schrödinger. Leipzig: Teubner; 1984. p. 97. DOI: 10.1007/978-3-322-92064-5

[14] Schrödinger E. Quantisierung als Eigenwertproblem (Erste Mitteilung.). Ann. Phys. 1926;**79**:361-376. DOI: 10.1002/andp.19263840404

[15] Klein O. Quantentheorie und fünfdimensionale Relativitätstheorie. Zeitschrift für Physik. 1926;**37**:895-906. DOI: 10.1007/BF01397481

[16] Fock V. Zur Schrödingerschen Wellenmechanik. Zeitschrift für Physik. 1926;**38**:242-250. DOI: 10.1007/BF01399113

[17] Gordon W. Der Comptoneffekt nach der Schrödingerschen Theorie. Zeitschrift für Physik. 1926;**40**:117-133. DOI: 10.1007/BF01390840

[18] Heisenberg W, Jordan P. Anwendung der Quantenmechanik auf das Problem der anomalen Zeemaneffekte. Zeitschrift für Physiotherapie. 1926;**37**:263-277. DOI: 10.1007/BF01397100

[19] Thomas LH. The motion of the spinning electron. Nature. 1926;**117**:514. DOI: 10.1038/117514a0

[20] Sokolov AA, Ternov IM, Kilmister CW. Radiation from

Relativistic Electrons. New York: American Institute of Physics; 1986. p. 312

[21] Berestetskii VB, Lifshitz EM, Pitaevski LP. Quantum Electrodynamics. Oxford: Butterworth-Heinemann; 2012. p. 667

[22] Madelung E. Mathematische Hilfsmittel des Physikers. 3rd ed. Berlin: Springer; 1936. p. 359

[23] Akhiezer AI, Berestetskii VB. Quantum Electrodynamic. New York: John Wiley & Sons; 1965. p. 875

[24] Dirac PAM. The current state of the relativistic theory of the electron. Proceedings of the Institute of History of Science and Technology (Trudy instituta istorii estestvoznaniya i tekhniki, in Russian). 1959;**22**:32-33

[25] The Birth of Particle Physics. Laurie B and Lillian H, edittors. Cambridge University Press; 1983. p. 432. Origin of quantum field theory, Paul A. M. Dirac, pp. 39-55. International Symposium on the History of Particle Physics, Fermilab, May 1980. "Some new relativistic equations are needed; new kinds of interactions must be brought into play. When these new equations and new interactions are thought out, the problems that are now bewildering to us will get automatically explained, and we should no longer have to make use of such illogical processes as infinite renormalization. This is quite nonsense physically, and I have always been opposed to it. It is just a rule of thumb that gives results. In spite of its successes, one should be prepared to abandon it completely and look on all the successes that have been obtained by using the usual forms of quantum electrodynamics with the infinities removed by artificial processes as just accidents when they give the right answers, in the same way as the successes of the Bohr theory are considered merely as accidents when they tum out to be correct"

[26] Mekhitarian V. The invariant representation of generalized momentum. Journal of Contemporary Physics. 2012;**47**:249-256. DOI: 10.3103/S1068337212060011

[27] Landau LD, Lifshitz EM. Mechanics. 3rd ed. Vol. 1. Oxford: Butterworth–Heinemann; 1982. p. 224

[28] Mekhitarian V. Canonical solutions of variational problems and canonical equations of mechanics. Journal of Contemporary Physics. 2013;**48**:1-11. DOI: 10.3103/S1068337213010015

[29] Mekhitarian V. Equations of Relativistic and Quantum Mechanics and Exact Solutions of Some Problems. Journal of Contemporary Physics. 2018; **53**:1-21. DOI: 10.3103/S1068337218020123

[30] Kratzer A. Die ultraroten Rotationsspektren der Halogenwasserstoffe. Zeitschrift für Physik. 1920;**3**:289-307. DOI: 10.1007/BF01327754

[31] Lennard-Jones JE. On the determination of molecular fields. Proceedings of the Royal Society of London A. 1924;**106**:463-477. DOI: 10.1098/rspa.1924.0082

[32] Morse PM. Diatomic molecules according to the wave mechanics. II. Vibrational levels. Physics Review. 1929; **34**:57-64. DOI: 10.1103/PhysRev.34.57

[33] Rosen N, Morse PM. On the vibrations of polyatomic molecules. Physics Review. 1932;**42**:210-217. DOI: 10.1103/PhysRev.42.210

[34] Mekhitarian V. The Faraday law of induction for an arbitrarily moving charge. Journal of Contemporary Physics. 2016;**51**:108-126. DOI: 10.3103/S1068337216020031

Chapter 8

Nature of Temporal (t > 0) Quantum Theory: Part I

Francis T.S. Yu

Abstract

It is our science governs the mathematics and it is "not" our mathematics governs our science. One of the very important aspects is that every science has to comply with the boundary condition of our universe; dimensionality and temporal (t > 0) or causality. In which I have shown that time is real and it is not an illusion, since every aspect within our universe is coexisted with time. Since our universe is a temporal (t > 0) subspace, everything within our universe is temporal. Science is mathematics but mathematics is not science, we have shown that any analytic solution has to be temporal (t > 0); otherwise, it cannot be implemented within our universe. Which includes all the laws, principles, and theories have to be temporal? Uncertainty principle is one of the most fascinated principles in quantum mechanics, yet Heisenberg principle was based on diffraction limited observation, it is not due to the nature of time. We have shown it is the temporal (t > 0) uncertainty that changes with time. We have introduced a certainty principle as in contrast with uncertainty principle. Of which certainty subspace can be created within our universe; which can be exploited for application. Overall of this chapter is to show that; it is not how rigorous the mathematics is, it is the physical realizable paradigm that we embrace.

Keywords: temporal universe, timeless space, physical realizable, uncertainty principle, certainty principle, quantum mechanics

1. Introduction

Strictly speaking every scientific solution has to be proven whether it is physical realizable before considering for experimentation, since analytical solution is mathematics. For example, if an elementary particle has proven not a temporal (t > 0) or a timeless (t = 0) particle, it has no reason to spend that big a budget for experimentally searching a timeless (t = 0) particle since timeless particle does not exist within our universe. Similarly, a mathematician discovers a 10-dimensional subspace, would not you want to prove that his 10-dimensional subspace is a temporal (t > 0) subspace, before experimentally search for it since mathematical solution is virtual.

Nevertheless at the dawn of science, scientists have been using a piece or pieces of papers; drawn models and paradigms in it and using mathematics as a tool analyzing for possible solution. But never occurs to them the back ground of that piece of paper represented a mathematical subspace that is "not" existed within our universe, for which practically all the laws, principles, and theories were developed from a piece or pieces of papers, which are timeless (t = 0) and strictly speaking are virtual.

Since science is mathematics but mathematics is "not" equaled to science, it is vitally important for us to understand what science really is. In order to understand science, firstly we have to understand what supported the science? For which the supporter must be the subspace within our universe. In other words, any scientific solution has to be proven existed within our universe; otherwise, it may be fictitious and virtual as mathematics is, since science is mathematics. In which we see that, our universe is a physical subspace that supports every physical realizable aspect within her space, "if and only if" the scientific postulation complies within the existent condition of our universe; dimensionality and causality or temporal (t > 0).

The essence of our temporal (t > 0) universe is that; if a mathematical solution is "not" complied within the temporal (t > 0) condition of our universe, it cannot exist within our universe. Since quantum mechanics is one of the pillars in modern science, I will start with one of the most intriguing principles in quantum mechanics; uncertainty principle. I will carry on the principle onto a newly found "certainty" principle. In which I will show Heisenberg's principle was based on diffraction limited observation, instead upon on "nature" of time, developing his principle. I will also show the mystery of coherence theory can be understood with principle of certainty. In which I will show that; certainty subspace can be created within our temporal (t > 0) universe. Samples as applied to synthetic aperture imaging and wave front reconstruction will be included.

2. Science and mathematics

There is a profound relationship between science and mathematics, in which we have seen that without mathematics there would be no science. In other words, science needs mathematics but mathematics does not need science. Although science is mathematics but mathematics is not science. For example, any mathematical solution if it cannot be proven it exists within our universe, then her solution is "not" a "physical realizable" solution that can be "directly" implemented within our temporal (t > 0) universe.

But this is by no means to say that; the solutions are not temporal (t > 0) or timeless (t = 0) solutions there are not science. In fact practically all the fundamental laws, principles, and theories are timeless (t = 0) or time-independent. And these timeless (t = 0) laws, principles, and theories were and "still" are the corner stone and foundation of our science, as I will call them timeless (t = 0) or time-independent science; a topic I will elaborate in a different occasion. For simplicity, let me take one of the simplest examples; Einstein's energy Eq. (1) as given by;

$$E = mc^2 \tag{1}$$

where E is the energy, m is the mass and c is the velocity of light. This equation is one of the most famous equations in science, yet it is timeless (t = 0). Although this equation has been repeatedly used and applied in practice, but strictly speaking; it cannot be directly implemented within our temporal (t > 0) universe, since it is not a time variable function. Let us transform Einstein's equation into a time variable equation as given by [1].

$$\frac{\partial E(t)}{\partial t} = -c^2 \frac{\partial m(t)}{\partial t}, \mathrm{t} > 0 \tag{2}$$

where $\partial E(t)/\partial t$ is the rate of increasing energy conversion, $-\partial m/\partial t$ is the corresponding rate of mass reduction, c is the speed of light, and $t > 0$ denotes a forward time-variable equation. In which we see Eq. (2) is a time-dependent equation exists at time $t > 0$, which represents a forwarded time variable function that only occurs after time excitation at $t = 0$. Incidentally, this is the well-known "causality" constraint (i.e., $t > 0$) [2] as imposed by our temporal ($t > 0$) universe.

Nevertheless in mathematical, a postulation is first needed to proof that there is solution existed before we search for the solution, although it is not guarantee that we can find it. But it seems to me it does not have a criterion to proof that a hypothetical science is existed within our universe, before we search for the science. For example, an analytically solution indicates that it exists an "angle particle" from a complicated mathematical analyses, will not you want to find out first is the solution existed within our temporal ($t > 0$) universe before experimentally to search for it. And this is precisely that we shall know first before experimentation is taken place, since it is a very costly in time and in revenue to find a physical particle.

Although science needs mathematics, but without simplicity mathematically approximation, science would be very difficult to learn and to facilitate. And this is precisely the reason practically all the fundamental laws are point-singularity approximated. In which we see precisely, science is a "law of approximation" and mathematics is "an axiom of certainty". Again we take Einstein's energy equation of Eq. (1) as an example, no dimension and size and it is a typical point-singularity approximated equation. It is discernible; if we include all the negligibly terms, "physical significances" of this equation would be over whelmed by the terms of mathematics. For which we see that an ounce of good approximation worth more than tons of mathematical calculation!

Let me stress that the essence of simplicity in science is that without the symbolic substitution and approximation, it will be extremely difficult or even impossible to develop science since science itself is already very complicated. Yet simplicity representation of science has also been misinterpreted as referred them as "classical and deterministic (i.e., classical physics)." The implication of deterministic or classical is a totally misled by our part, since our predecessors who developed those fundamental laws and principles were "precisely" understood the deficiency of approximation. Yet without the approximated presentation, how can we develop science? Instead of ignoring our predecessors' wisdom, turns around we had treated them "deterministic" or classical, which were "never" been our predecessors intention. Again without the point-singularity approximated science, please tell me how we can develop those simple and elegant laws, principles, and theories. Although those laws, principles, and theories were timeless ($t = 0$), most of them were and "still" are the foundation and corner stone of our science. Nevertheless, mathematics is a "symbolic" langue of science, but mathematics is not science.

Since all laws, principles, and theories were made to be broken or revised or even to replace, as science advances into sub-subatomic scale regime and moving closer to near real time processing, those timeless ($t = 0$) laws, principles, and theories could produce incomprehensible consequences; particularly as applied them directly confronting the temporal ($t > 0$) constraint of our universe. For example, as applying superposition principle to quantum computing and communication, since superposition is a timeless ($t = 0$) principle [3].

3. Temporal (t > 0) subspace

In this section, I will show several subspaces that have been used by the scientists, in the past as depicted in **Figure 1**. It is reasonable to stress that why subspace

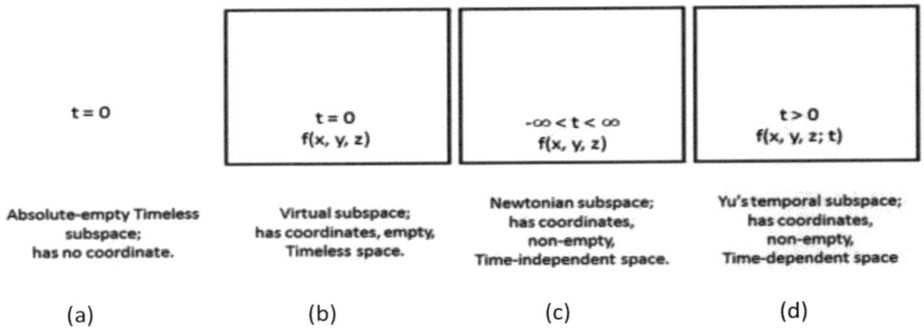

Figure 1.
(a) Shows an absolute-empty space, (b) a virtual mathematical space, (c) a Newtonian space, and (d) a temporal (t > o) space, respectively.

of a scientific model embedded is crucially important is that any analytical solution produced follows the "limitation" of the subspace, because it is the subspace dictates the science but "not" the mathematics changes the subspace.

For example, when you are designing a submarine, the subspace that the submarine is supposed to be situated within is vitally important; otherwise, your submarine will very "likely" not to survive thousands of feet underwater pressure. Therefore, it is necessary to know the subspace that a postulated science to be implementing into it; otherwise, the postulated science is very likely "cannot" be existed within the subspace.

In view of **Figure 1**, we see that; there is an absolute-empty space, a mathematical virtual space, a Newtonian's space [4], and a temporal ($t > 0$) space. An absolute-empty space or just empty space has no substance and has no time. A mathematical virtual space is an empty space which has no substance in it, but mathematicians and theoretical scientists can implant coordinate system in it, since mathematics is virtual and theoretical scientists are also mathematicians.

We note that mathematical virtual space has been used over centuries by scientists at the dawn of science, but this is a virtual space that does "not" exist within our temporal (i.e., $t > 0$) universe. The next subspace is known as Newtonian space [4]; it has substance and coordinates in it, but treated time as an "independent" variable, for which Newtonian and mathematical spaces are virtual the "same." Since Newtonian space is time independent, it "cannot" be exist within our temporal ($t > 0$) space since time and substance has to be "mutually coexisted" within our temporal ($t > 0$) universe. Yet scientists have been using Newtonian space for their analyses over centuries and not knowingly it is a virtual space.

The last subspace is known as temporal ($t > 0$) space [5], where time and substance are interdependently "coexisted" and time is a forward "dependent variable" runs at a "constant speed". We stress that this temporal ($t > 0$) subspace is currently "only" physical realizable space, where the space was created by Einstein energy Eq. (2).

Physical reality is that any scientific hypothesis that deviates "away" the boundary condition that imposed by our temporal ($t > 0$) universe is "not" a physically realizable solution. But this is by no means that the virtual mathematical empty space and Newtonian space are useless. The fact is that all the physical sciences were developed within timeless ($t = 0$) or Newtonian subspaces "inadvertently," at the dawn of science. Practically all the fundamental laws, principles, and theories were derived from a timeless ($t = 0$) subspace, which was from the background subspace of a piece of paper although not intentionally [6]. In which we see that practically all the laws, principle, and theories are timeless ($t = 0$).

Nevertheless what temporal (t > 0) space means is that any subspace is coexisted with time, where time is a forward dependent variable with respect to its subspace and its speed has been well settled when our universe was created. This means that before the creation of our temporal universe, there is a "larger" temporal space that our universe is embedded in; otherwise, our universe will "not" be existed. Nevertheless every subspace within our universe is a time varying stochastic [7] subspace, in which every substance or subspace changes with time. Strictly speaking our universe is a "temporal (t > 0) stochastic expanding subspace." For which we see that; any postulated law, principle, and theory has to comply with the temporal (t > 0) condition within our universe; otherwise, it is virtual as mathematics.

4. Timeless (t = 0) space

Let me show what mathematicians can do within a virtual subspace as depicted in **Figure 2**. Since quantum mechanists are also mathematicians, they can implant coordinate system within an empty space as they wishes, regardless whether the model is physically realizable or not.

The basic difference between **Figure 2(a)** and **(b)** is that there is a virtual coordinate system that has been added in **Figure 2(b)** by quantum mechanists. Once the coordinate system is implanted, dimensionality of the sub-atomic particles cannot be ignored. The reason is that for the atomic model to be existed within the subspace, the atomic model has to "comply" with the existence conditions within the subspace, since it is the subspace affects the solution and not the solution changes the subspace. In which we see that neither **Figure 2(a)** nor **Figure 2(b)** are "not" physical realizable paradigms. For which solutions obtained from these empty subspace models will be timeless (t = 0).

Aside the non-physical realizable paradigms of **Figure 2**, I will show what a timeless (t = 0) subspace can do for substances within the subspace. Let me assume we have three particles situated within an empty space, as normally do on a "piece of paper", shown in **Figure 3**.

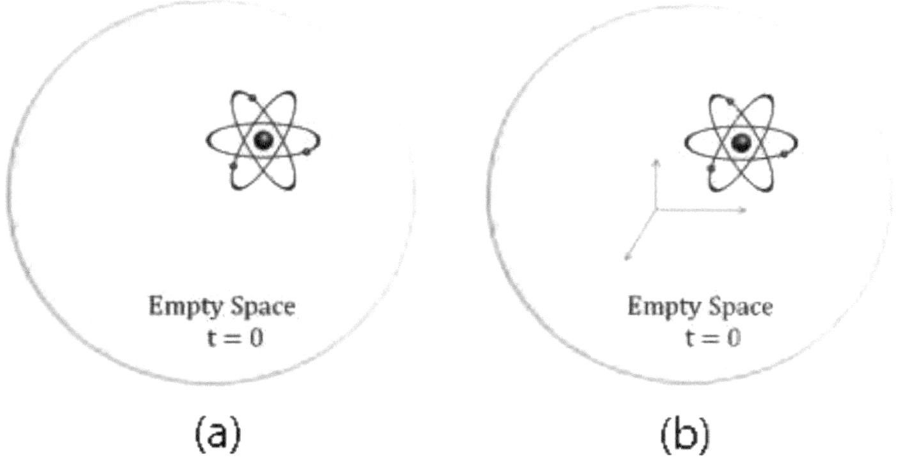

Figure 2.
A set of atomic models embedded within virtual empty subspaces. (a) shows a singularity approximated atomic model is situated within an empty space, which has no coordinate system. (b) shows an atomic model is embedded within empty space that has a coordinate system drawn into it.

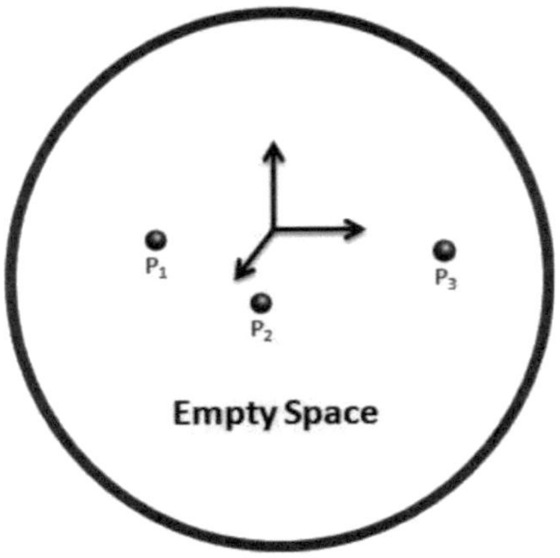

Figure 3.
A hypothetical scenario shows three particles are embedded within an empty subspace.

Since empty subspace has "no time," all particles within the subspace collapse or "superimposing" instantly all together at t = 0, because time is distance and distance is time. This is precisely the "simultaneous and instantaneous" superposition principle does in quantum mechanics [3]. The reason particles collapsed at t = 0, it is because the subspace has "no time." And the other reason that particles superimposed together, since within a timeless (t = 0) space, it has "no distance" or no space.

By virtue of energy conservation, we see that superimposed particles has a mass equals to the sum of entire superimposed particles, but it has "no size." In view of timelessness space, we see that the superimposed particles can be found everywhere within the entire timeless (t = 0) subspace, since timeless (t = 0) subspace has "no" distance, as depicted hypothetically in **Figure 4**. In which we see that Schrödinger's fundamental principle of superposition is existed within a virtual timeless (t = 0) subspace, and it cannot be existed within our temporal (t > 0) universe, since timeless and temporal are "mutually exclusive."

By the way, this is precisely the superposition principle that Einstein was objecting to, which he called it spooky. As I quote from a 1935 The New York Times' article (i.e., **Figure 5**), "Einstein and two scientists found quantum theory is

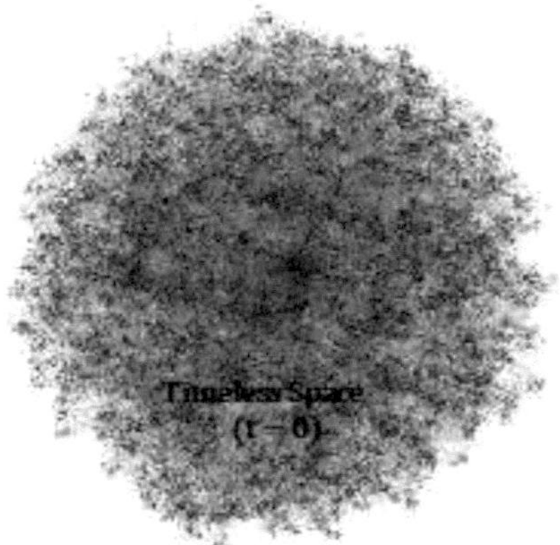

Figure 4.
Superimposed particle existed "simultaneously and instantaneously" all over the entire timeless (t = 0) subspace.

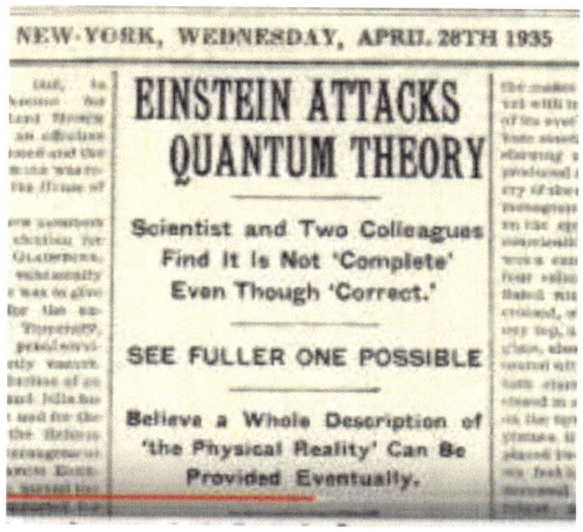

NEW-YORK, WEDNESDAY, APRIL 28TH 1935

EINSTEIN ATTACKS QUANTUM THEORY

Scientist and Two Colleagues Find It Is Not 'Complete' Even Though 'Correct.'

SEE FULLER ONE POSSIBLE

Believe a Whole Description of 'the Physical Reality' Can Be Provided Eventually.

Figure 5.
A 1935 New York times' article.

incomplete even though correct" [8]. In view of preceding illustration, we see that Schrödinger's superposition principle is "correct" but only within a timeless (t = 0) subspace and it is "incorrect" within our temporal (t > 0) space," since timeless space cannot exist within our temporal universe.

5. Time is not an illusion but real

As we accepted subspace and time are coexisted within our temporal (t > 0) universe, time has to be real and it cannot be virtual, since we are physically real. And every physical existence within our universe is real. The reason some scientists believed time is virtual or illusion is that; it has no mass, no weight, no coordinate, no origin, and it cannot be detected or even be seen. Yet time is an everlasting existed real variable within our known universe. Without time there would be no physical matter, no physical space, and no life. The fact is that every physical matter is associated with time which including our universe. Therefore, when one is dealing with science, time is one of the most enigmatic variables that ever presence and cannot be simply ignored. Strictly speaking, all the laws of science as well every physical substance cannot be existed without the existence of time. For which we see that time "cannot" be a dimension or an illusion. In other words, if time is an illusion, then time will be "independent" from physical reality or from our universe. And this is precisely that many scientists have treated time as an "independent" variable such as Murkowski's space [9], for which the space can be "curved" or time-space can be changed by gravity [10]. If time-space can be curved, then we can change the "speed" of time. In other words, is our universe exists with time, or time exists with universe? The answer is our universe exists with time, although space and time are interdependent but is not time exists with our universe.

As time is coexisted with subspace, we see that any subspace within our temporal (t > 0) universe cannot be empty and speed of time is the same everywhere within our universe. This means that the speed of time within a subspace is "relatively" with respect to the different subspaces, as based on Einstein's special theory of relativity [9]. For example, subspaces closer to the edge of our universe, their time runs faster "relative" to ours, but the speed of time within the subspaces near the edge as well within our subspace are the "same," which has been determined by the speed of light as our universe was created by a big bang theory using Einstein equation as given by [5];

$$\frac{\partial E(t)}{\partial t} = -c^2 \frac{\partial m(t)}{\partial t}, \mathbf{t > 0} \tag{3}$$

where $\partial E/\partial t$ is the rate of increasing energy conversion, $-\partial m/\partial t$ is the corresponding rate of mass reduction, c is the speed of light and $t > 0$ represents a forward time-variable. In which we see that it a "time-dependent" equation exists at time $t > 0$; a well-known causality constraint (i.e., $t > 0$) [2] as imposed by our universe. Similarly preceding equation can be written as:

$$\frac{\partial E}{\partial t} = -c^2 \frac{\partial m}{\partial t} = [\nabla \cdot S(v)] = -\frac{\partial}{\partial t}\left[\frac{1}{2}\varepsilon E^2(v) + \frac{1}{2}\mu H^2(v)\right], \mathbf{t > 0} \tag{4}$$

where ε and μ are the permittivity and the permeability of the deep space, respectively, υ is the radian frequency variable, $E^2(\upsilon)$ and $H^2(\upsilon)$ are the respective electric and magnetic field intensities, the negative sign represents the "out-flow" energy per unit time from an unit volume, $(\nabla\cdot)$ is the divergent operator, and S is known as the Poynting Vector or "Energy Vector" of an electro-magnetic radiator as can be shown by $S(\upsilon) = E(\upsilon) \times H(\upsilon)$ [11].

In view of this equation, we see how our universe was created as depicted by a composited diagram in **Figure 6**, in which we see that radian energy (i.e., radiation) diverges from the mass, as mass reduces with time. In which we see that our universe enlarges and her boundary expands at speed of speed of light.

Figure 7 shows a schematic diagram of our temporal $(t > 0)$ universe, which depicts approximately the behavior of subspace changes as her boundary expands with speed of light. In which we see that, subspace enlarges faster closer toward the boundary, but solid substance m (t) changes little within the subspace. We also see that the out-ward speed of particle (or subspace) increases "linearly" as boundary increases with light speed. For example; out-ward speed of particle 2 is somewhat faster than particle 1 (i.e., $v_2 > v_1$). For which we see that our universe is a dynamic temporal $(t > 0)$ "stochastic" universe that simple geometrical equation or mathematical abstract space can describe. One of the important aspects of our universe is that every subspace, no matter how small it is, "cannot" be empty and it has time.

For instance, in order for us to be existed within our planet, we must be temporal $(t > 0)$: that is we have time and must change with time; otherwise, we

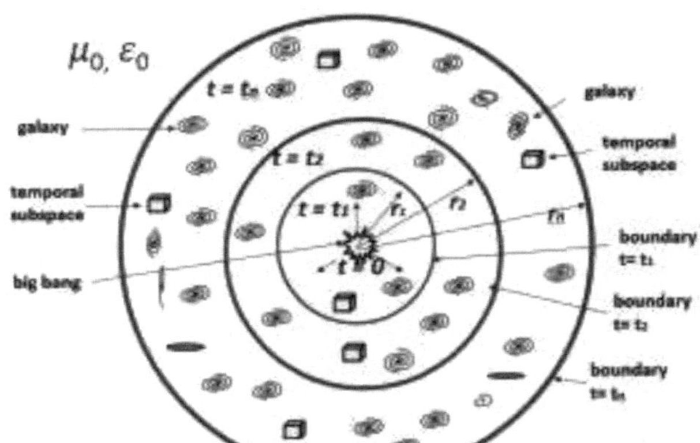

Figure 6.
Composite temporal $(t > o)$ universe diagrams. $r = ct$, r is the radius of our universe, t is time, c is the velocity of light, and ε_o and μ_o are the permittivity and permeability of the space.

DOI: http://dx.doi.org/10.5772/intechopen.93561

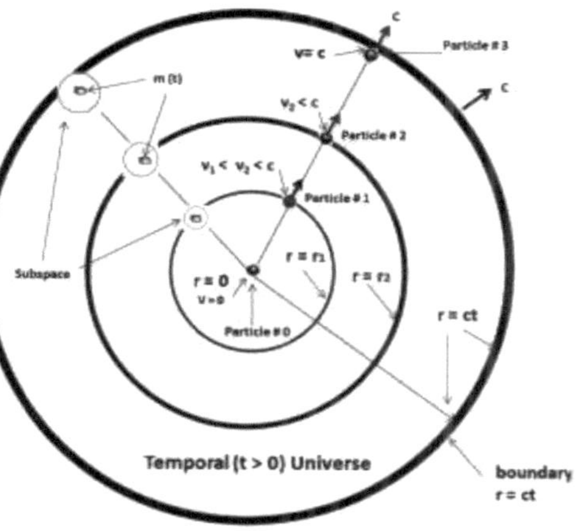

Figure 7.
A schematic diagram of our temporal (t > 0) universe. c is the speed of light, m(t) is the temporal mass, and v is the radial velocity.

cannot exist within our universe. In other words, our time is the same as our planet and the universe but the velocity of our planet is different from other subspaces. For example, subspaces near the edge of our universe are moving faster than us, for which it has "relative" speed of time between us and a subspace closer to the edge of our universe. On the other hand, if we assume that we are timeless (t = 0), we could "not" have existed within our universe, since time and timelessness are mutually exclusive.

I further note that any subspace within our universe cannot empty, since subspace is coexisted with time. Although subspace is coexisted with time, but time is neither equaled to subspace. Yet, space is time and time is space since time and space are mutually inclusive. For example, substance has dimension (or space), but time has no dimension and no mass. In which we see that time is "not" a dimension but it is "dependently" existed with respect to subspace. In which we stress that it is our universe governs the science and it is not the science changes our universe.

Once again, we have shown that time is "not "an illusion or virtual, time is physically real because everything existed within our living space is physical real; otherwise, it will not be existed within our temporal universe. In other words, everything within our universe is temporal (t > 0), of which I have discovered that practically all the laws, principles, theories, and paradoxes of science were developed from a timeless (t = 0) platform (i.e., a pieces or pieces of papers) for centuries, at the dawn of science "inadvertently" [6].

Nevertheless, one of the important aspects within our universe is that every subspace has a price, an amount of energy ΔE, and a section of time Δt to create (i.e., ΔE and Δt), and it is "not free." For example, a simple facial tissue takes a huge amount of energy ΔE and a section of time Δt to create. It is, however, a "necessary" but not sufficient condition, because it also needs an amount of information ΔI to make it happen (i.e., ΔE, Δt, and ΔI) [12].

In short, I would stress that if there is a beginning then there is an end. Since time and space are coexisted, then time and space have no beginning and no end. In which we see that time-space [or temporal (t > 0) space] is ever existed, since existence and non-existence are mutually exclusive. In other words, emptiness and non-emptiness are mutually excluded, then time "always" exists with space. Thus, time is real because the space is real, for which time-space has no beginning and has no end. And this must be the art of temporal (t > 0) universe.

6. Law of uncertainty

One of the most intriguing principles in quantum mechanics [13] must be the Heisenberg's Uncertainty Principle [14], as shown by the following equation:

$$\Delta p \ \Delta x \geq h \tag{5}$$

where Δp and Δx are the momentum and position errors, respectively, and h is the Planck's constant. As reference to "wave-particle dynamics," the momentum p of a "photonic particle" is presented by a "quanta" of energy $h\upsilon$ as given by:

$$p = h/\lambda = h\upsilon/c \tag{6}$$

where h is the Planck's constant, λ is the wavelength, υ is the frequency, and c is the velocity of light.

In which we see that Heisenberg's principle was based on "wave-particle duality" existed within an "empty space." The essence of the Heisenberg's uncertainty principle is that one cannot precisely determine the position x and the momentum p of a particle "simultaneously under observation", as illustrated in **Figure 8**. In which we see that; it is "independent" of time, since Heisenberg's principle was based on "observation" stand point which has nothing to do with changing naturally with time. Yet we know that if there is "no" time there is "no" uncertainty.

In view of **Figure 8**, Heisenberg principle was derived on an empty timeless (t = 0) subspace and it has "nothing to do or independent" with the "underneath subspace" that the particle is situated. Strictly speaking, it is "not" a physical realizable paradigm should be used in the first place, since particle and empty subspace are "mutual exclusive." Secondly, the position error Δx of Heisenberg was based on a "diffraction limited" microscopic observation, where the "spatial" ambiguity of Δx is given by [15]:

$$\Delta x = 0.6\,\lambda/\sin\,\alpha \tag{7}$$

where λ is the observation wavelength, $2(\sin\alpha)$ is the "numerical aperture" of the microscope and α is subtended half-angle of observation aperture. In which we

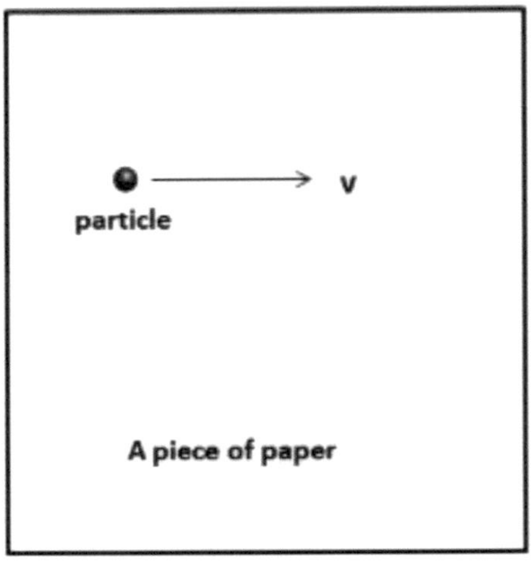

Figure 8.
A particle in motion within an "empty" subspace. v is the velocity. Note that background paper has been treated as an "empty" subspace for centuries.

see that the position error Δx is "not" due to particle in motion, but based on the diffraction limited aperture. This is precisely why Heisenberg's position error Δx has been interpreted as an "observation error" which is independent with time. But uncertainty changes naturally with time, since without time it has no uncertainty.

Secondly, the momentum error Δp as I quote [15]: after collision the particle being observed, the photon's path is only to lie within a cone having semi-vertical angle α in which momentum of the particle is uncertain by the amount as given by:

$$\Delta p = h(\sin \alpha)/\lambda \tag{8}$$

where λ is the wavelength of the quantum leap of $h\upsilon$. In which we see that; momentum error Δp is "not" due to band width $\Delta \upsilon$ of quantum leap since any physical radiator has to be band limited. In other words, the momentum error Δp of preceding Eq. (8) is a singularity approximated λ, which is "not" a band limited $\Delta\lambda$ of physical reality.

As we look back at the subspace that Heisenberg's principle developed from, it was an "inadvertently" timeless (t = 0) subspace as shown in **Figure 8**. Aside the timeless (t = 0) subspace, it is the uncertainty mainly due to diffraction limited observation, which is a "secondary cause" by human intervention, but not due to naturally change with time. This is similar to entropy theory of Boltzmann [16]: entropy increases naturally with time within an enclosed subspace. In which we see that uncertainty should be increasing with time, without human intervention. As I have noted, without time, there would be no entropy and no uncertainty.

Nevertheless, momentum error Δp and position error Δx are mutually "coexisted." In principle they can be traded. But the trading cannot without constraint, since time is a dependent forward variable. But Heisenberg uncertainty; Δp and Δx are "not" mutually dependent, since his position error Δx is due to diffraction limited observation, which is nothing to do with time. For which it poses a physical "inconsistency" within our universe, although Heisenberg principle has been widely used without any abnormality. But it is from the "physical consistency" standpoint, Heisenberg's position error Δx was based on diffraction limited observation has "nothing" to do with time. And also added and his momentum error Δp was based on singularity wavelength λ which is "not" a band limited reality.

Yet, uncertainty principle can be made temporal (t > 0), similar to entropy theory of Boltzmann. For which we have a "law of uncertainty" as stated: uncertainty of an isolated particle increases naturally with time. Or more specific: uncertainty of an isolated particle within an isolated subspace, increases with time and eventually reaches to a maximum amount within the isolated subspace. For which we see that there it exists a profound connection between uncertainty and entropy.

7. Temporal (t > 0) uncertainty

Since it is our universe governs the science and it is not the science governs our universe. Therefore, every principle within our universe has to comply with the temporal (t > 0) condition within our universe; otherwise, the principle cannot be existed within our universe. Which includes all the laws, principles and theories; such as Maxwell's Electro-Magnetic theory, Boltzmann's entropy theory, Einstein's relativity theory, Bohr's atomic model, Schrödinger's superposition principle, and others. Of which uncertainty principle cannot be the exception?

Let us now assumed a temporal (t > 0) particle m(t) is situated within a temporal (t > 0) subspace as depicted in **Figure 9**. Strictly speaking any particle existed within a temporal subspace must be a temporal (t > 0) particle; otherwise, the particle cannot be existed within our temporal (t > 0) universe.

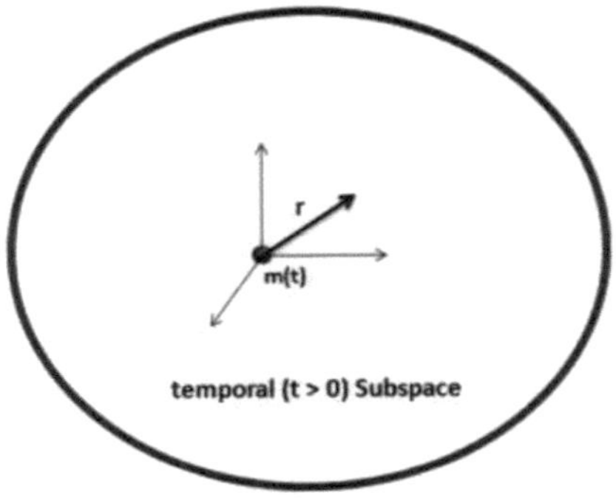

Figure 9.
A temporal (t > o) particle m(t) within a temporal (t > o) subspace. r is the radial direction. Note: it is a "physical realizable" paradigm, since a temporal particle m(t) is embedded within a temporal subspace.

For simplicity, we further assume m(t) has no time or "pseudo-timeless," after all science is a law of approximation. The same as Heisenberg's assumption, the particle is a photonic particle (i.e., a photon), as from wave particle-duality standpoint [17] momentum of a photon is given by:

$$p = h/\lambda = h\,\upsilon/c \tag{9}$$

where h is the Planck's constant, λ is the wavelength and υ is the frequency of the photonic particle. As I have mentioned earlier, within our universe any radiator has to be band limited. Thus the momentum error is naturally due changes of bandwidth Δυ, as given by;

$$\Delta p = h\,\Delta\upsilon/c \tag{10}$$

Instead of using a cone of light as Heisenberg had postulated. By virtue of time-bandwidth product Δυ Δt = 1, Δυ "decreases" with time. For which position error can be written as:

$$\Delta r = c\,\Delta t \tag{11}$$

where r is the radial distance, we have the following uncertainty relationship;

$$\Delta p\,\Delta r = [h\,\Delta\upsilon/c]\,[c\,\Delta t] = h\,\Delta\upsilon\,\Delta t \tag{12}$$

In which we see that; Δυ · Δt is the "time-bandwidth" product. As we imposed the optimum energy transfer criterion on time-bandwidth product [12], as given by:

$$\Delta\upsilon\,\Delta t \geq 1 \tag{13}$$

Since lower bound for a photonic particle is limited by Planck's constant, we have the following equivalent form as given by:

$$\Delta E\,\Delta t \geq h \tag{14}$$

Nevertheless, in view of Eq. (13), momentum uncertainty principle can be shown as:

$$\Delta p\,\Delta r \geq h, t > 0 \tag{15}$$

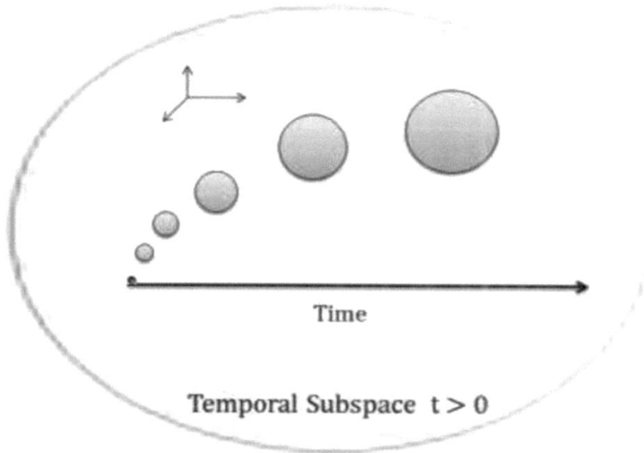

Figure 10.
Position error Δr (i.e., sphere of Δr) enlarges naturally with time within a temporal (t > 0) subspace: Δr represents a position error of the particle, at various locations as time moves constantly.

where (t > 0) denotes that uncertainty principle is complied with the temporal (t > 0) condition within our universe. In view of either conservation of momentum or energy conservation, we see that position error Δr increases naturally with time. Which shows that momentum error Δp "decreases" naturally with bandwidth Δυ, as in contrast with Heisenberg's assumption; momentum error Δp has "nothing" to do with the changes of Δυ. This is precisely the "law of uncertainty" as I have described earlier, uncertainty of an isolated particle increases naturally with time.

Since the increase in position error Δr is due to time, it must be due to the dynamic expansion of our universe [5]. For example, as the boundary of our universe constantly expanding at the speed of light, by virtue of energy conservation, it changes every dynamic aspect within our universe. As time moves on naturally, the larger the position error Δr increases with respect to that starting point, as illustrated in **Figure 10**.

Therefore we see that uncertainty is "not" a static process it is a temporal (t > 0) dynamic principle, as in contrast with Heisenberg's position error Δr is "independent" with time and his momentum error Δp is "independent" with Δυ. In which we see that if there is no time, there is no uncertainty and no probability. Nevertheless, each of the uncertainty unit or cell, such as (Δp, Δx), (ΔE, Δt) and (Δυ, Δt) is self-contained. In other words, ΔE and Δt are coexisted which they can be bilateral traded, but under the constraint of time as a forward moving dependent variable. In other words, if a section of Δt has been used, we cannot get the "same" section back, but can exchange for a different section of Δt. In which we see that we can trade for a narrower Δt with a wider ΔE or wider Δt with a narrower ΔE. But we "cannot" trade Δt for ΔE, since Δt is a real dependent variable has "no" substance to manipulate.

8. Certainty principle

One of the important aspects of "temporal uncertainty" is that subspace within our universe is a temporal (t > 0) uncertain "subspace." In other words, any subspace is a temporal (t > 0) stochastic subspace, such that the dynamic behavior of the subspace changes "dependently" with time. In which any change within our universe has a profound connection with the constant expanding universe. In which we have shown that uncertainty increases naturally with time, even though without any other perturbation or human intervention. Similar to the myth of Boltzmann's entropy theory [16], entropy increases naturally with time within an enclosed subspace, which has been shown is related to the expanding universe [5].

Similarly, there is a profound "connection" between coherence theory [18] and "certainty" principle as I shall address. Nevertheless, it is always a myth of coherence, as refer to **Figure 11**, where coherence theory can be easily understood by Young's experiment. In which degree of coherence can be determined by the "visibility" equation as given by:

$$\nu = \frac{\text{Imax–Imin}}{\text{Imax} + \text{Imin}} \tag{16}$$

where I_{max} and I_{min} are the maximum and minimum intensities of the fringes. But the theory does not tell us where the physics comes from. For which, it can be understood from "certainty principle," as I shall address.

It is trivial that if there is an uncertainty principle, it is inevitable not to have a certainty principle. This means that, as photonic particle we are looking for is "likely" to be found within a "certainty" subspace. Since "perfect certainty" (or absolute uncertainty) occurs at t = 0, which is a timeless (t = 0) virtual subspace not exist within our universe. Nevertheless, "certainty principle" can be written in the following equivalent forms;

$$\Delta p\ \Delta r < h, (t > 0) \tag{17}$$

$$\Delta E\ \Delta t < h, (t > 0) \tag{18}$$

$$\Delta \upsilon\ \Delta t < 1, (t > 0) \tag{19}$$

where (t > 0) denotes that equation is subjected to temporal (t > 0) constrain. In view of the position error Δr in Eq. (17), it means that it is "likely" the photonic particle can be found within the certainty subspace. Since the size of the subspace is limited by Planck constant h, it is normally used as limited boundary "not" to be violated. Yet within this limited boundary, certainty subspace had been exploited by Dennis Gabor for his discovery of wave front reconstruction in 1948 [19] and as well it was applied to synthetic aperture radar imaging in 1950s [20].

Since the size of certainty subspace is exponentially enlarging as the position error Δr increases, for which the "radius" of the certainty sub-sphere is given by:

$$\Delta r = c\ \Delta t = c/(\Delta \upsilon) \tag{20}$$

where c is the speed of light, Δt is the time error, and Δυ is the bandwidth of a light source or a quantum leap hυ. Thus we see that position error Δr is inversely proportional to bandwidth Δυ, as plotted in **Figure 12**.

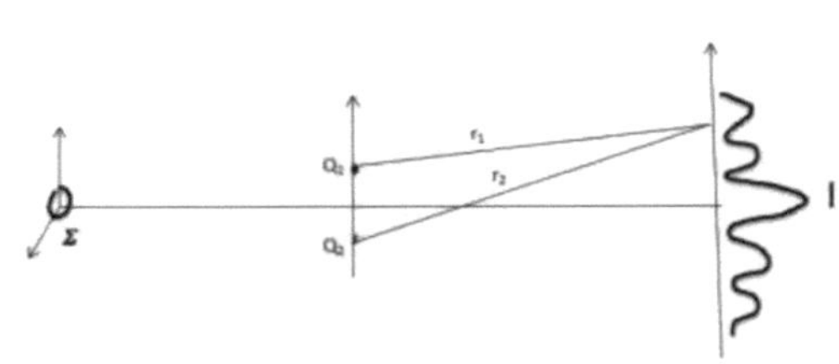

Figure 11.
Young's experiment. Σ represents an extended monochromatic source, Q_1 and Q_2 are the pinholes, and "I" represents the irradiance distribution.

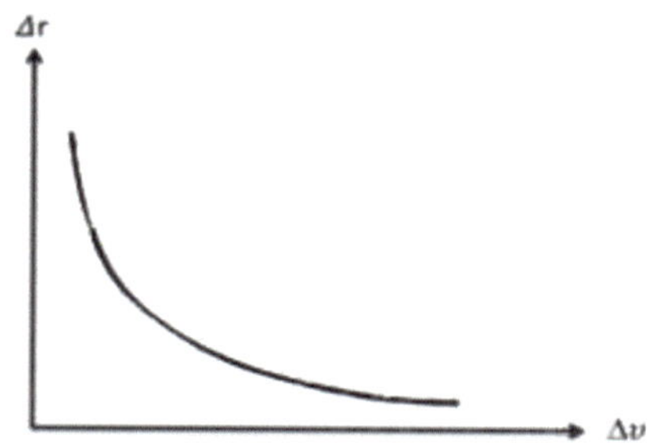

Figure 12.
A plot of position error Δr versus bandwidth Δυ.

In view of this plot, we see that when bandwidth Δυ decreases, a larger certainty subspace enlarges "exponentially" since the volume of the subspace is given by:

$$\text{Certainty subspace} = (4\pi/3)(\Delta r)^3 \tag{21}$$

In which we see that, a very "large" certainty subspace can be realized within our universe which is within limited Planck's constant h as depicted in **Figure 13**, where we see a steady state radiator A emits a continuous band limited Δυ electro-magnetic wave as illustrated. A "certainty subspace" with respect to an assumed "photonic particle" A for a give Δt can be defined as illustrated within r = c Δt, where Δt = 1/Δυ. In other words, it has a high degree of certainty to relocate particle A within the certainty subspace. Nevertheless, from electro-magnetic disturbance standpoint; within the certainty subspace provides a high "degree of certainty" (i.e., degree of coherence) as with respect to point A.

As from coherence theory stand point, any other disturbances away from point A but within the certainty subspace (i.e., within r < c Δt) are mutual coherence (i.e., certainty) with respect A; where r = c Δt is the radius of the "certainty subspace" of A. In other words, any point-pair within d < c Δt, where Δt = 1/Δυ, are "mutual coherence" within a radiation subspace. On the other hand, distance

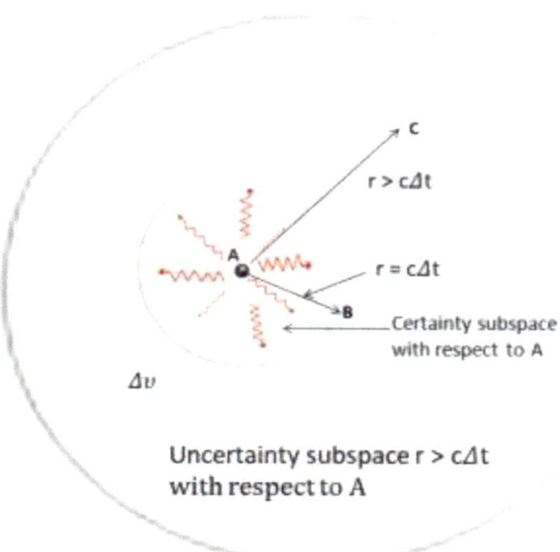

Figure 13.
A certainty subspace is embedded within uncertainty subspace. A is assumed a steady state photonic particle emits a band limited Δυ radiation, r is the radius with respect to the emitter A; and B represents the boundary of certainty subspace of A.

greater than $r > c\,\Delta t$ from point A is a mutual "uncertainty" subspace with respect to A. In other words, any point-pair distance is larger than $d > c\,\Delta t$ within the radiation space are mutually "incoherent." In which we see that; it is more "unlikely" to relocate a photonic particle, after it has been seen at point A, within a "certainty subspace."

Since certainty subspace represents a "global" probabilistic distribution of a particle's location as from particle physicists stand point, which means that it is "very likely" the particle can be found within the certainty subspace. In which we see that a postulated particle firstly is temporal ($t > 0$) or has time; otherwise, there is no reason to search for it. Then after it has been proven it is a temporal (i.e., $m(t)$) particle, it is more favorable to search the particle, within a certainty subspace.

The essence of "wave-particle duality" is a mathematical simplistic assumption to equivalence a package of wavelet energy as a particle in motion from statistical mechanics stand point, in which the momentum $p = h/\lambda$ is conserved. However one should "not" treated wave as particle or particle as wave. It is the package of wavelet energy "equivalent" to a particle dynamics (i.e., photon), but they are "not" equaled. Similar to Einstein's energy equation, mass is equivalent to energy and energy is equivalent to mass, but they are not equaled. Therefore as from energy conservation, bandwidth $\Delta \upsilon$ "decreases" with time is the physical reality instead of treating a package of wavelet as a particle (i.e., photon), which was due to the classical mechanics standpoint, treats quantum leap momentum $p = h/\lambda$. In which we see that photon is a "virtual" particle although many quantum scientists have been regarded photon as a physical particle?

We further note that any point-pair within the certainty subspace exhibits some degree of certainty or coherence, which has been known as "mutual coherence" [18]. And the mutual coherence can be easily understood as depicted in **Figure 14**, in which a steady state band limited $\Delta \upsilon$ electro-magnetic wave is assumed existed within a temporal ($t > 0$) subspace. As we pick an arbitrary disturbance at point B, a certainty subspace of B can be determined within $r \le c\,\Delta t$, as shown in the figure.

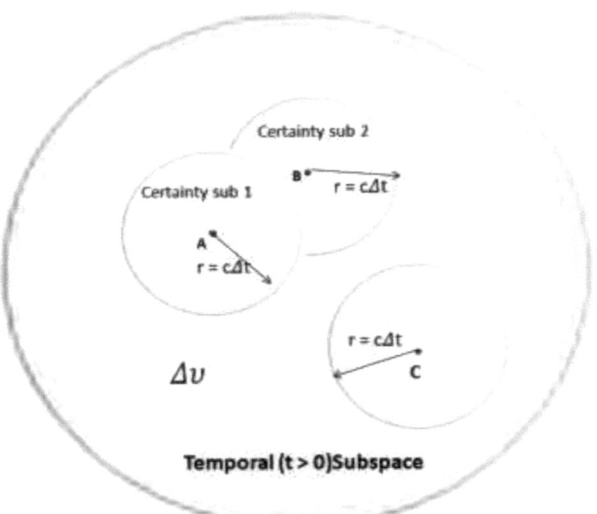

Figure 14.
Various certainty subspace configurations, as with respect to various disturbances within a steady state band limited $\Delta \upsilon$ electro-magnetic environment within a temporal ($t > 0$) subspace.

This means that any point disturbance within in the certainty subspace has a strong certainty (or coherence) with respect to point B disturbance. Similarly if we pick an arbitrary point A, then a certainty subspace of A can be defined as illustrated in the figure, of which we see that a portion is overlapped with certainty subspace of B. Any other disturbances outside the corresponding subspaces of certainty A, B, and C are the uncertainty subspace. It is trivial to see that a number of configurations of certainty subspaces can be designed for application. In which we see that multi wavelengths, such as $\Delta\upsilon_1$, $\Delta\upsilon_2$, and $\Delta\upsilon_3$, can also be simultaneously implemented to create various certainty subspace configurations, such as for multi spectral imaging or information processing application.

One of the commonly used for producing certainty subspaces for complex wave front reconstruction is depicted in **Figure 15** [21]. In which we see that a band limited $\Delta\upsilon$ laser is employed, where a beam of light is split-up by a splitter BS. One beam B_2 is directly impinging on a photographic plate at plane P and other beam B_1 diverted by a mirror and then is combined with beam B_2 at the same spot on the photographic plate P. It is trivial to know that if the difference in distances between these two beams is within the certainty subspace, then B_1 and B_2 are "mutually" coherence (or certainty); otherwise, they are mutually incoherence (or uncertain). In which we see that the distance between B_1 and B_2 is required as given by:

$$|d_1 - d_2| < c\,\Delta t = c/\Delta\upsilon \tag{22}$$

where d_1 and d_2 are the distances of bean B_2 and B2, respectively, from the splitter BS. In which we see that radius of certainty subspace of BS is written by;

$$\Delta r = |d_1 - d_2| < c\,\Delta t = c/\Delta\upsilon \tag{23}$$

where $|d_1 - d_2| = c/\Delta\upsilon$ is the "coherent length" of the laser. In which we see that by simply reducing the bandwidth $\Delta\upsilon$, a lager certainty subspace can be created within a temporal $(t > 0)$ subspace.

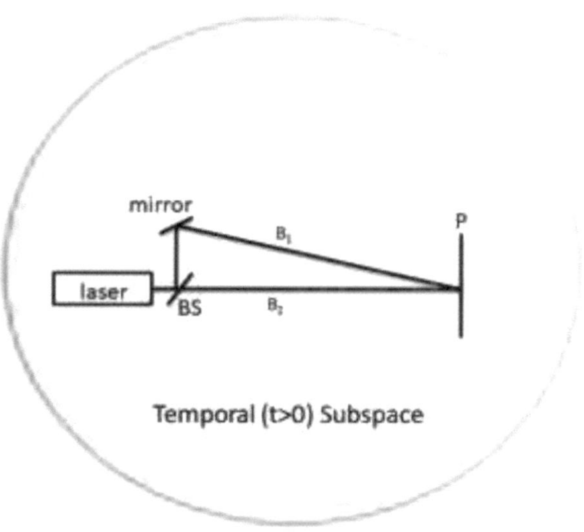

Figure 15.
An example of exploiting certainty subspace for wave front reconstruction. BS, beam splitter; P, photographic plate.

9. Essence of certainty principle

Since every substance or subspace within our universe was created by an amount of energy ΔE and a section of time Δt [i.e., (ΔE, Δt)], any changes of ΔE changes the size of certainty subspace Δr. This is a topic that astrophysicists may be interested. Similarly to particle physicists, subatomic particle has to be temporal (t > 0); otherwise, the particle must be a virtual particle cannot exist within our universe. Secondly, it is more "likely" a temporal (t > 0) particle to be found within its certainty subspace; otherwise, it will be searching a timeless (t = 0) particle "forever" within our temporal (t > 0) universe. In view of the certainty unit: ΔE and Δt are mutually coexisted in which time is a forward dependent variable. Any changes of ΔE can "only" happen with an expenditure of a section time Δt, but it "cannot" change the speed of time. Since the energy is "conserved," Δt is a section of time required to have the amount of ΔE within a certainty unit of (ΔE, Δt). In other words, ΔE and Δt can be traded; for example, a wider variance of ΔE is traded for a narrower Δt.

Nevertheless, time has been treated as an "independent" variable for decades, as normally assumed by scientists. But whenever a section of time Δt has been used, it is not possible to bring back the "original" moment of Δt, even though it is possible to reproduce the same section of Δt. This similar as we reconstructed a damaged car, but we cannot bring back the "original" car that has been crashed. And this is precisely the "price of time" to pay for everything within our universe. Then my question is that if time is a forward dependent variable with respect to its subspace, how can we "curve" the space with time? Similarly, we are coexisted with time, how can we get back the moment of time that has passed by?

Since certainty subspace changes with bandwidth Δυ as illustrated in **Figure 16**, in which we see that as bandwidth Δυ decreases a very large certainty subspace can be created within our universe as depicted in **Figure 16(a)–(c)**.

High resolution observation requires shorter wavelength but shorter wavelength inherently has broader bandwidth Δυ that creates a smaller certainty subspace, which can be used for high resolution wave front reconstruction [21]. On the other hand, for a larger certainty subspace, it required a narrower bandwidth of Δυ which has a larger certainty subspace for exploitation, such as applied to side looking radar imaging [20]. In which we see that the size of the certainty subspace can be manipulated by the bandwidth Δυ as will be shown in the following:

Since narrower bandwidth Δυ offers a huge certainty subspace that can be exploited for long distance communication, in which I have found that the certainty subspace is "in fact" the coherence subspace as I have discussed in the preceding. In other words, within a certainty subspace it exhibits a "point-pair certainty" or coherent property among them as illustrated in **Figure 17**. In other words, it has a

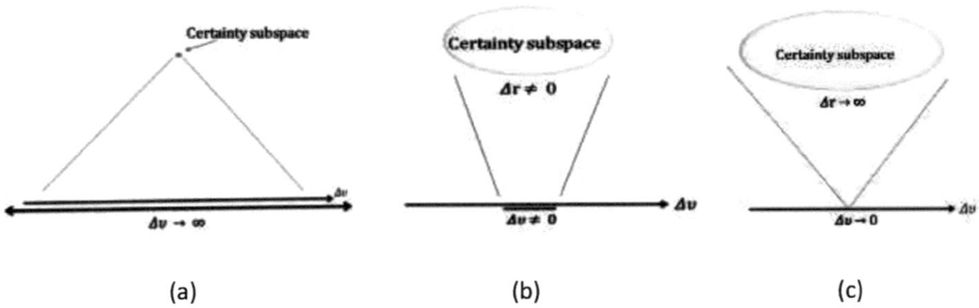

Figure 16.
Size of certainty subspace enlarges rapidly as band width Δυ narrows. (a) shows a very small size of certainty subspace as the result of Δυ approaching to very wide. (b) shows the size of certainty subspace reduces as Δυ continues to reduce. And (c) shows a huge size certainty subspace can be created as band width Δυ narrows.

high degree of certainty within a certainty subspace between points. This means that, if a photonic particle as it has been started at point u_1, then it has a high degree of certainty that the particle to be found at the next instantly Δt at u_2, since distance is time within a temporal (t > 0) subspace.

For example, given any two arbitrary complex disturbances $u_1(r_1; t)$ and $u_2(r_2; t)$, as long the separation between them is shorter than the radius Δr of the certainty subspace as given by:

$$d \leq c/(\Delta \upsilon) \tag{24}$$

the disturbances between $u_1(r_1; t)$ and $u_2(r_2; t)$ are "certainly" related (or mutually coherence). For which the "degree of certainty" (i.e., degree of coherence) between u_1 and u_2 can be determined by the following equation:

$$\gamma_{12}(\Delta t) = \frac{\Gamma_{12}(\Delta t)}{\Gamma_{11}(0)\Gamma_{22}(\Delta 0)} \tag{25}$$

where, "mutual certainty" (or mutual coherence) function between u_1 and u_2 can be written as:

$$\Gamma_{12}(\Delta t) = \lim_{T \to \infty} \frac{1}{T} \int_0^T u_1(t; r_1) u_2^*(t - \Delta t; r_1) dt \tag{26}$$

Similarly, the respective "self certainty" (or self coherence) functions are, respectively, given by:

$$\Gamma_{11}(\Delta t) = \lim_{T \to \infty} \frac{1}{T} \int_0^T u_1(t; r_1) u_1^*(t - \Delta t; r_1) dt \tag{27}$$

$$\Gamma_{22}(\Delta t) = \lim_{T \to \infty} \frac{1}{T} \int_0^T u_2(t; r_2) u_2^*(t - \Delta t; r_2) dt \tag{28}$$

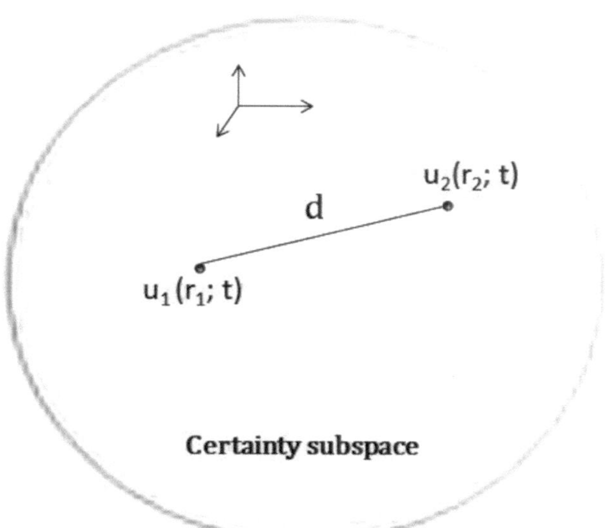

Figure 17.
Mutual certainty within a certainty subspace. $u_1(r_1; t)$ and $u_2(r_2; t)$ represent two arbitrary disturbances separated at distance d.

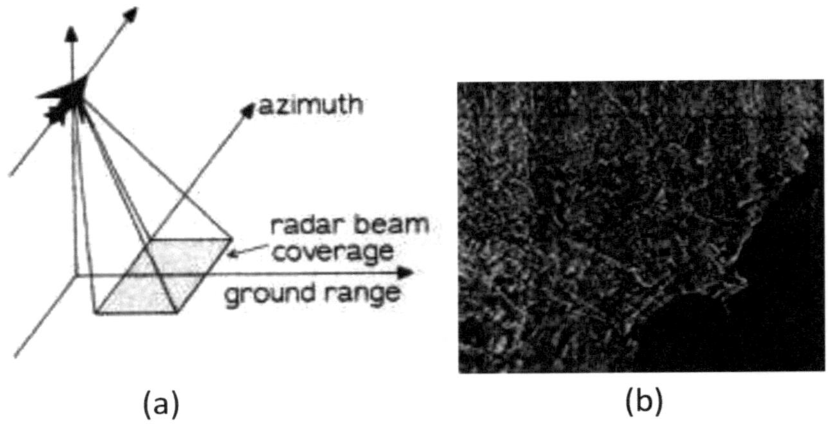

Figure 18.
Side-looking radar imaging within certainty subspace: (a) shows a side-looking radar scanning flight path; (b) shows an example of synthetic aperture radar imagery.

One of the interesting applications for certainty principle must be to synthetic aperture radar imaging as I have mentioned earlier is shown in **Figure 18**. In which we see an aircraft carried a side looking synthetic radar system shown in **Figure 18(a)**, emitting a sequence of radar pulses scanned across the flight path of the terrain. The returned pulses are combined with local radar pulses, which are "mutual coherence" (i.e., high degree of certainty), to construct a recording format that can be used for imaging the terrain, for which a synthetic imagery is shown in **Figure 18(b)**. In which we see a variety of scatters, including city streets, wooded areas, and farmlands and lake with some broken ice floes can also be identified on the right of this image. Since microwave antenna has a very narrow carrier bandwidth (i.e., $\Delta\upsilon$) and its certainty radius (i.e., $d = c \cdot \Delta t$) or the coherence length can be easily reached to hundreds of thousand feet. In other words, a very large certainty subspace for complex-amplitude imaging (or for communication) can be realized.

Finally I would address again within the certainty unite (Δp, Δr) [i.e., equivalently for (ΔE, Δt) and ($\Delta\upsilon$, t) unit] can be mutually traded. But it is the trading of Δp for Δr (or ΔE for Δt and $\Delta\upsilon$ for Δt) is physically visible, since time is not a physical substance but a forward constant dependent "variable" that we "cannot" manipulate. For which we see that the "section" of Δt that has been "used" cannot get it back. In other words, we can get back the same amount Δt, but "not" the same moment of Δt, that has been expensed. As I have shown earlier, everything within our universe has a price, an amount of energy ΔE, and a section of time Δt. Aside ΔE we can physically change, it is the moment of time Δt which has been expensed that is "preventing" us to get it back, because that moment of Δt is the "same moment" of time of our temporal ($t > 0$) universe that has been passed. And this is the "moment of time" Δt within our temporal ($t > 0$) universe, once the "moment" passes by and we can never able to get it back.

10. Conclusion

In conclusion, I would point out that quantum scientists used amazing mathematical analyses added with their fantastic computer simulations provide very convincing results. But mathematical analyses and computer animations are virtual and fictitious, and many of their animations are "not" physically real; for example such as the "instantaneous and simultaneous" superimposing principle for quantum computing is "not" actually existed within our universe. One of the important aspects within our universe is that, one cannot get something from nothing there is always a price to pay, an amount of energy ΔE and a section of time Δt. The important is that they are not free!

Since any science existed within our universe has time or temporal $(t > 0)$, in which we see that any scientific law, principle, theory, and paradox has to comply with temporal $(t > 0)$ aspect within our universe; otherwise, it may not be science. As we know that science is mathematics but mathematics is not equaled to science. In which we have shown that any analytic solution has to be temporal $(t > 0)$; otherwise, it cannot be implemented within our universe. Which includes all the laws, principles, and theories have to be temporal $(t > 0)$?

Since it is our universe governs our science and it is not our science changes our universe. In which we have shown every hypothetical science, law, principle, and theory has be temporal $(t > 0)$; otherwise, they are virtual and fictitious which cannot exist within our universe. Since time is a dependent variable coexisted with space, we have concluded that time is not an illusion but real, since we are real. As in contrast with most of the scientists, they believe that time is an independent variable and some of them even believe that time is an illusion?

Uncertainty principle is one of the most fascinating principles in quantum mechanics, yet Heisenberg principle was based on diffraction limited observation, it is not due to the nature of time or temporal $(t > 0)$ nature of our universe. We have shown uncertainty increases with time, as in contrast with Heisenberg's principle. We have also introduced a certainty principle, in which we have shown high degree of certainty within a certainty subspace can be exploited. For which we have shown that certainty subspace can be created within our temporal subspace for complex amplitude communication and imaging. Yet the important aspect of this chapter is that it is not how rigorous the mathematics is, but it is the physical realizably of science is, since mathematics is not science.

Author details

Francis T.S. Yu
Penn State University, University Park, PA, USA

*Address all correspondence to: fty1@psu.edu

References

[1] Yu FTS. From relativity to discovery of temporal (t > 0) universe. In: Origin of Temporal (t > 0) Universe: Correcting with Relativity, Entropy, Communication and Quantum Mechanics, Chapter 1. New York: CRC Press; 2019. pp. 1-26

[2] Bunge M. Causality: The Place of the Causal Principle in Modern Science. Cambridge: Harvard University Press; 1959

[3] Yu FTS. The fate of Schrodinger's cat. Asian Journal of Physics. 2019;**28**(1): 63-70

[4] Knudsen JM, Hjorth P. Elements of Newtonian Mechanics. Heidelberg: Springer Science & Business Media; 2012

[5] Yu FTS. Time: The enigma of space. Asian Journal of Physics. 2017;**26**(3): 143-158

[6] Yu FTS. What is "wrong" with current theoretical physicists? In: Bulnes F, Stavrou VN, Morozov O, Bourdine AV, editors. Advances in Quantum Communication and Information, Chapter 9. London: IntechOpen; 2020. pp. 123-143

[7] Parzen E. Stochastic Processes. San Francisco: Holden Day, Inc.; 1962

[8] Einstein Attacks Quantum Theory. Scientist and Two Colleagues Find It Is Not 'Complete' Even though 'Correct'. New York City: The New York Times; 1935

[9] Einstein A. Relativity, the Special and General Theory. New York: Crown Publishers; 1961

[10] Hawking S, Penrose R. The Nature of Space and Time. New Jersey: Princeton University Press; 1996

[11] Kraus JD. Electro-Magnetics. New York: McGraw-Hill Book Company; 1953. p. 370

[12] Yu FTS. Optics and Information Theory. New York: Wiley-Interscience; 1976

[13] Schrödinger E. An Undulatory theory of the mechanics of atoms and molecules. Physics Review. 1926;**28**(6): 1049

[14] Heisenberg W. Über den anschaulichen Inhalt der quantentheoretischen Kinematik und Mechanik. Zeitschrift für Physik. 1927; **43**(3–4):172

[15] Lawden DF. The Mathematical Principles of Quantum Mechanics. London: Methuen & Co Ltd.; 1967

[16] Boltzmann L. Über die Mechanische Bedeutung des Zweiten Hauptsatzes der Wärmetheorie. Wiener Berichte. 1866; **53**:195-220

[17] MacKinnon E. De Broglie's thesis: A critical retrospective. American Journal of Physics. 1976;**44**:1047-1055

[18] Yu FTS. Introduction to Diffraction, Information Processing and Holography, Chapter 10. Cambridge, Mass: MIT Press; 1973. pp. 91-98

[19] Gabor D. A new microscope principle. Nature. 1948;**161**:777

[20] Cultrona LJ, Leith EN, Porcello LJ, Vivian WE. On the application of coherent optical processing techniques to synthetic-aperture radar. Proceedings of the IEEE. 1966;**54**:1026

[21] Leith EN, Upatniecks J. Reconstructed wavefront and communication theory. Journal of the Optical Society of America. 1962;**52**:1123

Chapter 9

Nature of Temporal (t > 0) Quantum Theory: Part II

Francis T.S. Yu

Abstract

Since Schrödinger's quantum mechanics developed from Hamiltonian, I will show that his quantum machine is a timeless (t = 0) mechanics, which includes his fundamental principle of superposition. Since one of the most controversial paradoxes in science must be Schrödinger's cat. We will show that the myth of his hypothesis is "not" a physical realizable postulation. The most important aspect in quantum theory must be the probabilistic implication of science, a set of most elegant and simple laws and principles, which will be discussed. Since information and entropy have a profound connection, we will show that information is one of very important science in quantum theory, for which several significant aspects of information transmission will be stressed. Nevertheless, the myth of quantum theory turns out to be not Schrodinger's cat but the nature of a section of time Δt. Since time is a quantity that we cannot physically manipulate, we could change the section Δt but not the speed of time. Although we can squeeze a section of Δt, but we cannot squeeze Δt to zero. And this is the ultimate quantum limit of "instantaneous" response we can never be able to obtain. Since time traveling is one of the very interesting topics in science, I will show that time traveling is impossible even at the speed of light. Nevertheless, I will show quantum mechanics is a temporal (t > 0) physical realizable mechanics, and it should "not" be as virtual and timeless (t = 0) as mathematic does.

Keywords: quantum mechanics, Hamiltonian mechanics, timeless mechanics, temporal mechanic, temporal universe, timeless space, physical realizable, Schrödinger's cat

1. Introduction

Two of the most important discoveries in the twentieth century in modern science must be the Einstein's relativity theory [1] and Schrödinger's quantum mechanics [2]; one is dealing with very large objects and the other is dealing with very small particles. Yet they were connected by means of Heisenberg's uncertainty principle [3] and Boltzmann's entropy theory [4]. Yet, practically, all the laws, principles, and theories of science were developed from an absolute empty space, and their solutions are all timeless (t = 0) or time-independent. Since our universe is a temporal (t > 0) space, timeless (t = 0) solution cannot be "directly" implemented within our universe, because timeless and temporal are mutually exclusive.

IntechOpen

Although timeless laws and principles have been the foundation and cornerstone of our science, there are also scores of virtual solutions that are "not" physical realizable within our temporal ($t > 0$) space.

Yet, it is the major topic of the current state of science, fictitious and virtual as mathematics is. Added with very convincing computer simulation, fictitious science becomes "irrationally" real? As a scientist, I felt, in part, my obligation to point out where those fictitious solutions come from, since science is also mathematics.

Since Schrödinger's quantum mechanics is a legacy of Hamiltonian classical mechanics, I will first show that Hamiltonian was developed on a timeless ($t = 0$) platform, for which Schrödinger's quantum machine is also timeless ($t = 0$); this includes his quantum world as well his fundamental principle of superposition. I will further show that where Schrödinger's superposition principle is timeless ($t = 0$), it is from the adaption of Bohr's quantum state energy $E = h\upsilon$, which is essentially time unlimited singularity approximated. I will also show that nonphysical realizable wave function can be reconfigured to becoming temporal ($t > 0$), since we knew a physical realizable wave function is supposed to be. And I will show that superposition principle existed "if and only if" within a timeless ($t = 0$) virtual mathematical subspace but not existed within our temporal ($t > 0$) space.

When dealing with quantum mechanics, it is unavoidable not to mention Schrödinger's cat, which is one of the most elusive cats in science, since Schrödinger disclosed the hypothesis in 1935? And the interesting part is that the paradox of Schrödinger's cat has been debated by score of world renounced scientists such as Einstein, Bohr, Schrödinger, and many others for over eight decades, and it is still under debate. Yet I will show that Schrödinger's hypothesis is "not" a physical realizable hypothesis, for which his half-life cat should "not" have had used as a physical postulated hypothesis.

In short, the art of a quantum mechanics is all about temporal ($t > 0$) subspace, in which we see that everything existed within our universe; no matter how small it is, it has to be temporal ($t > 0$), otherwise it cannot exist within our universe.

2. Hamiltonian to temporal (t > 0) quantum mechanics

In modern physics, there are two most important pillars of disciplines: It seems to me one is dealing with macroscale objects of Einstein [1] and the other is dealing with microscale particle of Schrödinger [2]. Instead of speculating micro- and macro objects behave differently, they share a common denominator, temporal ($t > 0$) subspace. In other words, regardless of how small the particle is, it has to be temporal ($t > 0$), otherwise it cannot exist within our temporal ($t > 0$) universe.

As science progresses from Newtonian [5] to statistical mechanics [6], "time" has always been regarded as an "independent" variable with respect to substance or subspace. And this is precisely what modern physics has had been used the same timeless ($t = 0$) platform, for which they have treated time as an "independent" variable. Since Heisenberg [3] was one of the earlier starters in quantum mechanics, I have found that his principle was derived on the same timeless ($t = 0$) platform as depicted in **Figure 1**. And this is the "same" platform used in developing Hamiltonian classical mechanics [7]. Precisely, this is the reason why Schrödinger's quantum mechanics is "timeless ($t = 0$)" [8], since quantum mechanics is the legacy of Hamiltonian.

In view of **Figure 1**, we see that the background of the paradigm is a piece of paper, which represents a timeless ($t = 0$) subspace; it is "not" a physical realizable

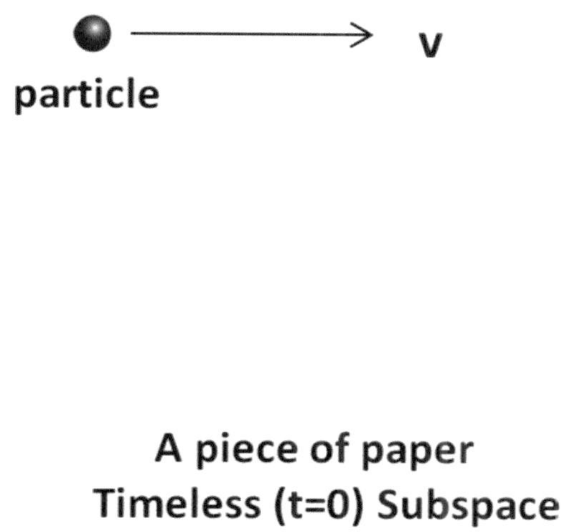

Figure 1.
A particle in motion within a timeless (t = o) subspace. v is the velocity of the particle.

model since particle and empty space are mutually exclusive. Notice that total energy of a "Hamiltonian particle" in motion is equal to its kinetic energy plus the particle's potential energy as given by [7];

$$\mathcal{H} = \mathrm{p}^2/(2\mathrm{m}) + \mathrm{V} \tag{1}$$

which is the well-known Hamiltonian equation, where p and m represent the particle's momentum and mass, respectively, and V is the particle's potential energy. Equivalently, Hamiltonian equation can be written in the following form as applied for a "subatomic particle";

$$\mathcal{H} = -\left[\mathrm{h}^2/\left(8\pi^2\mathrm{m}\right)\right]\, \nabla^2 + \mathrm{V} \tag{2}$$

where h is Planck's constant, m and V are the mass and potential energy of the particle, and ∇^2 is a Laplacian operator;

$$\nabla^2 = \frac{\partial^2}{\partial \mathrm{xi}\ \partial \mathrm{xj}}$$

We note that Eq. (2) is the well-known "Hamiltonian Operator" in classical mechanics.

By virtue of "energy conservation", Hamiltonian equation is written as

$$\mathcal{H}\psi = \{-\left[\mathrm{h}^2/(8\pi^2\mathrm{m})\right]\, \nabla^2 + \mathrm{V}\}\psi = \mathrm{E}\,\psi \tag{3}$$

where ψ is the wave function that remains to be determined and E and V are the energy factor and potential energy that need to be incorporated within the equation. And this is precisely where Schrödinger's equation was derived from; by using the energy factor E = hυ (i.e., a quanta of light energy) adopted from Bohr's atomic model [9], Schrödinger equation can be written as [7]

$$\frac{\partial^2\psi}{\partial x^2} + \frac{8\pi^2\mathrm{m}}{h^2}(E - V)\psi = 0 \tag{4}$$

In view of this Schrödinger's equation, we see that it is essentially "identical" to the Hamiltonian equation, where ψ is the wave function that has to be determined, m is the mass of a photonic-particle (i.e., photon), E and V are the dynamic quantum state energy and potential energy of the particle, x is the spatial variable, and h is Planck's constant.

Since Schrödinger's equation is the "core" of quantum mechanics, but without Hamiltonian's mechanics, it seems to me that we would "not" have the quantum mechanics. The "fact" is that quantum mechanics is essentially "identical" to Hamiltonian mechanics. The major difference between them is that Schrödinger used the dynamic quantum energy $E = h\upsilon$ as adapted from a quantum leap energy of Bohr's hypothesis, which changes from classical mechanics to quantum "leap" mechanics or quantum mechanics. In other words, Schrödinger used a package of wavelet quantum leap energy $h\upsilon$ to equivalent a particle (or photon) as from "wave-particle dynamics" of de Broglie's hypothesis [10], although photon is "not" actually a real particle. Nevertheless, where the mass m for a photonic particle in the Schrödinger's equation remains to be "physically reconciled", after all science is a law of approximation. Furthermore, without the adaptation of Bohr's quantum leap $h\upsilon$, quantum physics would not have started. It seems to me that quantum leap energy $E = h\upsilon$ has played a "viable" role as transforming from Hamiltonian classical mechanics to quantum mechanics, which Schrödinger had done to his quantum theory.

Although Schrödinger equation has given scores of viable solutions for practical applications, at the "same time", it has also produced a number of fictitious and irrational results which are not existed within our universe, such as his Fundamental Principle of Superposition, the paradox of Schrödinger's Cat [8], and others.

In view of Schrödinger's equation as given by Eq. (4), we see that it is a timeless ($t = 0$) or time-independent equation. Since the equation is the "core" of Schrödinger's quantum mechanics, it needs a special mention. Let me stress the essence of energy factor E in the Hamiltonian equation. Since Schrödinger equation is the legacy of Hamiltonian, any wave solution ψ emerges from Schrödinger equation depends upon the E factor. In other words aside the embedded subspace, solution comes out from Schrödinger equation whether is it a physical realizable; it depends upon the E factor that we introduced into the equation. As referring to the conventional Hamiltonian mechanics, if we let the energy factor E be a "constant" quantity that exists at time $t = t_0$, which is "exactly" the classical mechanics of Hamiltonian, this means that the Hamiltonian will take this value of E at $t = t_0$ and evaluates the wave function ψ as has been given by [7]:

$$\psi = \psi_0 \ \exp\left[-i\, 2\pi\, E(t - t_0)/h\right) \tag{5}$$

which is the Hamiltonian wave equation, where ψ_0 is an arbitrary constant, h is Planck's constant, and a constant energy factor $E(t - t_0)$ occurs at $t = t_0$. Although Hamiltonian wave equation is a time-variable function, it is "not" a time-limited solution, for which we see that it "cannot" be implemented within our temporal ($t > 0$) universe, since time unlimited solution cannot exist within our universe. This means that, wave solution ψ of Eq. (5) is "not" a physical realizable solution.

Then a question is being raised, why the Hamiltonian wave solution is time unlimited? The answer is trivial that Hamiltonian is mathematics and his mechanics was developed on an empty timeless ($t = 0$) platform as can be seen in **Figure 1**. Since it is the subspace that governs the mechanics, we see that particle-wave dynamics cannot exist within a timeless ($t = 0$) subspace. But Hamiltonian is mathematics and Hamilton himself is a theoretician; he could have had implanted a particle-wave dynamic into a timeless ($t = 0$) subspace, although timeless ($t = 0$) subspace and physical particle cannot coexist. Of which this is precisely all the scientific laws, principles, and theories were mostly developed on a piece or pieces of papers, since science is mathematics. This is by no means that timeless (t = 0) laws, principles, and theories were wrong [11], yet they were and "still" are the foundation and cornerstone of our science. However, it is their direct implementation within our temporal (t > 0) universe and also added a score of their solutions are irrational and virtual as "pretending" existed within our temporal (t > 0) subspace, for example, superposition principle of quantum mechanics, paradox of Schrödinger's cat, time traveling, and many others.

Nevertheless as we refer to **Figure 1**, immediately we see that it is "not" a physically realizable model that should be used in the first place. Secondly, even though we pretend that the particle in motion within can exist in an empty space, a question is being asked: how can a particle-wave dynamic propagate within an empty space? Thirdly, even though we assumed wave can be exited within an empty space, why it has to be time unlimited? From all these physical reasons, we see that time unlimited Hamiltonian wave equation of Eq. (5) is "not" a physically realizable solution, since it only existed within a timeless (t = 0) virtual mathematical space, which is similar within a Newtonian space, where time has been treated as an "independent" variable.

Since Schrödinger's mechanics is the legacy of Hamiltonian mechanics, firstly we see that Schrödinger's quantum "mechanics" is a solution as obtained from Hamiltonian's mechanics. Secondly, the reason why Schrödinger's quantum mechanics is timeless (t = 0) is the same reason as Hamiltonian, because its subspace is empty. Nevertheless, the major differences between Schrödinger's mechanics and Hamiltonian mechanics must be the name sake of "quantum", where comes Bohr's atomic quantum leap E = hυ, a quanta of light as shown in **Figure 2**, that Schrödinger has used for the development of his mechanics. This is precisely since Schrödinger's solution is very similar to Hamiltonian of Eq. (3) as given by [7],

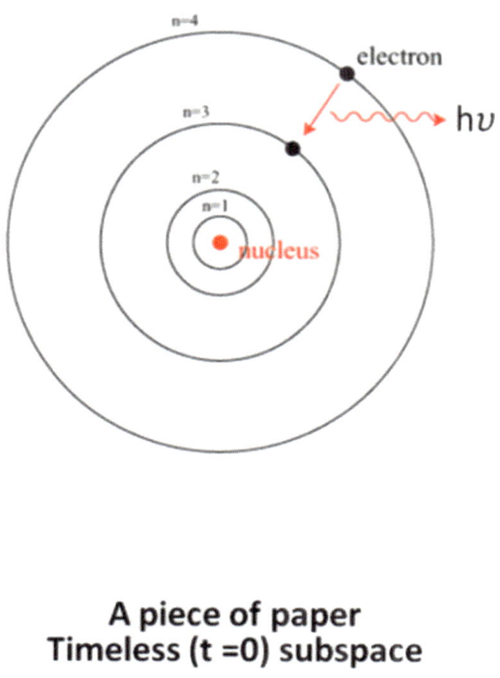

Figure 2.
Bohr atomic model embedded in a timeless (t = 0) platform (i.e., a piece of paper).

$$\psi(t) = \psi_0 \exp[-i\,2\pi\,\upsilon\,(t-t_0)/h] \quad (6)$$

which is the well-known Schrödinger wave equation, where ψ_0 is an arbitrary constant, $h\upsilon$ is the frequency of the quantum leap, and h is Planck's constant.

As anticipated, Schrödinger wave equation is also a "time unlimited" solution with "no" bandwidth. For the same reason as Hamiltonian, Schrödinger wave equation is "not" a physically realizable solution that can be implemented within our temporal ($t > 0$) universe, since any physically realizable wave equation has to be "time and band limited". Yet, many quantum scientists have been using this time unlimited solution to pursuing their dream for quantum supremacy computing [12] and communication [13] but "not" knowing the dream they are pursuing is "not" a physically realizable dream.

Since quantum mechanics is a "linear "system machine, similar to Hamiltonian mechanics, for a multi-quantum state energies atomic particle, the energy E factor to be applied in the Schrödinger's equation is a "linear" combination of those quantum state energies as given by

$$E = \Sigma h\upsilon_{n,}\ n = 1, 2, \ldots N \quad (7)$$

where υ_n is the frequency for the nth quantum leap, and h is Planck's constant. Therefore, the overall wave equation is a linear combination of all the wave functions as given by

$$\psi_N(t) = \Sigma\,\psi_{0n} \exp[-i\,2\pi\,\upsilon_n\,(t-t_{0n})/h], n = 1, 2, \ldots N \quad (8)$$

in which we see that all the wave functions are "super-imposing" together. This is precisely the Fundamental Principle of Superposition of Schrodinger. Yet, this is the principle that Einstein "opposed" the most as he commended as I quote: "mathematics is correct, but incomplete", published in The New York Times newspaper in 1935 [14]. And it is also the fundamental principle that quantum computing scientists are depending on the "simultaneous and instantaneous" superposition that quantum theory can offer to develop a quantum supremacy computer. But I will show that the superposition is a timeless ($t = 0$) principle and it does "not" exist within our universe.

Before I get started, it is interesting to show a hypothetical scenario of "superposition in life". If we assumed our life-expectancy can last for about 500 years, then we would have very good chance to coexist with Isaac Newton and possibly with Galileo Galilei somewhere in "time". Furthermore, if our universe is a "static" universe or timeless ($t = 0$), then we are also very likely to coexist with Galileo and Newton not only in "time" but superimposing with them everywhere in a timeless ($t = 0$) space. And this is precisely what "simultaneous and instantaneous" superposition can do for us, if our universe is timeless ($t = 0$) subspace.

As we understood from the preceding illustration, we know that any empty (i.e., timeless) subspace cannot be found within our universe. And we have also learned that within our universe, every quantum leap $h\upsilon$ has to be temporal ($t > 0$), that is time- and band-limited; otherwise it cannot be existed within our universe.

In view of Eqs. (7) and (8)), we see that they are time "unlimited" wave functions, and it is trivial to see that all of those wave functions, $\psi_N(t)$, n = 1, 2 ... N, are superimposing together at all times. Similar to an example that I had postulated earlier, if our life expectancy can be extended to 500 years, we would be coexist with Einstein and may be with Newton somewhere in time, although 500 years of life-expectancy is time limited. But again, time unlimited wave function is "not" a physical real function, since it cannot exist within our temporal ($t > 0$) universe.

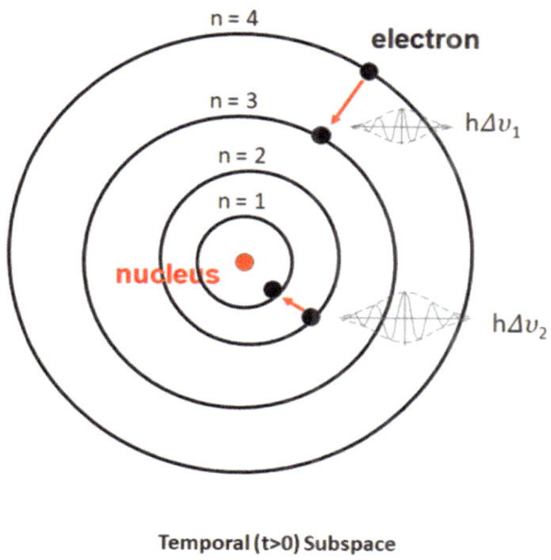

Figure 3.
A multi-quantum state atomic model embedded within a temporal subspace.

In order to mitigate the temporal (t > 0) requirement or the causality condition of those wave functions $\psi_N(t)$, we can "reconfigure" each of the wave function to becoming temporal (t > 0). In other words, we can reconfigure each of the wave function to "comply" with the temporal (t > 0) condition within our universe. For example, as illustrated in **Figure 3** we see that each of the quantum leap hΔυ is represented by "time limited" wavelets.

by which it can be shown that "reconfigured" wave functions are approximated by

$$\psi\ (t) = \Sigma\ \psi_{on}\ \exp\ \left[-\alpha_{on}\ (t - t_{on})^2\right]\ \cos\ (2\pi \upsilon_n t), t > 0, n = 1, 2, 3. \tag{9}$$

$$\psi\ (t) = 0, t \leq 0 \tag{10}$$

where t > 0 denotes equation is subjected to temporal (t > 0) condition, in words exited only in positive time domain. In view of these equations we see that the packages of quantum leaps are "likely" temporal separated, in which we see that all the wavelets are very "unlikely" to be "simultaneous and instantaneous" superposing together. Once again, we have proven that Schrödinger's fundamental principle of superposition "fails" to exist within our temporal (t > 0) universe.

3. Timeless (t = 0) space do to particles

On the other hand, if we take the preceding physical realizable wave functions of Eq. (9) and implement them within a timeless (t = 0) subspace, then it is trivial to see that how a timeless (t = 0) subspace can do to all the wave-particle dynamics within a timeless (t = 0) subspace. Since within a timeless (t = 0) space it has no time and no dimension, all wave-particles (i.e., package of wavelets) will be collapsed at t = 0, as can be seen in **Figure 4**.

Before this goes on, I would say that the wave-particle duality is a "nonphysical" reality assumption to "equivalence" a package wavelet of energy to a particle in motion, which is strictly from a statistical mechanics point of view, where momentum of a particle p = h/λ is conserved [7]. However, one should "not" be treated wave or a package of wavelet energy hΔυ as a particle or particle as wave. It is the package of wavelet energy "equivalent" to particle dynamics (i.e., photon), but they are "not" equaled [15]. Similar to Einstein's energy equation, mass is equivalent to energy and energy is equivalent to mass, but mass is not equal to energy and energy is not mass, for which quanta of light $h\Delta\upsilon$ or a "photon" is a "virtual" particle, in which we see that a photon has a momentum $p = h/\lambda$ but no mass, although many quantum scientists regard a photon as a physical real particle.

In view of **Figure 4** we see that within a timeless ($t = 0$) space, it has no time and no space; every particle exists anywhere within a timeless ($t = 0$) space but only exited at $t = 0$. This is precisely what the "simultaneous and instantaneous" superposition of Schrödinger's principle is anticipated for, since this is the fundamental principle that quantum scientists are aiming for, to build a quantum supremacy computer. This is as well applied to quantum entanglement communication, but unfortunately, the "simultaneous and instantaneous" superposition does "not" exist within our universe, of which we have had shown that superposition principle exists "if and only if" in a mathematical virtual timeless ($t = 0$) space, and it cannot exist within our temporal ($t > 0$) universe.

The reason that superposition principle "fails" to exist is coming from a nonphysical realizable paradigm used in the analysis, which can be traced back to the development of Hamiltonian mechanics, since quantum mechanics is an extension of Hamiltonian. I have found that it is the background subspace (i.e., a piece of paper) used in quantum mechanical analysis. Since the background represents an "inadvertently" empty timeless ($t = 0$) subspace, where a photonic particle in motion was embedded, it is also that piece of paper that Bohr's atomic model was used, added his quantum state energy $h\upsilon$ is not a time limited physical reality.

Aside the substance and emptiness are mutually excluded; it is the subspace that governs the behavior of each wave functions $\psi_n(t)$. In which within a timeless ($t = 0$) subspace, we have shown all the wave functions $\psi_N(t)$, regardless time limited or time unlimited, collapse all together at $t = 0$. In other words, all the quantum state wavelets superimposed at a "singularity" $t = 0$. This is the reason that superposed quantum state energies can be found anywhere and everywhere within a virtual mathematical timeless ($t = 0$) space, since a timeless ($t = 0$) space has no distance.

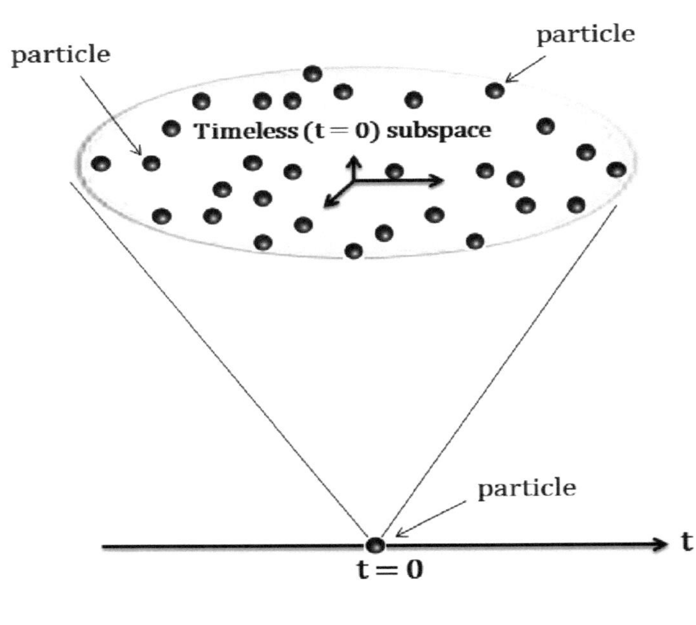

Figure 4.
All the particles within a timeless (t > 0) subspace actually have done; converges all the particles at t = 0.

DOI: http://dx.doi.org/10.5772/intechopen.93562

From this illustration, we have shown once again that; it is not how rigorous the mathematics is, it is the physical realizable paradigm determines her analytical solution is physical realizable or not? For which we see that; the wave functions as obtained from Schrödinger equation is virtual as mathematics is, because Schrödinger's quantum mechanics was developed on an empty subspace platform, the same platform as Hamiltonian classical mechanics.

4. Schrödinger's cat

When we are dealing with quantum mechanics, it is inevitable not to mention Schrödinger's cat since it is one of the most elusive cats in the science since Schrödinger's disclosed it in 1935 at a Copenhagen forum. Since then his half-life cat has intrigued by a score of scientists and has been debated by Einstein, Bohr, Schrödinger, and many others as soon as Schrödinger disclosed his hypothesis. And the debates have been persisted for over eight decades, and still debating. For example, I may quote one of the late Richard Feynman quotations as: "After you have leaned quantum mechanics, you really "do not" understand quantum mechanics ... ".

It is however not the art of the Schrödinger's half-life cat; it is the paradox that quantum scientists have treated it as a physical "real paradox". In other words, many scientists believed the paradox of Schrödinger's cat actually existed within our universe, without any hesitation. Or literally "accepted" superposition is a physical reality, although fictitious and irrational solutions have emerged; it seems like looking into the Alice wonderland. In order to justify some of their believing some quantum scientists even come up with their believing; particle behaves weird within a microenvironment as in contrast within a macro space. Yet, some of their potential applications such as quantum computing and quantum entanglement communication are in fact in macro subspace environment. Nevertheless, I have found many of those micro behaviors are "not" existed within our universe; and the paradox of Schrödinger's cat is one of them, as I shall discuss briefly in the following:

Let us start with the Schrödinger's box as shown in **Figure 5**; inside the box we have equipped a bottle of poison gas and a device (i.e., a hammer) to break the

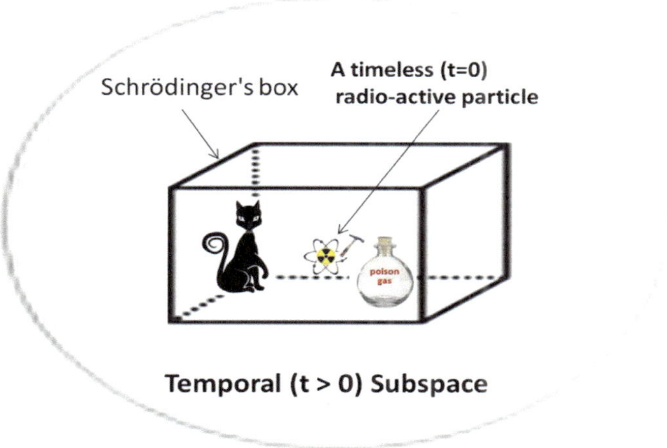

Figure 5.
Inside the box we equipped a bottle of poison gas and a device (i.e., hammer) to break the bottle, triggered by the decaying of a radio-active particle, to kill the cat.

bottle, triggered by the decaying of a radio-active particle, to kill the cat. Since the box is assumed totally opaque of which no one knows that the cat will be killed or not, as imposed by the Schrödinger's superposition principle until we open his box.

As we investigate Schrödinger's hypothesis of **Figure 5**, immediately we see that; it is "not" a physical realizable postulation at all, since within the box it has a timeless ($t = 0$) or time independent radioactive particle in it. As we know that; any particle within a temporal ($t > 0$) subspace has to be a temporal ($t > 0$) particle or has time with it, otherwise the proposed radioactive particle cannot be existed within Schrödinger's temporal ($t > 0$) box. It is therefore, the paradox of Schrödinger's cat is "not" a physical realizable hypothesis and we should "not" have had treated Schrödinger's cat as a physically real paradox.

Since every problem has multi solutions, I can change the scenarios of Schrödinger's box a little bit, such as allow a small group of individuals take turn to open the box. After each observation, close the box before passing on to the next observer. My question is that how many times the superposition has to collapse? With all those apparent contradicted logics, we see that Schrödinger's cat is "not" a paradox after all! And the root of timeless ($t = 0$) superposition principle as based on Bohr's quantum leap $h\upsilon$, represents a time "unlimited" radiator, which is a singularity approximated wave solution. For which we should "not" have treated quantum leap $h\upsilon$ a physical real radiator, since any quantum leap has to be time and band limited within our universe.

Finally I would address that; all the laws, principles, theories and paradoxes were made to be broken, revised and replaced, it is not they were all approximated, because they all changes with time or temporal ($t > 0$). Yet, without approximated science, then there would be no science in which we have shown that a simple hypothetical paradox takes decades to resolve! And this is the nature of quantum mechanics and is all about temporal ($t > 0$) subspace.

5. Nature of Δt

Since our universe was assumed created with a huge energy explosion with time situated within a "non-empty" space. Every subspace "no" matter how small is created by an amount of energy ΔE and a section of time Δt for which every subspace is temporal ($t > 0$) (i.e., existed with time).

In view of modern science, there is a set of simple, yet elegant laws and principles that are profoundly associated with a unit of (ΔE Δt). The objective of this section is to explore the relationship between these laws and principles as related with the unit of (ΔE, Δt). Since time is a dependent forward variable moves at a constant speed, we see that Δt is one of the most "esoteric" variable existed within our universe. We will show that once a moment of Δt is used, we "cannot" get it back although ΔE and Δt can be traded. In which I will show that; there it is a physical limit for Δt to approaching to "none" (i.e., $\Delta t \longrightarrow 0$), that "prevents" us to reach; even though we have the all the price to pay. And this must be the nature of Δt?

Nevertheless, there is a set of "simple and elegant" laws and principles that are profoundly associated with a section of time Δt. These are laws and principle of entropy of Boltzmann [4], information of Shannon [16], uncertainty of Heisenberg [3], relativity of Einstein [1] and temporal ($t > 0$) universe [17]. Each of them has associated with a section of time Δt which changes naturally with time. And all these evidences tell us science has to be temporal ($t > 0$) and dynamics, which cannot be "static" or timeless ($t = 0$). In other words, if there has no time, then there has no science. Nevertheless, science is a law of "approximation", as in contrast with mathematics, which is an axiom of "certainty", of which I state these laws and principles "approximately" as follows:

Law of entropy; entropy within an enclosed subspace increases naturally "with time" or remains constant.

Theory of information; the higher the amount the information, the more uncertain the information is.

Principle of uncertainty; uncertainty of an isolated particle increases naturally "with time".

Theory of special relativity; when a subspace moves faster "relatively" than the other subspace; there is a "relativistic" time speed between them, although time speed within the subspaces remains the same.

Nature of universe; every isolated subspace was created by amount of energy ΔE and a section of time Δt and it is a dynamic temporal (t > 0) stochastic subspace changes naturally with time.

Nevertheless, it is easier to facilitate these laws and principles in mathematical forms, since mathematics is a "language", as given by

$$S = -k \ln p \tag{11}$$

$$I = -\log_2 p \tag{12}$$

$$\Delta E \ \Delta t \geq h \tag{13}$$

$$\Delta t' = \frac{\Delta t}{\sqrt{1 - v^2/c^2}} \tag{14}$$

$$U : \Delta E \, \Delta t \geq (\Delta mc^2)\Delta t, \ \Delta E \, \Delta t \geq h \tag{15}$$

where S, I, and U are entropy, information and universe respectively. k is the Boltzmann's constant, h is the Planck's constant, p is the probability, Δt is a section of time, Δt' is the dilated section of time, v is the velocity, m is the mass and c is the speed of light. $k = 1.38 \times 10^{-16}$ ergs per degree centigrade and $h = 6.624 \times 10^{-27}$ erg-second.

In this we see that our universe was created by means a "huge" amount of energy ΔE and a "long" section of time Δt. And Δt is "still" extending rapidly, since the boundary of our universe is still expanding at the speed of light [17].

In view of these laws and principles, they must be the most "elegant and simple" science equations that existed today in which these equations either attached or associated with a section of time Δt, except Eq. (12) since information theory is mathematics. But as soon information is recognized as related to entropy, information is equivalent to an amount of entropy; this makes an amount of information a physical quantity which is acceptable in science. For which we will show that a section of Δt will be associated with the theory of information, otherwise information will be very difficult to apply in science. Since Δt is coexisted with ΔE, we will further see that; every bit of information takes an amount of energy ΔE and a section of time Δt to transmit, to create, to process, to store, to process and to "tangle".

As we got back from Eq. (11) to Eq. (15) we see that; they are all point-singularity approximated; otherwise it will be very difficult to write in simple mathematical forms. As the laws and principles stated, there are all associated with time, by which they are all space-time variable laws and principles, since time is space and space is time within our temporal (t > 0) universe. In short, they are all connected to a unit of (Δt, ΔE) which is the basic building blocks of our universe. For which I envision that; every existence within our universe has a beginning and has an end. But it is time; it has "no" beginning and has "no" end!

Since our temporal ($t > 0$) universe was created based on a commonly accepted Big Bang Theory [17], we see that our universe as is a temporal ($t > 0$) dynamic "stochastic" subspace [18]. The boundary of our universe increases at the speed of light, we see that; every subspace within our universe is a "nonempty" temporal ($t > 0$) stochastic subspace. By the way, any one or two dimensional subspaces "cannot" be existed within our universe, since one or two dimensional subspaces are volume-less for which any independent Euclidian subspace "cannot" be simply applied to describe a temporal ($t > 0$) subspace. Because all the dimensional coordinates (e.g., x y z coordinates) of a temporal space are all "interdependent" with time, where time is a forward variable with respect to the subspace. In other words, every substance no matter how small it is, has to have time and temporal ($t > 0$).

In view of the time dilation of Einstein's relativity of Eq. (14) and Heisenberg's uncertainty principle of Eq. (13); we see that they are associated with a section of time Δt; which represents a "temporal ($t > 0$)" subspace, as given by;

$$\Delta r = c\ \Delta t$$

where r is the radius of a spherical subspace and c is the velocity of light. In which we see that subspace enlarges rapidly as Δt increases is given by

$$V = (3/4)\ \pi\ (c\ \Delta t)^3 \tag{16}$$

This shows precisely our universe is expanding with a section of time Δt. Since ΔE is a physical quantity equivalent to a subspace that "cannot" be empty and coexisted with Δt, then every unit (ΔE, Δt) is a temporal ($t > 0$) subspace, in which we see that time and space "cannot" be separated. In other words, time and space are "interdependent" although ΔE is a physical quantity but Δt is an invisible "real" variable.

6. Entropy and information

As we look back at Boltzmann entropy Eq. (11), we see that it is a typical timeless ($t = 0$) point-singularity approximated equation. But the law described; entropy increases with "time", implies that entropy is associated with a section of time Δt, although it is "not" shown in the equation. Nevertheless, law of entropy is essentially identical to the law of information as can be seen by their logarithmic expressions of Eq. (11) and Eq. (12), for which we have the following relationships as given by [19];

$$S = k\, I\ \ln\ 2 \tag{17}$$

where I is an amount of information in "bit" and k is Boltzmann's constant in which we see that, "every bit" of information is equal to an amount of entropy ΔS which is given by

$$\Delta S = k\ \ln\ 2, \text{ per bit of information} \tag{18}$$

Although an amount of information can be "traded" for a quantity of entropy, but entropy is a "cost" in energy "equivalents" to an amount of information, but "not" the "actual" information. In other words, it is a "necessary cost" of an amount of entropy to pay for an amount of information in bits. For example, if an amount of entropy ΔS is equivalent to 1000 bits of information of a specific book. Then how many books have the same 1000 bits or how many different items has also 1000 bits? Similarly, an amount of information in bits is not given us the actual information, but it is a "necessary cost" but "not sufficient" to obtain the precise information. In which we see that; the amount of entropy ΔS is a "necessary cost" needed to obtain an equivalent number of information in bits.

Since entropy is a "physical quantity" similar to energy, as given by

$$\Delta S = \Delta E/T = h\Delta\upsilon/T \tag{19}$$

where $\Delta E = h\Delta\upsilon$ is the quantum leap energy and $T = C + 273$ is the absolute temperature in Kelvin, C is the temperature in degree Celsius. In which we see that; higher the thermal noise requires higher energy to transmit a of bit information.

$$\Delta E = T\,k\,\ln 2 \tag{20}$$

Thus, we see that an amount of entropy is equivalent to an amount of information, but it is "not" the information. But an amount of information is equivalent to an amount of entropy that makes information a very "viable" physical quantity can be applied in science. In which we see that; information and entropy can be simply traded as given by

$$\Delta S \Longleftrightarrow \Delta I \tag{21}$$

Nevertheless, we have shown that; either information or entropy has to be a temporal $(t > 0)$ or time dependent law, as given by respectively:

$$I(t) = -\log_2 p(t), t > 0 \tag{22}$$

$$S(t) = -k\,\ln\,p(t), t > 0 \tag{23}$$

where k is the Boltzmann's constant. In which we see that either information or entropy "increases" with time, and $(t > 0)$ denotes imposition by temporal $(t > 0)$ constraint. The amount of entropy for $I(t)$ bits of information can be written as

$$S(t) = k\,I(t)\ln 2, t > 0 \tag{24}$$

where $I(t)$ is in bits and k is the Boltzmann's constant. In view of preceding equation, it shows that entropy increases as amount information increases. In which we see that "every bit" of information ΔI takes an amount of energy ΔE and a section of time Δt to "create" or to transmit as given by

$$\Delta I \sim \Delta E\ \Delta t = h, \text{per bit of information} \tag{25}$$

Since "every bit" of information is equivalent to an amount of entropy ΔS,

$$\Delta S = k\,\ln\,2, \text{per bit of information} \tag{26}$$

Thus, every quantity of entropy ΔS is "equivalently" equaled to an amount of energy ΔE and a section of time Δt to produce as shown by

$$\Delta S = \Delta E/T \tag{27}$$

where $T = C + 273$ is the absolute thermal noise temperature in Kelvin, C is the temperature in degree Celsius, h is the Planck's constant. Since ΔE is "coexisted" with Δt, it is reasonable to say that; every ΔS is also associated with a section of time Δt as given by

$$\Delta S \sim E\,\Delta t/T = h/T, \text{per bit of information} \quad (28)$$

In which we see that information is connected with the law of uncertainty, where "every bit" of information is profoundly associated with ΔE and Δt.

Since every subspace within our universe is created by an amount of energy ΔE and a section of time Δt, we see that; Boltzmann's entropy, Shannon's information, Heisenberg's uncertainty and Einstein's relativity has a profound association with a section of Δt and of ΔE since they are coexisted. In other words, all the laws, principles, and theories as well the paradoxes have to comply with the "coexistence" of ΔE and Δt, otherwise those laws and principles cannot guarantee to be existed within our universe.

Nevertheless, increasing entropy is regarded as a "degradation" of energy by Kelvin [19], although entropy was originated by Clausius [19]. But he might have intended it to be used as a "negative" of entropy (i.e., neg-entropy) in which we see that as entropy or amount of information increases means that there is "energy degradation". This is also meant that entropy or amount of information "degrades with time". Let me stress again "energy degradation" within our universe is due to boundary expansion of our universe at the speed of light [17]. For which I see it entropy increases with time is "no longer" a myth, as most scientists believed it is.

Since all the laws and principles are attached with a price-tag of (ΔE, Δt), but it is the $\Delta t \longrightarrow 0$ that "cannot" be reached, even though we assumed having all the energy of ΔE to pay for! This is precisely the "physical" limit of a temporal ($t > 0$) subspace, by which the "instantaneous" moment of time (i.e., $t = 0$) can be approached but can "never" be able to attend, regardless how much of energy ΔE we willing to pay for. And this is the nature of Δt!

7. Uncertainty and information

Every substance or subspace has a piece of information which includes all the elementary particles, basic building blocks of the subspaces, atoms, papers, our planet, solar system, galaxy, and even our universe! In other words, the universe is flooded with information (i.e., spatial and temporal), or information fills up the whole universe. Strictly speaking, when one is dealing with the origin of the universe, the aspect of information has never been absence. Then, one would ask: What would be the amount of information, aside the needed energy ΔE, is required to create a specific substance? Or equivalently, what would be the "cost" of entropy to create it? To answer this question is to let me start with the law of uncertainty, in equivalent form, as given by

$$\Delta \upsilon\ \Delta t = 1 \quad (29)$$

where υ is the bandwidth, in which there exists a profound relationship of an "information cell" [20], as illustrated in **Figure 6**. In which we see that, the shape of ($\Delta \upsilon$, Δt) or equivalently (ΔE, Δt) can be "mutually" exchanged. Since every bit of information can be efficiently transmitted, if and only if it is transmitting within the constraint of the uncertainty principle (i.e., $\Delta \upsilon \cdot \Delta t \geq 1$). This relationship implies that the signal bandwidth should be either equal or smaller than the system bandwidth (i.e., $1/\Delta t \leq \Delta \upsilon$). In which we see that Δt and $\Delta \upsilon$ can be "traded".

It is however the unit region but not the shape of the information cell that determines the limit, as illustrated in **Figure 6**, we see that; within each unit cell, that is ($\Delta \upsilon$, Δt) [or equivalently (ΔE, Δt)] can be mutually traded. But it is from $\Delta \upsilon$

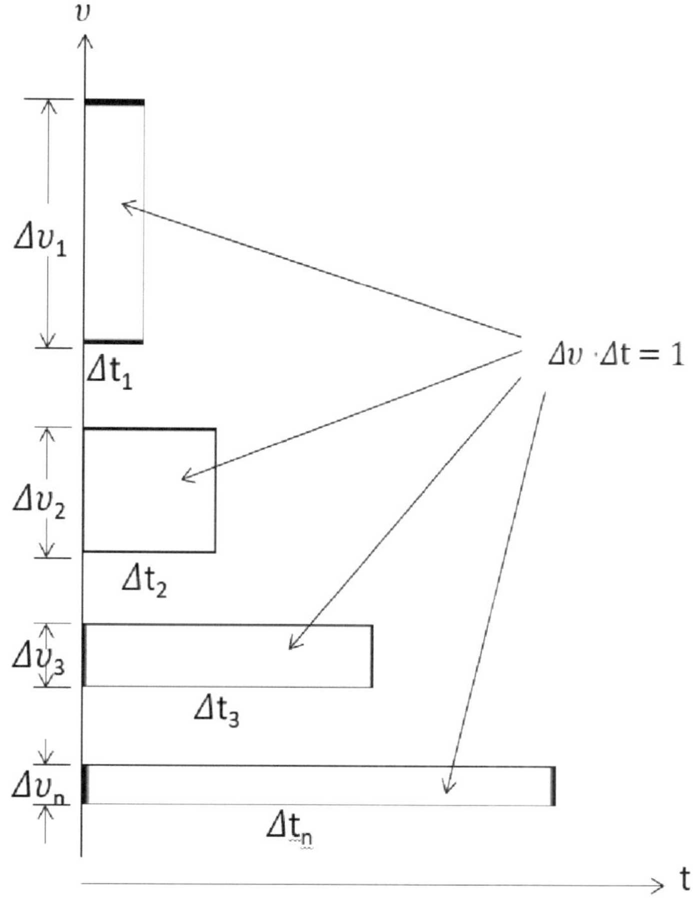

Figure 6.
Various (Δυ, Δt) information cells, where $\Delta\upsilon_n$ and Δt_n are the bandwidths and time-limited sections, and $\upsilon_1 > \upsilon_2 > \upsilon_3 > \ldots > \upsilon_n$ are the frequencies.

to Δt or from ΔE to Δt, since Δυ and ΔE are physical quantities. For which we see that; once a section of Δt is "used", we "cannot" get back the same moment of Δt, although we can create the same section of Δt, since time is a forward dependent variable.

Nevertheless, there are basically two types of information transmission; one is limited by uncertainty Principle and the other is constrained within the "certainty subspace". And the boundary between these two regimes is given by $\Delta\upsilon \cdot \Delta t = 1$ (or $\Delta E \cdot \Delta t = h$) as I called this limit a Quantum Unit [21]. In which we see that Δυ can be traded for Δt. But under uncertainty regime, information is carried by means of intensity (i.e., amplitude square) variation. Yet, information can also be transmitted within the certainty regime, such as applied to complex-amplitude communication [22, 23]. As limited by the law of uncertainty, a quantum unit subspace QLS, for (ΔE, Δt) and (Δυ, Δt), are shown in **Figure 7** for reference.

Since every subspace within our universe is a temporal (t > 0) subspace, the radius of any subspace can be described by a time-dependent variable as given by

$$r = c \cdot \Delta t \tag{30}$$

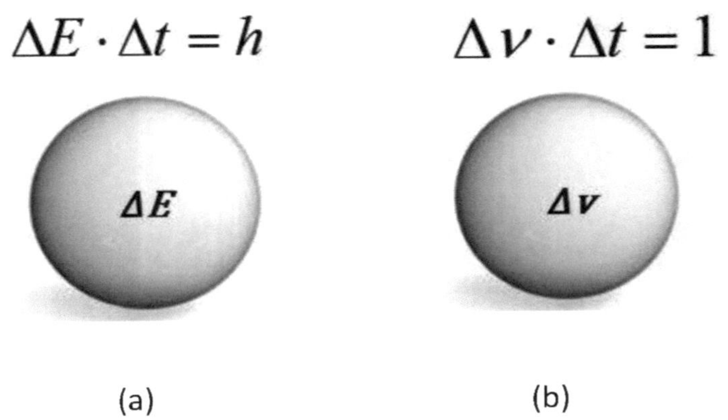

Figure 7.
A set of quantum limited subspaces (QLS). (a) Shows a ΔE limited subspace; (b) Shows a Δv limited subspace.

where c is the speed of light, and Δt represents a section of time. In which we see that the size of the subspace enlarges rapidly as Δt increases as given by

$$V = (3/4)\, \pi\, (c\, \Delta t)^3 \tag{31}$$

Since the carrier bandwidth Δυ and time resolution Δt are exchangeable, we see that the size of the QLS enlarges as the carrier bandwidth Δυ decreases. In other words, narrower the carrier bandwidth Δυ has the advantage of having a larger quantum limited subspace for complex-amplitude communication as depicted in **Figure 8**.

In this we see that it is possible to create a temporal ($t > 0$) subspace within a temporal ($t > 0$) space (i.e., our universe) for communication. We stress that; it is "not" possible to create any time independent or timeless ($t = 0$) subspace within our temporal universe, since timeless ($t = 0$) or time independent "cannot" be existed within temporal universe. And this timeless ($t = 0$) or the "instantaneous limit" (or the causal condition) is the fact of physical limit (i.e., $\Delta t \longrightarrow 0$) within our universe. This limit can only be approached with huge amount of energy ΔE, but we can "never" be able to reach it?

Furthermore, let me note that; timeless ($t = 0$) or time independent subspace is "not" an "inaccessible" space as some scientists claimed, since inaccessible implies it existed within our universe. Nevertheless, one of the apparent aspects of using large

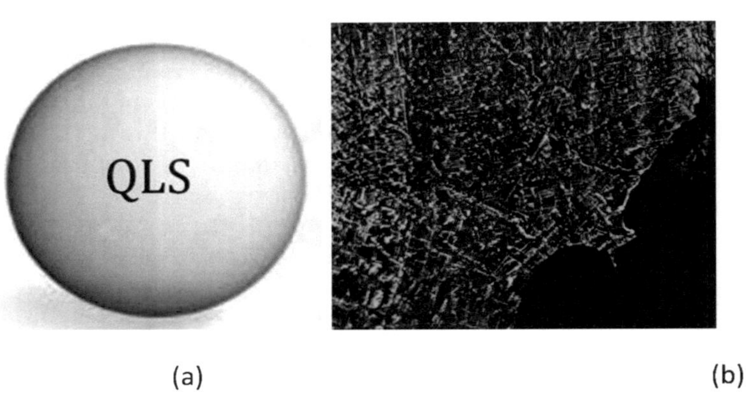

Figure 8.
A "very large" quantum limited subspace as depicted in (a) can be realized in practice within our temporal (t > o) space, for example, such as applied to synthetic aperture radar imaging shown in (b).

quantum limited subspace is for complex information transmission, for example, as applied to complex wave front construction (i.e., holographic recording) [23], complex-match filter synthesis [24], as well as synthetic aperture radar imaging [22]. But there is an apparent price paid for using a "wider" section of time Δt; which "deviates" further away from real-time transmission.

8. Reliable communication

One of the important aspects of information transmission is that "reliable" information can be transmitted, such that information can be reached to the receiver with a "high degree of certainty". Let me take two key equations from information theory, "mutual information" transmission through a "passive additive noise channel" as given by [19]

$$I(A;B) = H(A) - H(A/B) \tag{32}$$

and

$$I(A;B) = H(B) - H(B/A) \tag{33}$$

where $H(A)$ is the information provided by the sender, $H(A/B)$ is the information loss (or equivocation) through transmission due to noise, $H(B)$ is information received by the receiver, and $H(B/A)$ is noise entropy of channel.

However, there is a basic distinction between these two equations: one is for "reliable" information transmission and the other is for "retrievable" information. Although both equations represent the mutual information transmission between sender and receiver; but their objectives are rather different. Example; using Eq. (32) is purposely designed for "reliable information transmission" in which the transmitted information has a high degree of "certainty" to reach the receiver. While Eq. (33) is purposely designed to "retrieve information" from "unreliable" information" by the receiver. For which we see that; for "reliable" information transmission, one can simply increase the signal to noise ratio at the transmitting end. While for "unreliable" information transmission is to extract information from ambiguous information. In other words, one is to be sure information will be reached to the receiver "before" information is transmitted, and the other is to retrieve the information "after" information has been received.

In communication, basically there are two orientations: one by Norbert Wiener [25, 26] and the other by Claude Shannon [16]. But there is a major distinction between them; Wiener's communication strategy is that; if the information is corrupted through transmission, it may be recovered at the receiving end, but with a "cost" mostly at the receiving end. While Shannon's communication strategy carries a step further by encoding the information before it is transmitted such that, information can be "reliably" transmitted, also with a "cost" mostly at the transmitting end. In view of the Wiener and Shannon information transmission orientations; mutual information transfer of Eq. (32) is kind of Shannon type, while Eq. (33) is kind of Wiener type. In which we see that; "reliable" information transmission is basically controlled by the sender; It is to "minimize" the noise entropy $H(A/B)$ (or equivocation) of the channel, as shown by

$$I(A;B) \approx H(A) \tag{34}$$

One simple way to do it is by increasing the signal to noise ratio, with a "cost" of higher signal energy (i.e., ΔE).

On the other hand, to recovering the transmitted information is to "maximize" H(B/A) (the channel noise). Since the entropy H(B) at the receiving end is "larger" than the entropy at the sending end; that is H(B) > H(A), we have,

$$I(A; B) = H(B)–H(B/A) \approx H(A) \tag{35}$$

Eq. (35) essentially shows us that; information can be "recovered" after being received, again with a price; ΔE and Δt. In view of these strategies; we see that the cost paid for using Weiner type for information transmission is "much higher" than the Shannon type; aside the cost of higher energy of ΔE it needs extra amount of time Δt for "post processing". Thus, we see that Wiener communication strategy is effective for a "none cooperating" sender, for example, applied to radar detection, and others. One the other hand, Shannon type provides a more reliable information transmission, by simply increasing the signal to noise so that every bit of information can be "reliably transmitted" to the receiver.

Therefore, we see that quantum entanglement communication [13] is basically using Wiener communication strategy. The price will be "much higher and very inefficient", such as post processing is one thing. And it is "illogical" to require the received signal be "more equivocal" (i.e., uncertain); the better the information recovery it can be received at the receiving end. In which quantum entanglement communication is designed for extracting information as Weiner type communication. However, it is "not" the purpose for reliable information-transmission of Shannon.

9. Relativistic transmission

One of most esoteric aspects in time must be Einstein's special theory of relativity [1] as stated approximately as follows: when a subspace moves faster than the other, there is a "relative" time speed between them, although time speed within the subspaces is the "same". In this we see that the "relativistic time" within a vast cosmological space may not be the same. Let me start with the relativistic time dilation as given by

$$\Delta t' = \frac{\Delta t}{\sqrt{1 - v^2/c^2}} \tag{36}$$

where $\Delta t'$ is the relativistic time window, as compared with the time window Δt of a standstill subspace; v is the velocity of a moving subspace; and c is the velocity of light. In which we see that time dilation $\Delta t'$ within a moving subspace, "relative" to the time duration of the standstill subspace Δt, appears to be wider as velocity increases.

In view of law of uncertainty limit as given by

$$\Delta E \; \Delta t = h \tag{37}$$

we see that every subspace is limited by ΔE and Δt. In other words, it is the h region, but not the shape of that determines the boundary of (ΔE, Δt). For example, the shape can be either elongated or compressed, as long as it is equaled to h region, as can be seen depicted in **Figure 6**.

Incidentally, the uncertainty limit of Eq. (37) is also the limit of "reliable" bit information transmission [16]. Nonetheless the connection with the special theory of relativity is that; subspaces near the edge of our universe will receive a "narrower" section of relativistic time (Δt') with respect to an standstill subspace, since relativistic dilation time window is wider Δt' > Δt. In which we see that; "relativistic" uncertainty within the moving subspace, as with respect to a standstill subspace, can be shown as given by

$$\Delta E\ \Delta t'[1-(\nu/c)^2]^{1/2} = h \tag{38}$$

Or equivalently we have,

$$\Delta \upsilon\ \Delta t'[1-(\nu/c)^2]^{1/2} = 1 \tag{39}$$

In which we see ΔE energy is "conserved". Thus a "narrower" time-window Δt can be squeeze as with respect to standstill subspace. This is precisely physically possible to exploit for "time-domain" digital communication, as from ground station to satellite information transmission.

One the other hand, as from satellite to ground station digital-transmission, we might want to use digital-bandwidth (i.e., Δν). This is a "frequency-domain" information transmission strategy, as in contrast with time-domain, which has "not" fully exploited yet. In which the "relativistic" uncertainty relationship within the standstill subspace as with respect to the moving subspace can be written as

$$\frac{\Delta E\ \Delta t}{\sqrt{1-\left(\frac{v}{c}\right)^2}} = h \tag{40}$$

Or equivalently we have,

$$\frac{\Delta v\ \Delta t}{\sqrt{1-\left(\frac{v}{c}\right)^2}} = 1 \tag{41}$$

In this we see that a narrower bandwidth Δv can be used for "frequency domain" digital communication.

Nevertheless, the essence of ΔE Δt = h (or Δυ Δt = 1) shows that ΔE and Δt or Δυ and Δt can be mutually traded. Again, trading from ΔE for Δt or equivalently from Δυ for Δt is physically viable, since ΔE and Δυ are physical quantities and Δt is "not". Since Δt is coexisted with ΔE (or equivalently with frequency Δυ), we can change Δt, but we "cannot" change the speed of time. In other words, it is time dictates the science but "not" science changes or "curves" the speed of time. In which we have shown that in principle, we can "squeeze" Δt as small as we wish with a huge price of ΔE, but we can "never" able to squeeze Δt to zero (i.e., Δt = 0). In which we see that; it is "not" possible to transmit a "bit" of information "instantaneously" (i.e., t = 0) within our temporal (t > 0) universe.

Since digital communication requires a "narrower" Δt for rapid transmission and complex amplitude communication needs a "wider" Δt for transmission, this is what communication between satellites and ground stations can do with the "relativistic" uncertainty principle. For example, using digital transmission from ground station to satellite stations has the advantage to squeeze the relativistic Δt somewhat at receiving satellite station. On the other hand, from a satellite station to ground stations, one might use wider relativistic Δt for digital frequency signal transmission. Wider Δt also offers a lager "certainty" communication space for complex wave front transmission [22].

Let me assume a "relativistic" communication scenario as depicted in **Figure 9**, in which we assume Q1 and Q2 satellite stations situated within two distinct

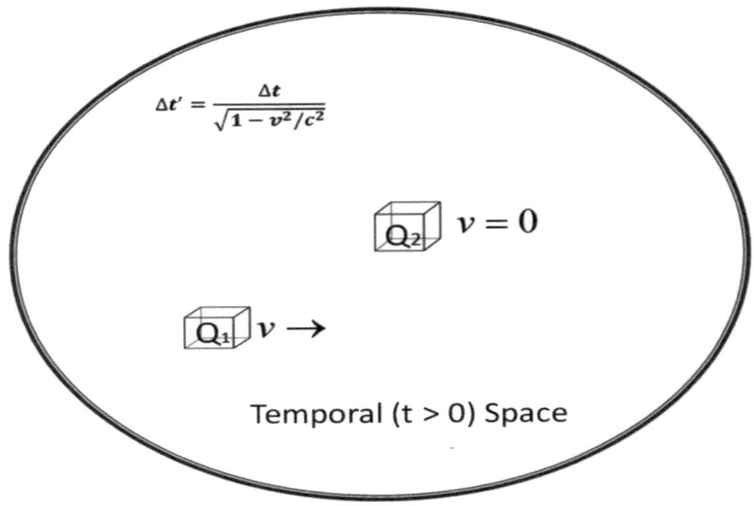

Figure 9.
Relativistic digital transmission within temporal subspace.

subspaces, one travels at a velocity v and the other is stand still. In view of this figure, we see that the hypothetical scenario is a "physical realizable" paradigm, since these two subspaces are embedded within a temporal (t > 0) space.

Now, if we let Q_1 station transmits a pulse signal with a duration Δt to Q_2 station. Assuming without any significant time delay, the digital pulse as received by Q_2 station appeal "wider" due to relativistic dilation as can be seen from Eq. (41). For instance, if we assume the time-dilation from Q_1 station relatively with respect to Q_2 station is two time wider (i.e., $\Delta t' = 2\Delta t$), then $\Delta t'$ is two times wider as received by Q_2; to complete for a bit" of information transmitted from Q_1, as depicted in **Figure 10**, where we see that the transmitted ΔE is "conservation". Needless to say that if the received pulse of Δt is transmitted back to Q_1 in motion; the receiving pulse width will be 2 time broader, as can be seen in the figure. In which we see that; one can exploits faster "time-digital" transmission from a static station Q_2 to a moving station Q_1. From Q_1 to Q_2 static station, one can take advantage for larger communication subspace, such as synthetic aperture radar imaging [22].

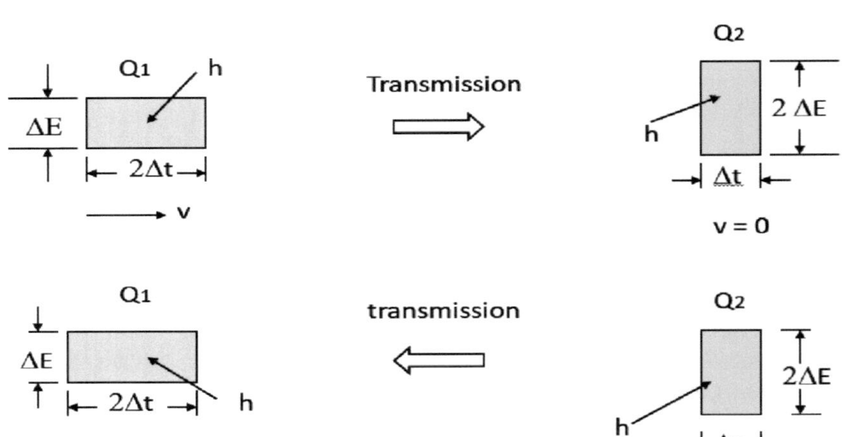

Figure 10.
A relativistic digital information transmission, Δt' = 2Δt.

As I see it; it is our universe governs the science and it is not the science dictates our universe. Within our universe every subspace is created by an amount of energy ΔE and a section of time Δt. Once a section of Δt has been used, it cannot bring it back, although we can create the same Δt at a different time. Although ΔE can be traded for Δt, but it is "impossible" to squeeze Δt equals to zero (i.e., t = 0), and this is the "temporal limit" of our universe. In this we see that there is "no" substance that can travel instantly (i.e., t = 0) within our universe. Even someday we may discover substance that travels beyond the speed of light, this is by "no" means that the substance can travel instantly (i.e., t = 0) within our universe.

Nevertheless, the nature of a section of time Δt is all about our temporal (t > 0) universe, in which time is space and space is time. I have shown that within our universe every subspace takes an amount of energy ΔE and a section of time Δt to "tangle"; by which ΔE and Δt cannot be separated. Although ΔE and Δt can be mutually traded, it is trading ΔE for Δt, or Δυ for Δt, but not trading for ΔE or Δt for Δυ since Δt is a real variable but "not" a physical quantity. But we cannot trade Δt for ΔE; once a section of Δt has been used, it cannot bring it back since time is a forward dependent variable. It is however, in principle, possible to trade ΔE (or Δυ) for a smallest Δt, but it is "not" possible to squeeze Δt to zero, no matter how much energy ΔE that one is willing to pay. Since Δt = 0 is the "instantaneous" response that "cannot" be reached within a temporal (t > 0) subspace, in which we see that Δt is lower bounded by Δt =0. But Δt = 0 exists only within a timeless (t = 0) space but not within our universe.

In view of the laws of entropy, information, uncertainty, relativity, and universe as given by

$$\Delta \mathrm{I} \sim \Delta \mathrm{E}\ \Delta \mathrm{t} = \mathrm{h}, \text{per bit of information} \tag{42}$$

$$\Delta \mathrm{S} \sim \mathrm{E}\ \Delta \mathrm{t}/\mathrm{T} = \mathrm{h}/\mathrm{T}, \text{per bit of information} \tag{43}$$

$$\Delta \mathrm{E}\ \Delta \mathrm{t} \geq \mathrm{h} \tag{44}$$

$$\Delta t' = \frac{\Delta t}{\sqrt{1 - v^2/c^2}} \tag{45}$$

$$\mathrm{U} : \Delta \mathrm{E}\, \Delta \mathrm{t} \geq (\Delta \mathrm{mc}^2)\Delta \mathrm{t},\ \Delta \mathrm{E}\, \Delta \mathrm{t} \geq \mathrm{h} \tag{46}$$

Notice that law of universe in Eq. (46) has a set of equations; one is for an isolated mass m and the other is for isolated photonic-particle, since photon is a "virtual" particle has no mass. Nevertheless, these laws and principles are profoundly associated with (ΔE, Δt), where unit (ΔE, Δt) is the "necessary" cost within our universe. We have shown that it is possible to "squeeze" Δt by widening ΔE. This corresponds to a higher energy of shorter wavelength λ. But it is "impossible" to trade for infinitesimal small section of Δt (i.e., t ≈ 0), which is physical limited as imposed by our temporal (t > 0) universe.

10. Time traveling?

One of the most interesting topics in science must be time traveling for which I assume a photonic traveler (i.e., a photon) is situated within subspace 1 at the center of our temporal universe in **Figure 11**. In view of this figure, the outward speed of subspaces 3 moves somewhat faster than subspace 2 (i.e., $v_3 > v_2$) toward the boundary of our universe since subspace 3 is closer to the boundary.

Now we let the photonic traveler start his voyage with a "narrow pulse" width Δt' from subspace 1 (i.e., very closed to our planet earth) to a distant subspace 3,

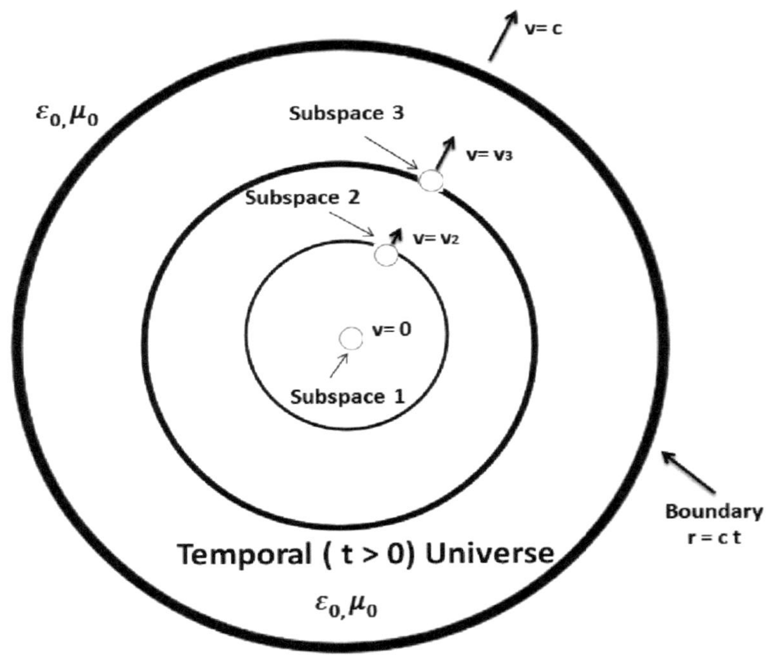

Figure 11.
A schematic diagram of our expanding universe. It shows our universe is a temporal (t > 0) dynamic stochastic universe; time and space are "coexisted." (μ_o, ε_o) are the permeability and permittivity of space.

which has an outward velocity of v_3. If the "relativistic" time dilation Δt' between these two subspaces is "two" times wider than the static sunspace 1 (i.e., Δt = 2 Δt'). Then velocity of subspace 3 can be calculated by means Einstein's special theory of relativity as given by

$$\Delta t' = \frac{\Delta t}{\sqrt{1 - v^2/c^2}} \tag{47}$$

For which the outward velocity V_3 is given by

$$V_3 = 0.87\ c = 0.87 \times 186,000 = 161,820 \text{ miles/s}$$

With reference to Hubble space telescopic observation [27], the boundary of our universe is about 15 billion light years away from subspace 1; for which Subspace 3 is estimated about 13 billion light years away from the center of our universe. Which will take the photonic traveler a 13 billion light-years and possible added another 13 billion light-years to catch-up to subspace 3, since subspace 3 has moved away as traveler's voyage started. For which the traveler will take about 26 billion light-years to reach subspace 3, at speed of light.

Nevertheless, as arrived at subspace 3, the traveler's pulse pulse-width reduces to about 1/4 the size. Which has a 3/4 "gain" in relative time-duration with respect to the static subspace 1 and the gain can be translated into "duration" of time that has been taken during the voyage. Since it took about a total 26 billion light-years journey to reach subspace 3, there is a "net gain" **of** about 19.5 billion light-years ahead "relatively" to the time duration that has gone by at the subspace 1. In other words; there is a total 19.5 billion light-years "relatively ahead" of subspace 1, after a total 26 billion light-years journey to subspace 3, as illustrated in **Figure 12**.

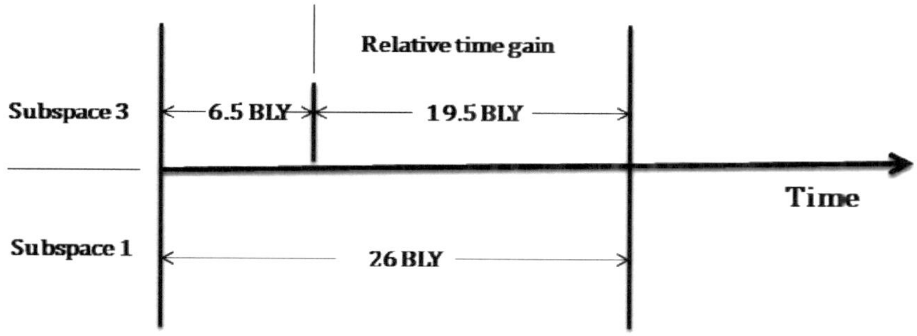

Figure 12.
The "relative" time gain as the traveler reached subspace 3. BLY represents billion light-years.

After the long journey arrived at subspace 3, the traveler is contemplating when he should return back. The "dilemma" is that if he waited too long, he may not be able to return home soon enough to enjoy some of his time-gained, since subspace 3 is moving even faster closer to the speed of light. For which he has decided to return right away, since is a longer journey of "more" than 26 billion light-years to cover, in view of an outward velocity of subspace 3 to overcome.

But as I see it; all the "relative" time-gained will be used up on his journey back home; it turns out the traveler will be home at precisely the same time of subspace 3 "without" any time gain. This part I will let you to figure out, since you have all the mathematics to play with. Yet the worst scenario is that; the traveler "cannot" find his home, since his home had been gone a few billion light-years ago after he had departed from subspace 1 to subspace 3.

On the other hand, if the traveler is "not" a cruising photonic particle, then the kinetic energy to reach a velocity of V_3 = 161,820 miles/s can be calculated as

K.E. = $\frac{1}{2} m v^2 = \frac{1}{2} m (161{,}820)^2$.

which is a price that "nobody" can afford, even just for one-way trip to subspace 3, where m is the mass of the traveler, in which we see that "time traveling" to the future is "unlikely", even assume we can travel at the speed of light.

Nevertheless, every subspace within our universe is always attached a price; a section of time Δt and an amount of energyΔE, although the unit (ΔE, Δt) is a "necessary" cost. For example the "cost" to create a golf ball; it need a huge amount of energy ΔE and a section of time Δt, but without an amount of information ΔI (or equivalent an amount of ΔS) it will not make it happen.

Another scenario is that traveling within "empty" space as depicted in **Figure 13**, as normally assumed. in spite it is a nonphysical paradigm; we see that traveler can reach subspace 3 instantly and return back as he wishes, since within a timeless ($t = 0$) space it has "no" distance and no time, although the diagram shows it has. And this is precisely a virtual mathematical paradigm do to science, even though the subspace has no time and yet appears it has. For which I have found; practically all the laws, principles and theories of science were developed from the same empty space, which is "not" a physical realizable subspace.

Since science is a "principle" of logic, in which we see that a simple logic worth more than tons of mathematics. For example, as illustrated in **Figure 14**, if a time-traveler able to remove himself from current moment of 2020 and searching for last year of the same moment of 2019. The question is can he find it? The apparent answer is that; last year of our universe has been departed. Similarly, the traveler is wishing to visit next year 2021, but next year of our universe has not arrived yet.

In short, I remark that it is physical realizable science that "directs" the mathematics, but not the virtual mathematics that leads science, although science needs

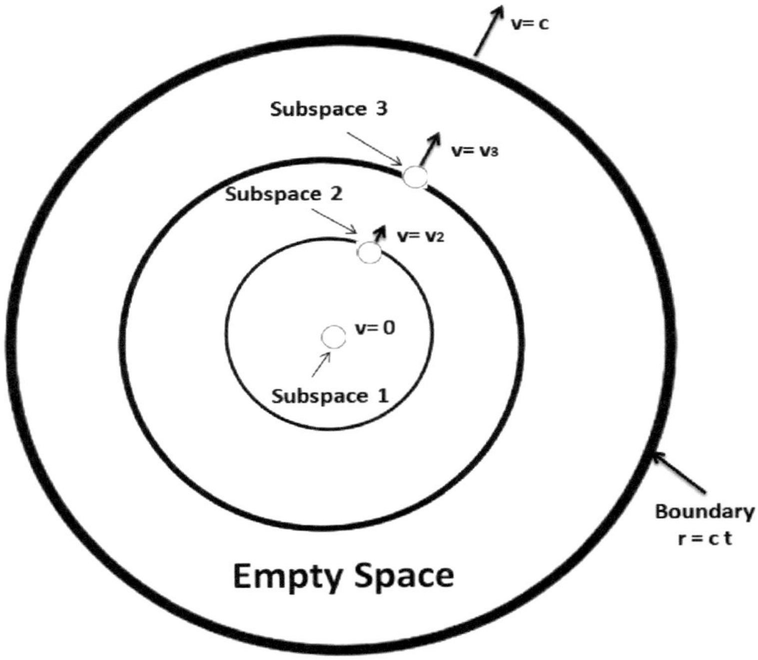

Figure 13.
Our universe model embedded within an empty space. This is a subspace that normally used since the dawn of science.

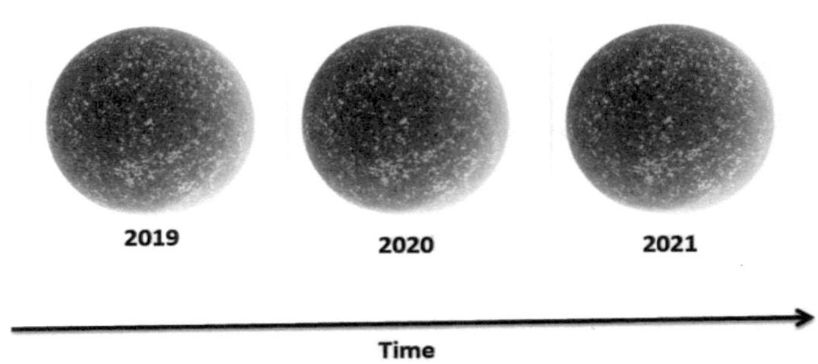

Figure 14.
A composited temporal universe as function of time. Notice that these temporal universes cannot be "simultaneous" existed as superposition principle of quantum mechanics.

mathematics. In which I note that; it is "not" how rigorous the mathematics is, it is the physical realizable science we embrace. Otherwise more and more virtual sciences will continuingly emerge. In view of relativity, we can "relatively" slow down the time somewhat, but we can "never" change the speed of time. It is you walk with time, and it is "not" time walks with you.

11. Conclusion

In conclusion, I would point out that quantum scientists used amazing mathematical analyses added to their fantastic computer simulations that provide very convincing results. But mathematical analyses and computer animations are virtual and fictitious, and many of their animations are "not" physically real, for example, the "instantaneous and simultaneous" superimposing principle for quantum computing did "not" actually exist within our universe. One of the important aspects within our universe is that one cannot get something from nothing there is always a price to pay an amount of energy ΔE and a section of time Δt. The important is that they are not free!

Any science that existed within our universe has time or temporal ($t > 0$), in which we see that any scientific law, principle, theory, and paradox has to comply with temporal ($t > 0$) aspect within our universe, otherwise it may not be a physically realizable science, as we know that science is mathematics but mathematics is not equal to science. In this we have shown that any analytic solution has to be temporal ($t > 0$), otherwise it cannot be implemented within our universe, which includes all the laws, principles, and theories.

Since Schrödinger's quantum mechanics is a legacy of Hamiltonian classical mechanics, we have shown that Schrödinger's mechanics is a timeless ($t = 0$) machine since Hamiltonian mechanics is timeless ($t = 0$). This includes Schrödinger's fundamental principle of superposition which is "not" a physically realizable principle. Since Schrödinger's cat is one of the most controversial paradoxes in modern history of science, we have shown that the paradox of Schrödinger's cat is "not "a physically realizable paradox, which should not have been postulated!

The most esoteric nature of our universe must be time, for which every fundamental law, principle, and theory is associated with a section of time Δt. We have shown that it is the section of Δt that we have used cannot bring it back. And this is the section of Δt that a set of most elegant laws and principles are associated with. In this we have shown that we can squeeze Δt approaches to zero, but it is "not" possible to reach zero even though we have all the energy ΔE to pay for it, in which we see that we can change the section of Δt, but we cannot change the speed of time.

Information is a very important aspect in science, since everything is a piece of information. Nevertheless, without the connection with entropy, information would be very difficult to apply in science. Since entropy is in energy form, but this is by "no" means that entropy is conserved implies that information is conserved since entropy is equivalent to an amount of information. We have shown; information has two major orientations; Shannon transmission is for "reliable" information while Weiner communication is for information "retrieval", for which we see that every bit of information takes an amount of energy ΔE and a section of time Δt to transmit, and it is not free.

Nevertheless, time traveling is a very interesting topic for all scientists, in which I have shown it is physically "not" realizable; it is simply we cannot "curve" a temporal ($t > 0$) space, since time in a "dependent" forward variable with space. It is science can change a section of time Δt but "not" change the speed of time. In other words, we walk on the street and it is not the street that walks on us. However, time traveling is possible if our universe is embedded within an empty space. But emptiness is a timeless ($t = 0$) space which is "not" exited within our temporal universe. And this is precisely most of the scientists uses this empty space for over a few centuries since the dawn of science. And this is precisely why all the laws, principles and theories are timeless ($t = 0$) or time-independent.

Overall, this chapter is to show that it is not how rigorous the mathematics is, it is the physically realizable paradigm that produces viable solution. If one used a nonphysical realizable model, it is very "likely" one will get a nonphysical realizable solution, virtual and fictitious as mathematics is.

Finally, I would stress that the nature of temporal ($t > 0$) quantum mechanics is all about the temporal ($t > 0$) universe, in which we have seen that it is our universe that governs our science; it is not our science that "curves" our universe. Although we can change a section of time Δt, we cannot change the speed of time. In short, it is the physically realizable science we value, but not the fancy mathematical solution we adored.

Author details

Francis T.S. Yu
Penn State University, University Park, PA, USA

*Address all correspondence to: fty1@psu.edu

References

[1] Einstein A. Relativity, the Special and General Theory. New York: Crown Publishers; 1961

[2] Schrödinger E. An undulatory theory of the mechanics of atoms and molecules. Physics Review. 1926;**28**(6): 1049

[3] Heisenberg W. Über den anschaulichen Inhalt der quantentheoretischen Kinematik und Mechanik. Zeitschrift für Physik. 1927; **43**(3–4):172

[4] Boltzmann L. Über die Mechanische Bedeutung des Zweiten Hauptsatzes der Wärmetheorie. Wiener Berichte. 1866; **53**:195-220

[5] Knudsen JM, Hjorth P. Elements of Newtonian Mechanics. Heidelberg: Springer Science & Business Media; 2012

[6] Tolman RC. The Principles of Statistical Mechanics. London: Dover Publication; 1938

[7] Lawden DF. The Mathematical Principles of Quantum Mechanics. London: Methuen & Co Ltd.; 1967

[8] Yu FTS. The fate of Schrodinger's cat. Asian Journal of Physics. 2019;**28**(1): 63-70

[9] Bohr N. On the constitution of atoms and molecules. Philosophical Magazine. 1913;**26**(1):1-23

[10] MacKinnon E. De Broglie's thesis: A critical retrospective. American Journal of Physics. 1976;**44**:1047-1055

[11] Yu FTS. What is "wrong" with current theoretical physicists? In: Bulnes F, Stavrou VN, Morozov O, Bourdine AV, editors. Advances in Quantum Communication and Information, Chapter 9. London: IntechOpen; 2020. pp. 123-143

[12] Bennett CH. Quantum information and computation. Physics Today. 1995; **48**(10):24-30

[13] Życzkowski K, Horodecki P, Horodecki M, Horodecki R. Dynamics of quantum entanglement. Physical Review A. 2001;**65**:1-10

[14] Einstein Attacks Quantum Theory. Scientist and Two Colleagues Find It Is Not 'Complete' Even though 'Correct'. New York City: The New York Times; 1935

[15] Yu FTS. Aspect of particle and wave dynamics. In: Origin of Temporal (t > 0) Universe: Correcting with Relativity, Entropy, Communication and Quantum Mechanics, Appendix. New York: CRC Press; 2019. pp. 145-147

[16] Shannon CE, Weaver W. The Mathematical Theory of Communication. Urbana, IL: University of Illinois Press; 1949

[17] Yu FTS. From relativity to discovery of temporal (t > 0) universe. In: Origin of Temporal (t > 0) Universe: Correcting with Relativity, Entropy, Communication and Quantum Mechanics, Chapter 1. New York: CRC Press; 2019. pp. 1-26

[18] Parzen E. Stochastic Processes. San Francisco: Holden Day, Inc.; 1962

[19] Yu FTS. Optics and Information Theory. New York: Wiley-Interscience; 1976

[20] Gabor D. Communication theory and physics. Philosophical Magazine. 1950;**41**(7):1161

[21] Yu FTS. Information transmission with quantum limited subspace. Asian Journal of Physics. 2018;**27**(1):1-12

[22] Cultrona LJ, Leith EN, Porcello LJ, Vivian WE. On the application of

coherent optical processing techniques to synthetic-aperture radar. Proceedings of the IEEE. 1966;**54**:1026

[23] Leith EN, Upatniecks J. Reconstructed wavefront and communication theory. Journal of the Optical Society of America. 1962;**52**:1123

[24] Yu FTS. Introduction to Diffraction, Information Processing and Holography, Chapter 10. Cambridge, Mass: MIT Press; 1973. pp. 91-98

[25] Wiener N. Cybernetics. Cambridge, MA: MIT Press; 1948

[26] Wiener N. Extrapolation, Interpolation, and Smoothing of Stationary Time Series. Cambridge, MA: MIT Press; 1949

[27] Zimmerman R. The Universe in a Mirror: The Saga of the Hubble Space Telescope. Princeton, NJ: Princeton Press; 2016

Section 3

Applications

Chapter 10

Analysis of Quantum Confinement and Carrier Transport of Nano-Transistor in Quantum Mechanics

Aynul Islam and Anika Tasnim Aynul

Abstract

Quantum mechanics is the branch of physics that consists of laws explaining the physical properties of the nature of nano-particles and their characteristics on an atomic scale. The study of nano-particles significantly challenges our current perception of the universe and the fabric of reality itself. Quantum particles have both wave-like and particle-like characteristics. The fundamental equation that predicts the physical behaviour of a quantum system is the Schrödinger equation and the Poisson equation using Monte Carlo simulations. This gives rise to the *wavefunction, electron and hole densities, energy levels and band structure* of the system which contains all the measurable information about the particle such as time and position, where position is represented using probabilities. This is because particles do not have one definite position during the time before measurement. In fact, they exist as a fuzzy distribution of all possible states where the likelihood of finding the particle in some states is more probable than others. This is known as being in a *superposition* of all states. When the quantum system is observed, however, its wavefunction *collapses* so it consequently falls into one specific position. Moreover, in this chapter we present the simulation results of conduction band profile, electron density (classical and quantum mechanical), eigenstate and eigenfunctions for Si, SOI and III-V MOSFET structures at bias voltage 1.0 V using 1D Poisson-Schrödinger solver.

Keywords: nano-devices, nano-particles, MOS, SOI and III-V structures, 1D Poisson-Schrödinger solver, conduction and valence band profile, carrier density and wavefunctions in the potential well, wave-particle duality

1. Introduction

In this chapter, a connection between the band structure and quantum confinement effects with device characteristics in nano-scale devices is established. Three different devices are presented: a 25 nm gate length Si MOSFET, a 32 nm SOI MOSFET and a 15 nm $In_{0.3}$ $Ga_{0.7}As$ channel MOSFET. We use a 1D Poisson-Schrödinger solver across the middle of the gate along the channel of the devices. The goal is to obtain the calculations of an energy of bound states and associated carrier wavefunctions which are carried out self consistently with electrostatic potential.

The obtained wavefunctions are then used to calculate a carrier density which allows to obtain a sheet density across the structure at given bias.

One of the architectures seriously considered the Silicon-On-Insulator (SOI) transistor as a replacement for bulk MOSFETs. SOI transistors have many advantages compared with the conventional bulk MOSFET architecture. One of the most important is a better electrostatic integrity. SOI devices tolerate thicker gate oxides and low channel doping, allowing scaling to sub-10 nm channel lengths without substantial loss of performance.

However, the transition to this new device architecture and the eventual introduction of new materials in order to further boost device performance is a challenging task for the industry. However, the simulation of UTB transistors has to consider the impact of quantum confinement effects on the device electrostatics. The confinement effects can be induced into classical device simulation approaches using various approximations which include the density gradient approach [1, 2], the effective potential approach [3] and 1D Poisson-Schrödinger solver acting across the channel [4].

This idea leads onto another fundamental quantum superpower called quantum tunnelling. Quantum tunnelling causes particles to simply pass through physical barriers. If a particle was trapped in a well where it has not got enough kinetic energy to escape the well, it would stay in the well as one would expect, however, there is a slight difference. There will also be an exponentially decaying probability that the particle is found outside the well (under specific conditions, that is)! This has to do with the fact that the particles have a 'wave' of probable locations it can be in which extends beyond the well.

2. Quantum mechanics in a semiconductor

The purpose of this chapter is to understand the behaviour and properties of the particles of semiconductor material and devices. In order to get a conception of conduction band and valence band profile, drift velocity, energy, characteristics of the electrical field, wavefunction, carriers density and the mobility of carriers, we need to have an idea on the behaviour of carriers and then proper analysis about semiconductor materials which is related to the different potential. For more understanding about the particles in the theory of semiconductor physics we need to increase our knowledge extensively on the area of quantum mechanical wave theory. Furthermore, we will get idea about the physical behaviour of the materials in semiconductor physics whose electrical properties are related to the behaviour of the carriers in the crystal lattice structure. We will do an analysis of these carriers with the formulation of quantum mechanics, so called 'wave mechanics'. One of the most important parts to describing wave mechanics is 'Schrödinger wave equation'. The gradient of Poisson equation describes about the carriers density of the materials in the semiconductor devices. More details about the characteristics and behaviour of carriers of the semiconductor materials related with the quantum mechanical behaviour are described in this chapter.

2.1 Action of quantum mechanics

Generally, in quantum mechanics, we need to know the basic idea about the principle of tiny energy behaviour (photon), the wave-particle duality, wavefunction behaviour in the potential well, Heisenberg uncertainty principle, and the Schrödinger and Poisson equation.

2.1.1 Wave and photon energy

In general, a wave is a 'perturbation' from the surrounding or 'collision' between particles that travel from one position to another position over time. As we know a wave like classically or electromagnetic wave, which carries momentum and energy during changing the position, while in quantum mechanically a wavefunction can be applied to find out probabilities. We can say the equation for such wavefunction, so called Schrödinger wave equation, which will be described and developed mathematically with more details in the next section [1, 2].

We will now explore the physics behind the photoelectric effect. If light (monochromatic) falls on a smooth and clean surface of any material, then at some specific conditions, electrons are emitted from the surface. According to classical physics, the high intensity of light where the work function of the material will be overcome and an electron will be emitted from the surface, does not depend on the incident frequency, which is not observable. The observed effect is that, at a constant intensity of the incident light, the kinetic energy of the photoelectron increases linearly with frequency, start at specific frequency ν_0, below this frequency we did not observe any emission of photoelectron, as shown in **Figure 1(a)** and **(b)**.

After heating the surface, from that surface, thermal radiation will be emitted continuously (Planck), which form in discrete packets of energy called 'quanta'. The energy of these quanta is generally described by $E = h\nu$, where ν is the frequency of radiation and h is a Planck's constant. However, Einstein explained that the energy in a light wave also contains photon or quanta, whose energy is also given by $E = h\nu$. A photon with high energy can emit an electron from the surface of the material. The required energy to emit an electron is equal to the work function of the material, and rest of the incident photon energy can be converted into the kinetic energy of the photoelectron [1–3]. The maximum kinetic energy of the photoelectron can be written below as in the equation form:

$$T_{max} = \frac{1}{2} mv^2 = h\nu - h\nu_0 \tag{1}$$

2.1.2 Wave-particle duality

As we know the light waves in the photoelectric effect behave like particles. In the Compton effect experiment, an X-ray beam was incident on a solid - individual photons collide with single electrons that are free or quite loosely bound in the atoms of matter, as a result colliding photons transfer some of their energy and momentum of electrons.

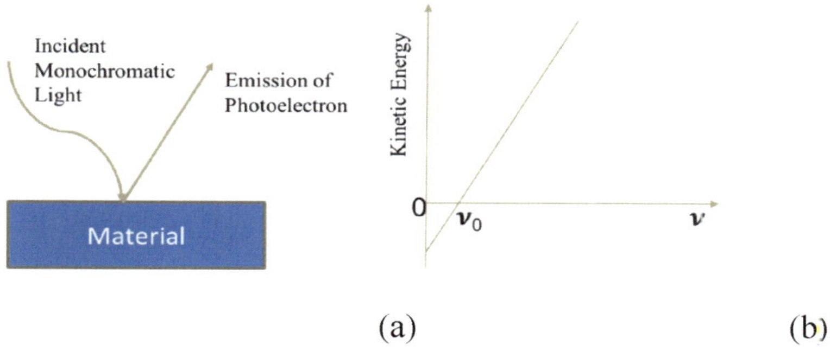

Figure 1.
(a) The photoelectric effect and (b) the kinetic energy of the photoelectron as a function of the incident frequency.

A portion of the X-ray beam was deflected and the frequency of the deflected wave had shifted compared to the incident wave as shown in **Figure 2**. The observed change in frequency and the deflected angle corresponded exactly to the collision between an X-ray (photon) and an electron in which both energy and momentum are conserved [1, 2].

In 1924, de Broglie observed that just as the waves exhibit particle-like behaviour, the particles also show wave-like characteristics. So, the assumption of the de Broglie was the existence of a wave-particle duality principle. The momentum of a photon is given by:

$$\boldsymbol{p} = \frac{\boldsymbol{h}}{\lambda}; = \frac{\boldsymbol{h}}{\boldsymbol{p}}, \tag{2}$$

where $\boldsymbol{p}$ is the momentum of the particle and λ is known as the de Broglie wavelength of the matter wave. In general, electromagnetic waves behave like particles (photons), and sometimes particles behave like waves. This wave-particle duality principle quantum mechanics applies to small particles such as electrons, protons and neutrons. The wave-particle duality is the basis on which we will apply wave theory to explain the motion and behaviour of electrons in a crystal.

2.1.3 Uncertainty principle

The uncertainty principle describes with absolute accuracy the behaviour of subatomic particles, which makes two different relationships between conjugate variables, including position and momentum and also energy and time [1, 2].

For the case 1, it is impossible to simultaneously explain with accuracy the position and momentum of a particle. If the uncertainty in the momentum is $\Delta \boldsymbol{p}$, and the uncertainty in the position is $\Delta \boldsymbol{x}$, then the uncertainty principle is stated as

$$\Delta \boldsymbol{p} \Delta \boldsymbol{x} \geq \frac{\hbar}{2} \; \boldsymbol{or} \; \hbar, \tag{3}$$

where $\hbar$ is defined as $\hbar = \frac{h}{2\pi} = 1.054 \times 10^{-34}$ J-s and is called international Planck's constant.

For the case 2, it is impossible to simultaneously describe with accuracy the energy of a particle and the instant of time the particle has this energy. So, if the uncertainty in the energy is given by $\Delta \boldsymbol{E}$ and the uncertainty in the time is given by $\Delta \boldsymbol{t}$, then the uncertainty principle is stated as

$$\Delta \boldsymbol{E} \Delta \boldsymbol{t} \geq \hbar \tag{4}$$

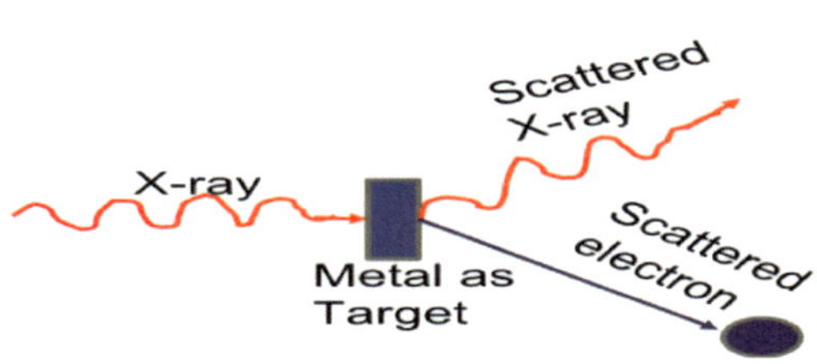

Figure 2.
Compton scattering diagram showing the relationship of the incident photon and electron initially at rest to the scattered photon and electron given kinetic energy.

One consequence of the uncertainty principle is that we cannot, for example, determine the exact position of an electron. We, instead, will determine the probability of finding an electron at a particular position [5, 6].

2.2 Basic principle of Schrödinger and Poisson equation

Generally, the Schrödinger equation description depends on the physical situation. The most common form is the time-dependent Schrödinger equation which gives an explanation of a system related with time and also predicts that wave functions can form standing waves or stationary states. The stationary states can also be explained by a simpler form of the Schrödinger equation, the time-independent Schrödinger Equation [7–9]. We will explain here the motion of electrons in a crystal by theory, which is described by Schrödinger wave equation.

2.2.1 Time dependent and time independent Schrödinger wave equation and the density probability function

The Schrödinger equation is one of the fundamental tools for the understanding and prediction of nano-scaled semiconductor devices. For the case of one dimension the wave vector and momentum of a particle can be considered as scalars, so relating the de Broglie equation, we can write as

$$E = \hbar\omega \; p = \hbar k \tag{5}$$

We use these equation and properties of classical waves to set up a wave equation, known as the Schrödinger wave equation. We solve this equation for the particles which are confined to a potential well, and also to find the solution for particular discrete values of the total energy. However, we develop a theory by considering a particle, moving under the influence of a potential, V (x, t). For this case, the total energy E is equal to the sum of the kinetic and potential energies which can be written as,

$$E\psi = H\psi = \left(\frac{p^2}{2m} + V\right)\psi \tag{6}$$

As we know the momentum operator, and energy are given by

$$p = -i\hbar\frac{\partial}{\partial x}, E = i\hbar\frac{\partial\Psi\,(\mathbf{x},\mathbf{t})}{\partial t} \tag{7}$$

After solving Eqs. (6) and (7), we can develop the one-dimensional time-dependent Schrödinger equation, which can be written as

$$i\hbar\frac{\partial\psi\,(x,t)}{\partial t} = -\frac{\hbar^2}{2m}\frac{\partial^2\psi(x,t)}{\partial x^2} + V(x)\psi(x,t), \tag{8}$$

where $\psi(x,t)$ is the wave function, which describes the behavior of an electron in the device and V(x) is the potential function assumed to be independent of time, and m is the mass of the particle. Assume that the wave function can be written in the form

$$\psi(x,t) = \psi(x)\phi(t), \tag{9}$$

where $\psi(x)$ is a function of the position x only and $\phi(t)$ is a function of time t only. Substituting this form in the Schrödinger wave Eq. (8), we get

$$i\hbar\, \psi(x) \frac{\partial \phi(t)}{\partial t} = -\frac{\hbar^2}{2m} \phi(t) \frac{\partial^2 \psi(x)}{\partial x^2} + V(x)\, \psi(x)\, \phi(t) \tag{10}$$

Now, if we divide both sides of the Eq. (10) by the wave function, $\psi(x)\, \phi(t)$, we get

$$i\hbar\, \frac{1}{\phi(t)} \frac{\partial \phi(t)}{\partial t} = -\frac{\hbar^2}{2m} \frac{1}{\psi(x)} \frac{\partial^2 \psi(x)}{\partial x^2} + V(x) \tag{11}$$

After solving the Eq. (11) (using differential equation), the solution of the above equation can be written as

$$\phi(t) = e^{-i\,\left(\frac{E}{\hbar}\right)t} \tag{12}$$

We notice the solution of Eq. (12) is the classical exponential form of a sinusoidal wave. As we see from Eq. (11) on the left hand side with function of time, which is equal to the constant of total energy of the particle, and the right hand side is a function of the position x only. After simplification of the above equation the time-independent portion of the Schrödinger wave equation can now be written as

$$-\frac{\hbar^2}{2m} \frac{1}{\psi(x)} \frac{\partial^2 \psi(x)}{\partial x^2} + V(x) = E$$

$$\frac{\partial^2 \psi(x)}{\partial x^2} + \frac{2m}{\hbar^2} (E - V(x))\, \psi(x) = 0 \tag{13}$$

Now, the total wave function can be written as in the form of product of the position or time independent function and the time dependent function,

$$\psi(x,t) = \psi(x)\, \phi(t) = \psi(x) e^{-i\,\left(\frac{E}{\hbar}\right)t} \tag{14}$$

According to the Max Born, the function $|\psi(x,t)|^2 dx$ is the probability of finding the particle between x and $x + dx$ at a given time, we can express also $|\psi(x,t)|^2\, dx$ as a probability density function.

$$|\psi(x,t)|^2\, dx = \psi(x,t) \cdot \psi^*(x,t), \tag{15}$$

where $\psi\ \ (x,t)$ is the complex conjugate function. Following Eq. (14), we can rewrite:

$$\psi^*(x,t) = \psi^*(x) e^{i\,\left(\frac{E}{\hbar}\right)t} \tag{16}$$

Finally, we can develop the density of the probability function using Eq. (14), and Eq. (16), which is independent of time.

$$|\psi(x,t)|^2\, dx = \psi(x) \cdot \psi^*(x) = |\psi(x)|^2 \tag{17}$$

The main difference between classical and quantum mechanics is that in classical mechanics, the position of a particle can be determined precisely, whereas in quantum mechanics the position of a particle is related in terms of probability [1, 2].

2.2.2 Wave function behaviour: finite square well, infinite square well, and tunnelling behaviour

In quantum mechanics, finite square well is an important invention to explain the particle wave function behaviour in the crystal. It is a further development of the infinite potential well, in which particle is confined in the square well. The finite potential well, there is a probability to find the particle outside the box. The idea in quantum mechanics is not like the classical idea, where if the total energy of the particle is less than the potential energy barrier of the walls it is not possible to find the particle outside the box. Alternatively, in quantum mechanics, there is a probability of the particle existing outside the box even if the particle energy is not enough by comparing the potential energy barrier of the walls [1, 8, 9].

We apply here the time independent Schrödinger equation for the case of an electron in free space. Consider the potential function $V(x)$ will be constant and energy must have the condition $E > V(x)$. For analysis, we assume that the potential function $V(x) = 0$ for the region II inside the box, as shown in **Figure 3**, and then the time-independent wave equation can be written as from Eq. (13) as

$$\frac{\partial^2 \psi(x)}{\partial x^2} + \frac{2mE}{\hbar^2}\psi(x) = 0 \tag{18}$$

Letting $k = \sqrt{2mE}/\hbar$ or $E = \hbar^2 k^2/2m$, then Eq. (18) leads to

$$\frac{\partial^2 \psi(x)}{\partial x^2} = -k^2 \psi(x) \tag{19}$$

After solving the Eq. (19) using differential equation, the general solution becomes

$$\psi(x) = A \sin(kx) + B \cos(kx),$$

where A and B are complex numbers, and k is any real number.

Now, for the region I and region III, outside the box, where the potential assumed to be constant, $V(x) = V_0$, and Eq. (13) becomes

$$\frac{\partial^2 \psi(x)}{\partial x^2} + \frac{2m}{\hbar^2}(E - V(x))\,\psi(x) = 0$$

$$-\frac{\hbar^2}{2m}\frac{\delta^2 \psi(x)}{\delta x^2} = (E - V_0)\psi(x) \tag{20}$$

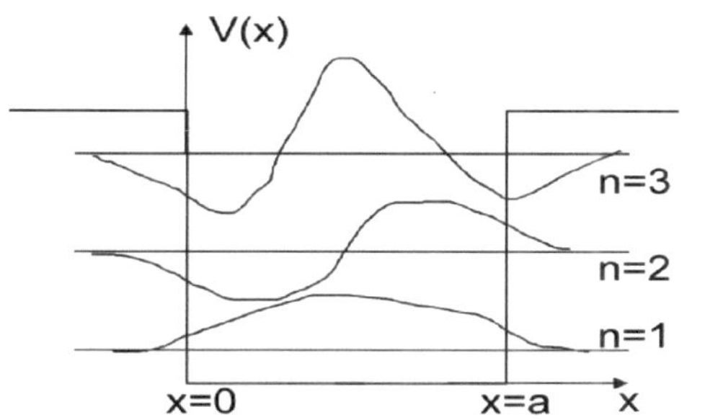

Figure 3.
Potential function of the finite potential well for different regions along the x-direction, with three discrete energy levels and corresponding wave function.

We will get here two possible solutions, depending on energies, where $\boldsymbol{E}$ is smaller than $\boldsymbol{V_0}$, that means the particle is bound the potential and $\boldsymbol{E}$ is greater than $\boldsymbol{V_0}$, that means the particle is moving in free space, which is represented by travelling wave (shown in **Figure 3**).

The potential $V(x)$ as a function of the position is shown in **Figure 4**. The particle is assumed to exist in region II so the particle is contained within a finite region of space. The time-independent Schrödinger wave equation can be written as

$$\frac{\partial^2 \psi(x)}{\partial x^2} + \frac{2m}{\hbar^2}(E - V(x))\,\psi(x) = 0, \tag{21}$$

where E is the total energy of the particle. If E is finite, the wave function must be zero, or $\psi(x) = 0$, in both regions I and III. A particle cannot penetrate these infinite potential barriers, so the probability of finding the particle in regions I and III is zero.

In the quantum mechanics, the particle in a box (also known as the infinite potential well) describes a particle free to move in a small space surrounded by impenetrable barriers (shown in **Figure 4**). In classical systems, for example, a particle trapped inside a large box can move at any speed within the box and it is no more likely to be found at one position than another. However, when the well becomes very narrow (on the scale of a few nanometers), quantum effects become important. The particle may only occupy certain positive energy levels [1, 9, 10].

The energy of the incident particle ($\boldsymbol{E} > \boldsymbol{V}$) in region I and transmitted particle ($\boldsymbol{E} > \boldsymbol{V}$) in region III through the potential barrier ($\boldsymbol{E} < \boldsymbol{V}$) in region II, where the tunnelled particle is the same but the probability amplitude is decreased. There is a finite probability that a particle impinging a potential barrier will penetrate the barrier and will appear in region III (shown in **Figure 5**). This quantum mechanical tunnelling phenomenon can be applied to semiconductor devices.

Quantum tunnelling is the quantum mechanical phenomenon where a subatomic particle's probability disappears from one side of a potential barrier and appears on the other side without any probability appearing inside the well. Quantum tunnelling is not predicted by the laws of classical mechanics where surmounting a potential barrier requires enough potential energy [9, 10].

2.2.3 Maxwell's equations: Poisson equation

To develop the Poisson equation we need to describe the famous Maxwell's equations in their differential form. In mathematics, Poisson's equation is a partial differential equation, which describe the potential field caused by a given charge distribution [2, 8]. Our goal is to find the density of the electron in the crystal of the

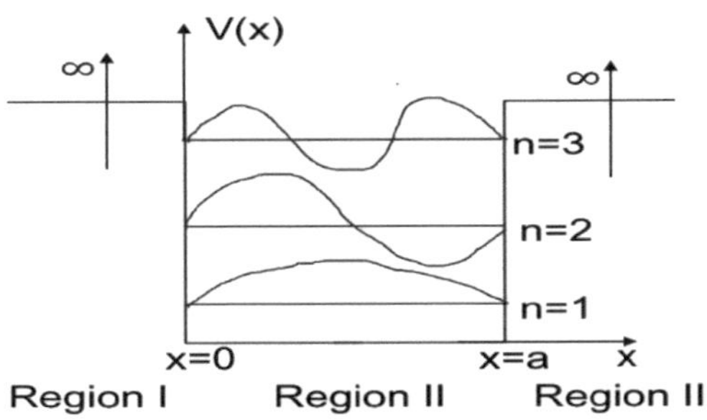

Figure 4.
Potential function of the infinite potential well for different regions along the x-direction, with three discrete energy levels and corresponding wave function in the box or potential well.

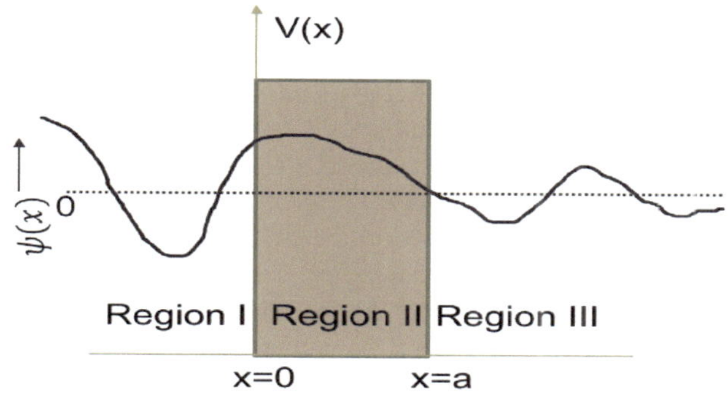

Figure 5.
Quantum tunnelling: The wavefunctions through the potential barrier, a significant tunnelling effect can be seen in three different regions.

semiconductor device classically. As we know the Gauss's law is $\nabla \cdot \boldsymbol{D} = \boldsymbol{\rho}$, where ∇ is the divergence operator, **D** is the electric displacement law ($\boldsymbol{D} = \boldsymbol{\varepsilon E}$), $\boldsymbol{\rho}$ is the free charge density, $\boldsymbol{\varepsilon}$ is the permittivity of the medium, and **E** is the electric field ($\boldsymbol{E} = -\nabla \mathbf{V}$). Maxwell's four equations describe the electric and magnetic fields arising from distributions of electric charge and currents, and how those fields change in time. Maxwell's equations described as follows [2]:

$$\nabla \cdot \boldsymbol{E} = \frac{\rho}{\epsilon_0}, \tag{22}$$

$$\nabla \cdot \boldsymbol{B} = \mathbf{0} \tag{23}$$

$$\nabla \times \boldsymbol{E} = -\frac{\delta B}{\delta t}, \tag{24}$$

$$\nabla \times \boldsymbol{B} = \mu_0 j + \mu_0 \epsilon_0 \frac{\delta E}{\delta t} \tag{25}$$

We can substitute the value of electric displacement in the basic equation of Gauss's law, which can be rewritten as $\nabla \cdot \boldsymbol{E} = \frac{\rho}{\epsilon_0}$, so called Eq. (22). In electrostatic, we suppose that there is no magnetic field, then Eq. (24) can be rewritten as $\nabla \times \boldsymbol{E} = 0$, The electric field as the gradient of a scalar function **V**, is called electrostatic potential. Thus we can write $\boldsymbol{E} = -\nabla \mathbf{V}$, and the minus sign is chosen so that **V** is introduced as the potential energy per unit charge. Finally, we can develop the derivation of Poisson's equation using Eq. (22), and Eq. (24), which leads to

$$\nabla \cdot \boldsymbol{E} = \nabla \cdot (-\nabla \mathbf{V}) = \frac{\rho}{\epsilon_0}$$

$$\nabla^2 \mathbf{V} = -\frac{\rho}{\epsilon_0} \tag{26}$$

2.3 Contribution of Schrödinger and Poisson equation in nano-particles

In this section, a connection between the bandstructure and quantum confinement effects with device characteristics in nano-scale devices is established. Three different devices are presented: a 25 nm gate length Si (Silicon) MOSFET (Metal Oxide Semiconductor Field Effect Transistor), a 32 nm Silicon-on-insulator (SOI) MOSFET and a 15 nm implant free (IF) $In_{0.3}Ga_{0.7}As$ (Indium Gallium Arsenide) channel MOSFET. We use a 1D Poisson-Schrödinger solver across the middle of the gate along the channel of the devices. The goal is to obtain the calculations of an energy of bound states and associated carrier wavefunctions which are carried out self consistently with electrostatic potential. The obtained wavefunctions are then used to calculate a carrier density which allows to obtain a sheet density across the structure at given bias. We have chosen the SOI MOSFET for comparison because it is considered for low power applications. The SOI technology is developing now into the commercial area and is included in the ITRS. The SOI based MOSFETs have a silicon channel made of a narrow layer of less than 10 nm grown on a relatively thick SiO_2 layer. Such strongly confined device channel creates an ultra-thin body (UTB) which provides enhanced carrier transport and, therefore, this transistor architecture is better to be referred as UTB SOI. The FD (fully depleted) SOI MOSFET has superior electrical characteristics and a threshold control from a bottom gate compared to the bulk CMOS device, which are described as follows [11, 12]:

1. decrease in a power dissipation and faster speed due to reduced junction area,

2. steep subthreshold slope,

3. negligible floating body effects,

4. increased channel mobility,

5. reduced short-channel effects and an excellent latchup immunity.

From a point of simplicity, we will first consider the UTB SOI transistor architecture because it is quite illustrative for quantum-mechanical calculations of a confined structure. We will consider a semiconductor material with a small energy gap sandwiched between energy barriers from a material with a larger energy gap. In this way, a quantum well is formed between the barriers which introduce a potential well with discrete energy levels, where particles are confined in one dimension and move free in other two directions as shown in **Figure 6** [13].

We will now focus exclusively on the calculation of quantum states related to electrons. The calculation of quantum states related to holes or any other particles or quasi-particles are equivalent. In the calculations, we will determine the conduction band profile, electron density, energy levels (eigenstates), wavefunctions (eigenfunctions) and electron sheet density in the semiconductor device structure under external potential. In this case, both Schrödinger and Poisson equations have to be solved self-consistently. The one-dimensional, time independent Schrödinger's wave equation for a particle in a potential distribution is a second order ordinary differential equation, which is given by [14].

$$\frac{\partial^2\psi(x)}{\partial x^2}+\frac{2m}{\hbar^2}(E-V(x))\,\psi(x)=0$$

$$\left(-\frac{\hbar^2}{2}\frac{\delta}{\delta x}\frac{1}{m(x)}+V(x)\right)\psi(x)=E\psi(x) \tag{27}$$

$$\boldsymbol{H}\psi(x)=E\psi(x),$$

where

$$\boldsymbol{H}=-\frac{\hbar^2}{2}\frac{\delta}{\delta x}\frac{1}{m(x)}+V(x), \tag{28}$$

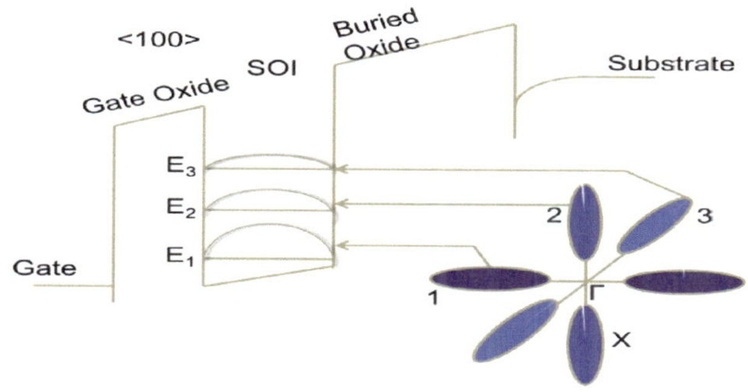

Figure 6.
Schematics of conduction band structure of silicon in <100> oriented narrow channels [assumed (100) plane for the Si-SiO_2 interface]. The energy levels are also shown in the silicon quantum well.

where ψ is the wave function, E is the energy eigenvalue, $V(x)$ is the potential energy assumed to be independent of time, $\hbar$ is Planck's constant divided by 2π, $m \equiv m(x)$ is the effective mass of an electron which is a position dependent, $\boldsymbol{H}$ represents the Hamiltonian operator associated with the sum of the kinetic and potential energies of the system.

The potential distribution $\phi(x)$ in the semiconductor can be determined from a solution of the 1 D Poisson Eq. (26), which is given by

$$\frac{d}{d(x)}\left(\varepsilon_S(x)\frac{d}{d(x)}\right)\phi(x) = \frac{-\rho(x)}{\varepsilon_0} \tag{29}$$

The charge density ρ is given by

$$\rho = N_D(x) - N_A(x) + p(x) - n(x). \tag{30}$$

Eq. (25) can be written as

$$\frac{d}{d(x)}\left(\varepsilon_S(x)\frac{d}{d(x)}\right)\phi(x) = \frac{-q[N_D(x) - N_A(x) + p(x) - n(x)]}{\varepsilon_0}, \tag{31}$$

where ϕ is the electrostatic potential, ε_S is the semiconductor dielectric constant, ε_0 is the permittivity of free space. In the static behaviour, N_D and N_A are called the ionised donor and acceptor concentrations and, in the case of dynamic behaviour, n and p are known as electron and hole density distributions. When dealing with n-type majority carriers semiconductor devices, we can ignore the holes contribution due to their slow movement compared to the electron dynamics. Then only the electrons and donors are considered. As a result the above Eq. (31) can be written as

$$\frac{d}{d(x)}\left(\varepsilon_S(x)\frac{d}{d(x)}\right)\phi(x) = \frac{-q[N_D(x) - n(x)]}{\varepsilon_0} \tag{32}$$

The potential energy V in the Hamiltonian is related to the electrostatic potential ϕ as follows [15]:

$$V(x) = q\phi(x) + \Delta E_C(x), \tag{33}$$

where ΔE_C is the pseudopotential energy due to the band offset at the heterointerface. The wavefunction $\psi(x)$ in Eq. (27) and the electron density $n(x)$ in Eq. (32) are related by

$$n(x) = \sum_{k=1}^{M} \psi_k^*(x)\psi_k(x)n_k, \tag{34}$$

where the summation runs over all the subbands, M is the number of bound states, and n_k is the electron occupation for each state. The electron occupation of a state k is given by the Fermi-Dirac distribution:

$$n_k = \frac{m}{\pi\hbar^2}\int_{E_k}^{\infty} \frac{1}{1 + exp\,(E - E_F/k_BT)}dE, \tag{35}$$

where E_k is the eigenenergy, is the E_F Fermi energy, and k_BT is the thermal energy. An iteration procedure is employed to obtain self-consistent solutions for Eqs. (27) and (32). Starting with a trial potential (x), the wave functions, and their corresponding eigenenergies, E_k are used to calculate the electron density distribution n_x using Eqs. (34) and (35).

2.4 Simulation results: wavefunctions behaviour of particles in the semiconductor devices: 1D Poisson-Schrödinger solver

The previous method of solving the Schrödinger-Poisson equations (see Section 2.3) has been applied to calculate the conduction band profile, electron concentration, energy levels (eigenstates) and wavefunctions (eigenfunctions) in a cross-section placed in middle of the gate of Metal-Oxide-Semiconductor (MOS) structure.

2.4.1 Si MOS structure

The MOS structure consists of a Metal-Oxide-Semiconductor capacitor, which is in the heart of the MOSFET. **Figure** 7 shows the ideal MOS structure for p-type silicon in the flat band condition. The MOS structure is called the flat-band condition if the two following conditions are met [16]:

1. The work function of metal and silicon are equal, which implies that in all the materials, all energy levels in both the silicon and oxide are flat. When there is no applied voltage between the metal and silicon, their Fermi levels line up.

2. There exists no charge, the electric field is zero everywhere in the oxide and at the Si-SiO_2 interface.

The energy bands in the semiconductor near the oxide-semiconductor interface bend as a voltage is applied across the MOS capacitor [13]. We will assume three different bias voltages. One is below the threshold voltage, V_T, the second is just above the threshold voltage, and third one is at an on-current condition. The threshold voltage is defined as the applied gate voltage required to create the inversion layer charge and is one of the important parameters of MOSFETs. For enhancement mode, n-type MOS structure, the accumulation is for $V_G < 0$, the depletion for $V_T > V_G > 0$, the inversion for $V_G \sim V_T$ and the strong inversion for $V_G >> 0$ [14].

An accumulation layer of holes occurs in the oxide-semiconductor junction typically for negative voltages when the negative charge on the gate attracts holes from the substrate to the oxide-semiconductor interface. The induced space charge region is created for positive voltages. The positive charge on the gate pushes the

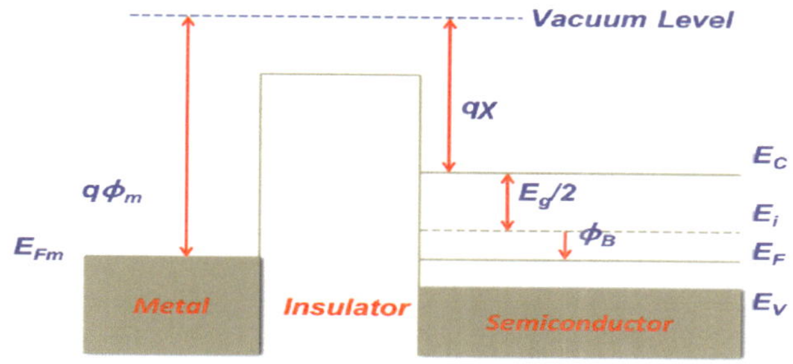

Figure 7.
Ideal metal-oxide-semiconductor (MOS) structure with p-type silicon substrate in a flat band condition.

mobile holes into the substrate. Therefore, the semiconductor is depleted of mobile carriers and a negative charge occurs at the interface because the fixed ionised acceptor atoms are in fixed positions [16].

We will investigate a Si MOS structure at a cross-section in the middle of a gate of the 25 nm gate length Si MOSFET. The structure shown in **Figure 8(a)** has a p-type silicon substrate, oxynitride (ON) gate oxide with a thickness of 1.6 nm, a dielectric constant of ε_{ON} = 7 and a metal gate.

The conduction band profile in a MOS structure of bulk silicon, biased at gate voltage of VG = 1.0 V is shown in **Figure 8(b)**. The ground state energy rises from the conduction band edge as shown in **Figure 8(b)**. This phenomenon is called surface quantization by applied higher gate voltage. The surface quantization is often expressed by a triangular well approximation and the potential near the interface has almost a triangular shape because the potential barrier of SiO_2 is relatively high in silicon MOS structure [13]. **Figure 8(b)** shows also the classically calculated electron density which will peak at the interface and predicts a much larger electron density and higher energy level when compared to the lower gate voltage [14, 16]. The quantum-mechanically calculated electron density is smaller and a displacement of the charge from the interface occurs when compared to the classical calculation.

2.4.2 Silicon-on-insulator (SOI) MOS structure

The investigated 32 nm gate length silicon-on-insulator (SOI) MOSFET is grown on a silicon (Si) substrate. The SOI structure has a layer of silicon dioxide (SiO_2)

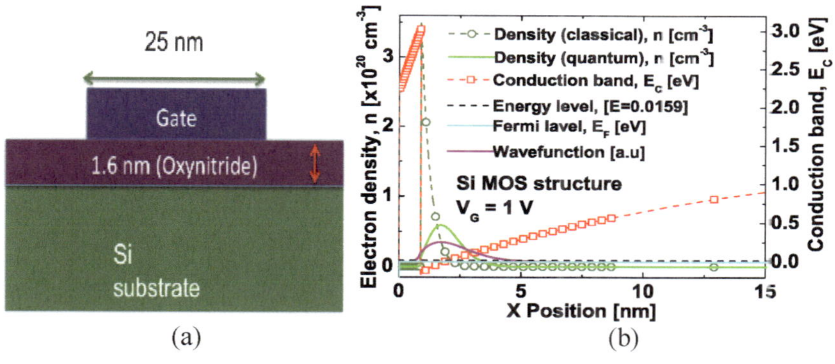

Figure 8.
(a) A schematic metal-oxide-semiconductor (MOS) structure for the 25 nm gate length MOSFET with p-type silicon substrate, (b) conduction band, electron density (classical and quantum-mechanical), energy level, Fermi energy level and wavefunction under an applied bias of V_G = 1.0 V across the channel for the MOS structure of the 25 nm gate length Si MOSFET, where T = 300 K.

with a thickness of 20 nm, which is called buried oxide (BOX) and is fabricated on a Si substrate, and a silicon body (which creates a device channel) with a thickness of 8 nm. A Hafnium Oxide (HfO_2) layer is deposited above the silicon body as a gate oxide with a thickness of 1.19 nm and a dielectric constant of $\varepsilon_{HfO_2} = 20$ and a top metal contact referred to as a gate as shown in **Figure 9(a)**. The metal gate will be able to bend the semiconductor bands with the application of a gate potential [13].

We have investigated the specified 32 nm gate length SOI MOSFET using, again, a self-consistent solution of 1D Schrödinger and Poisson equations. **Figure 9(b)** shows the electron conduction band and density profiles, which are obtained along a slice taken through the middle of a SOI MOS structure, from the surface to the substrate biased at V_G = 1.0 V at room temperature. In this structure, the potential energy creates a square quantum well, because the potential difference between the front interface and the back interface is small and the potential barriers are very high. Electrons are therefore confined in the ultra-thin Si body, which is sandwiched between the gate oxide and the BOX. The electron energy in the perpendicular direction is quantized and the energy of the ground state rises [14, 16] when compared to the conduction band. We can find two discrete energy levels in the quantum well. The classically calculated electron density will again peak at the interface of oxide and semiconductor. The quantum-mechanically calculated electron density will peak away from the oxide-semiconductor interface due to displacement of the charge from the interface [13].

2.4.3 MOS structure for an InGaAs channel transistor

We have selected an $In_{0.3}Ga_{0.7}As$ channel because of its optimal electron mobility and low effective mass. We investigate the effect of a confined channel in the implant free (IF) $In_{0.3}Ga_{0.7}As$ channel MOSFET with a gate length of 15 nm aimed for the future sub-22 nm Si technology. The IF MOSFET is derived from a HEMT structure which has

1. an oxide layer to prevent gate tunnelling,

2. a δ-doping layer placed below the channel. This placement allows the metal gate to maintain a good control of carrier transport in the channel, and

3. an ultra-thin body channel to the heterostructure used in a transistor design.

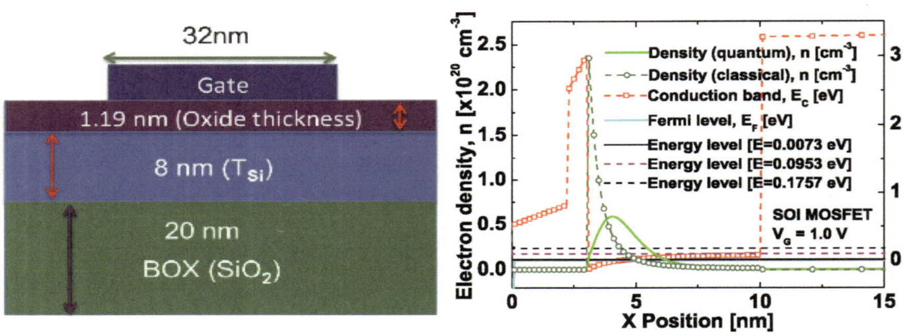

(a) (b)

Figure 9.
(a) A schematic for silicon on insulator (SOI) structure with the 32 nm gate length. (b) Electron density (classical and quantum-mechanical) distribution, conduction band and energy levels under an applied bias of V_G = 1.0 V.

We have used, again, the 1D self-consistent solution of the Poisson-Schrödinger equation to obtain the conduction band profile, energy levels, wavefunctions and electron density in a confined body of this heterostructure MOSFET.

The III-V MOSFET consists of $In_{0.3}Ga_{0.7}As$ channel with thickness of 5 nm, high-ε dielectric layer of Gadolinium Gallium Oxide (GdGaO) as a gate dielectric with a thickness of 1.5 nm and whose dielectric constant is $\varepsilon_{GGO} = 20$. The $In_{0.3}Ga_{0.7}As$ channel is located between an $Al_{0.3}Ga_{0.7}As$ layer with a thickness of 1.5 nm and an $Al_{0.3}Ga_{0.7}As$ layer of a thickness of 3 nm. The δ-doping layer is placed below the channel with a concentration of 7×10^{12} cm^{-2}. The $Al_{0.3}Ga_{0.7}As$ layer at the bottom of the structure is grown as a thick buffer layer of 50 nm as shown in **Figure 10(a)**. The whole device is grown on a GaAs substrate [13, 14].

Figure 10(b) shows the conduction band, five discrete energy levels and electron concentration (classical and quantum mechanical) across the channel for the 15 nm gate length $In_{0.3}Ga_{0.7}As$ MOSFET biased at V_G = 1.0 V.

At V_G = 1.0 V, we obtain three discrete energy levels in the quantum well with corresponding wavefunction for these energy levels shown in **Figures 11(a)** in SOI MOSFET. We summarise that in future technology the bulk MOSFET will be replaced by an ultra-thin-body (UTB) silicon-on-insulator (SOI) on the basis of better electrostatistic integrity, low channel doping to get high mobility, high dielectric material to prevent gate leakage and metal gate [17–19].

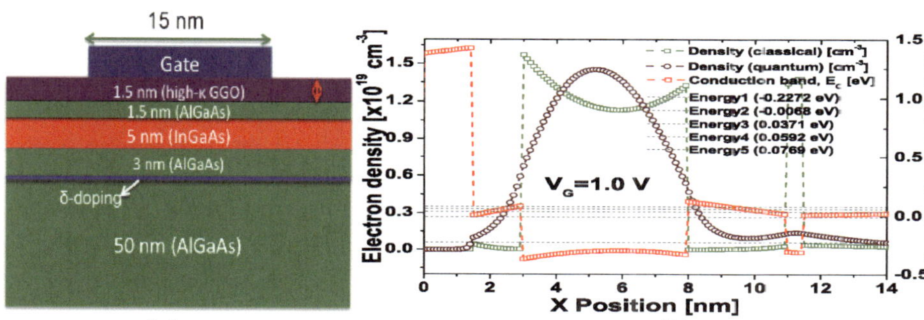

(a) (b)

Figure 10.
(a) Cross-section of the 15 nm gate length $In_{0.3}Ga_{0.7}As$ channel MOS structure with a high- ε dielectric layer which is located below the metal gate, and (b) conduction band, electron density and discrete energy levels under an applied bias of VG = 1.0 V, across the channel for a MOS structure of the 15 nm gate length $In_{0.3}Ga_{0.7}As$ MOSFET.

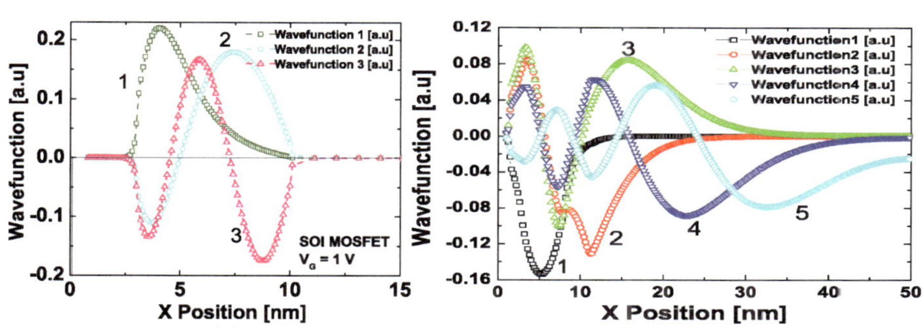

(a) (b)

Figure 11.
(a) Electron wave functions under an applied bias of V_G = 1.0 V across the channel for a MOS structure of the 32 nm gate length SOI MOSFET. (b) The wavefunctions under an applied bias of V_G = 1.0 V, across the channel for a MOS structure of the 15 nm gate length $In_{0.3}Ga_{0.7}As$ MOSFET.

Figure 10(b) shows the conduction band and electron concentration for the 15 nm gate length In0.3Ga0.7As MOSFET biased at V_G = 1.0 V. Five discrete energy levels can be observed at this high bias with the corresponding wave functions in the quantum well shown in **Figure 11(b)** [20].

3. Conclusions

We have been carried out using a self-consistent solution of 1D Poisson-Schrödinger equation to determine conduction band profiles, electron density, energy levels (eigenstates) and wavefunctions (eigenfunctions) in the Si, SOI and InGaAs MOS structures under external potential. We have afterwards simulated the electron sheet density as a function of the applied gate bias and made a comparison among the three device structures, the 25 nm gate length bulk Si, 32 nm UTB SOI, and 15 nm gate length InGaAs MOSFETs.

We have investigated the effect of electron confinement in nanoscaled transistor channels using 1D simulation through cross-sections of the devices. These investigations have been carried out using a self-consistent solution of 1D Poisson-Schrödinger equation to determine conduction band profiles, electron density, energy levels (eigenstates) and wavefunctions (eigenfunctions) in the Si, SOI and $In_{0.3}Ga_{0.7}As$ MOS structures under external potential. We have afterwards simulated the electron sheet density as a function of the applied gate bias and made a comparison among the three device structures, the 25 nm gate length bulk Si MOSFET, the 32 nm UTB SOI Si MOSFET, and the 15 nm gate length IF $In_{0.3}Ga_{0.7}As$ MOSFET [20].

I can envisage that my future work could be related to the investigations of the new physical phenomena present in the UTB MOSFET architectures. As explained previously, the planar and non-planar UTB device architectures are preferred solutions for future technology nodes because the conventional bulk MOSFETs suffer from a poor electrostatic behaviour when scaled to sub-22 nm gate lengths exhibiting unsatisfactory short channel effects. These short channel effects can be summarised as follows:

1. reduced carrier mobility at high channel doping, hampering the device performance,

2. band-to-band drain leakage current [19],

3. and high gate tunnelling current and poor electrostatic control despite employment of metal/high-ϙ gate stacks, etc. [1, 3].

The UTB MOSFET architectures do not show such severe short channel effects because they have superior electrostatic integrity thanks to the electron confinement of their channel region.

Author details

Aynul Islam[1,2] and Anika Tasnim Aynul[3]*

1 Bangor College, Bangor University, United Kingdom

2 Central South University Forestry and Technology, Hunan, China

3 Department of Physics and Astronomy, University College of London (UCL), London, United Kingdom

*Address all correspondence to: anika_tasnim@live.co.uk

References

[1] Donald Neamen A. Semiconductor Physics and Devices, Chapter 2. 3rd ed. New Delhi/New York: University of New Mexico; 2007

[2] Rae AM, Napolitano J. Quantum Mechanics, Chapter 3 and 5. 6th ed. London/New York: CRC Press/Taylor and Francis Group; 1986

[3] Ridley BK. Quantum Processes in Semiconductors. London: Oxford; 1982

[4] Frank DJ, Dennard R, Nowak E, Solomon P, Taur Y, Wong H-S. Proceedings of the IEEE. 2001;**89**: 259-288

[5] Winstead B, Ravaioli U. IEEE Transactions on Electron Devices. 2003; **50**(2):440-446

[6] Lundstrom M. Fundamentals of Carrier Transport. 2nd ed. Cambridge, UK: Cambridge University Press; 2000

[7] Kazutaka T. Numerical Simulation of Submicron Semiconductor Devices, Chapter 2. New York: Artech House, Inc; 1993. p. 102

[8] Jacoboni C, Lugli P. The Monte Carlo Method for Semiconductor Device Simulation. Vienna, Austria: Springer-Verlag; 1989. p. 114

[9] Griffiths David J. Introduction to Quantum Mechanics. 2nd ed. Edinburgh Gate, Harlow: Prentice Hall; 2004. ISBN 978-0-13-111892-8

[10] Physicist Erwin Schrödinger's Google doodle marks quantum mechanics work. The Guardian. 13 August 2013 [Accessed: 25 August 2013]

[11] Schrödinger E. An undulatory theory of the mechanics of atoms and molecules. Physical Review. 1926;**28**(6): 10491070

[12] Laloe F. Do We Really Understand Quantum Mechanics. New York: Cambridge University Press; 2012. ISBN: 978-1-107-02501-1

[13] Aynul I, Kalna K. Nano-Transistor Scaling and their Characteristics Using Monte Carlo. Moldova, UK: LAP LAMBERT Academic Publishing; 2018. p. 60. ISBN-13: 978-613-9-94747-8. Available from: https://www.lap-publishing.com/

[14] Aynul I, Kalna K. Analysis of electron transport in the nano-scaled Si, SOI and III -V MOSFETs: Si/SiO_2 interface charges and quantum mechanical effects. In: IOP Conf. Series: Materials Science and Engineering. Vol. 504. UK: IOP Publishing; 2019. p. 012021. DOI: 10.1088/1757-899X/504/1/012021

[15] Shankar R. Principles of Quantum Mechanics. 2nd ed. New York/London: Kluwer Academic/Plenum Publishers; 1943. ISBN: 978-0-306-44790-7

[16] Aynul I. "Monte Carlo Device Modelling of Electron Transport in Nanoscale Transistors" Doctor of Philosophy. Wales, United Kingdom: College of Engineering, Swansea University Swansea SA2 8PP; 2012

[17] Jacoboni C, Lugli P. The Monte Carlo Method for Semiconductor Device Simulation. Vienna, Austria: Springer-Verlag; 1989. p. 90

[18] Oda S, Ferry DK. Silicon Nano-electronics. London/New York: Technology and Engineering; 2006. pp. 89-95

[19] Frank DJ, Laux SE, Fischetti MV. Monte Carlo simulation of a 30 nm dual-gate MOSFET: How short can Si go. In: Technical Digest - International Electron Devices Meet; 1992. pp. 553-556

[20] Sekigawa T, Hayashi Y. Solid-State Electronics. 1984;**27**:827-828

Chapter 11

Development of Supersymmetric Background/Local Gauge Field Theory of Nucleon Based on Coupling of Electromagnetism with the Nucleon's Background Space-Time Frame: The Physics beyond the Standard Model

Aghaddin Mamedov

Abstract

A new reformulated gauge field theory comprising discrete super symmetry matrixes **U (1) = SU (2) + SO (3)** has been developed which explains why all the elementary particles appear in three families with very similar structures. The three families' performance is the product of discrete conservation of energy—momentum eigenvalue **Es = 1/2Ea** within space–time frame which appears to be the genetic code of new physics. A new supersymmetric gauge field theory of photon was developed, which describes fundamental conservation laws through invariant translation of the discrete symmetries of nature. A new gauge theory describes all the fundamental laws through isomorphism of the discrete space–time **SU (2)** frame and energy-momentum **SO (3)** symmetry group. Coupling of space and time phases of energy conservation generates the background gauge field, which in conjugation with the local gauge field mediates discrete performance of three fractional proton-neutron families of baryon structure. The presented theory requires to have a new look to our understanding of symmetry and conservation laws.

Keywords: supersymmetric theory, discrete double gauge field, discrete space–time symmetry, strong interactions, matter–antimatter symmetry

1. Introduction

From the beginning, I would like to show that there is no matter–antimatter asymmetry in nature and the matter–antimatter asymmetry would eliminate existence of our world in cyclic mode, moving to the randomness. Matter–antimatter, holding discrete supersymmetric genetic code **2Es = Ea**, appear in different phases of energy conservation with the change of frequency. The phenomenon called symmetry breaking is the discrete supersymmetric invariant translation of background gauge symmetry to the local gauge symmetry with the invariant inverse.

Nature does not distinguish a difference between the laws, describing different scale events, and selects very simple principle, which holds symmetry with the perfect conservation laws. The main problem of classic, relativistic, and quantum mechanics theories is the application of mathematical models, which describe a break of *continuous symmetry, associated with the continuous energy conservation*. Due to the consumption of energy in dynamical processes with the discrete energy portions, application of continuous functions, such as Lagrangian and Hamiltonian, for differentiation of change leads to the runaway of the energy solutions to infinity. Artificial renormalization of Lagrangian/Hamiltonian dynamical equations leads to the approximate symmetry; therefore we cannot use these linear differential equations to get correct fundamental laws of nature. Due to these problems, some authors, for example, Weinberg, suggest [1] that nature is approximately simple and Yang-Mill symmetry naturally should produce approximate symmetry. Weinberg sand Glashow suggested that [1, 2] nuclear interactions have spontaneous symmetry breaking that is why these interactions may produce only approximate symmetry.

The theories, describing spontaneous breaking of continuous symmetry, with application of renormalization approach, do not provide proper mathematics of energy conservation, associated with the symmetry. Presently there is no theory, which may describe conservation of action during "change" of an event at small space and time intervals. An action is the product of energy consumption, and due to the discrete consumption of energy, the outcome product of an action has to produce discrete action formulation. However, Lagrangian continuous action principle (Hamiltonian as well) does not hold this requirement.

Our opinion is that the fundamental laws of nature cannot appear in differential formulations in correct way, if these equations do not include dynamical superposition origin and describe interactions through change of continuous function. Without involvement of the original position to the differential equation and using the continuous function, we cannot conserve energy at the origin and remove renormalization groups, adding to the physical theories.

Feynman showed [3] that you could describe an event in the Hamiltonian in the form of differential equation, which describes how the function changes in term of operator. We may provide our comment that such a task is not realizable with the Hamiltonian, because it is a continuous function and does not involve dynamical initial position. The action is the discrete space–time function; therefore, the continuous outcome of a discrete action without relation to dynamic local position is uncertain.

Feynman [4] applied renormalized Lagrangian action to quantum mechanics but even in Feynman's renormalized Lagrangian action is not conserved. The main problem of Lagrangian action and Feynman approach is the application of linear continuous action-response relation.

Therefore, we need an entirely new theory to describe reversible fundamental laws of nature, combining discrete conservation of energy with the boundary-mapped space–time, which would combine all kind of fields and forces within discrete background symmetry. Application of discrete energy conservation law and discrete symmetry as the product of this law may change drastically our present knowledge on the nature of forces and their roles in fundamental interactions. It is possible that nature and performance of forces, merging at the background symmetry, will be different, and reversibility of dynamical laws at discrete symmetry may change completely the role and classification of forces.

In our previous studies [5–7], we showed that wrong description of dynamical laws through simple continuous displacement in space–time structure using only intervals and linearity in differential equations might be the reason for appearance of uncertainty problems of quantum mechanics and infinity in classic physics formulations. Similarly Nobel Laureate Hoof't showed [8] that "when we send distances and time intervals to zero we do assume that the philosophy of differential equations works."

In the present paper, we will discuss a new reformulated gauge theory, combining "twin brothers" of background/local gauge fields, conjugated with the new energy-momentum rotational symmetry group and associated with the discrete space–time symmetry. We will describe fundamental change of physical laws when we shift from the theories based on continuous energy conservation law to the discrete energy-momentum invariant translation, carried within the discrete space–time frame. Simply, we will present an invariant action-response exchange parity, relative to the response, which drastically changes features of classic, relativistic, and quantum theories. *The paper will describe why all the elementary particles come in three families with very similar structure. The three families' performance is the product of energy conservation within energy-momentum exchange interaction, which became the genetic code of our existence and new physics. We believe that our theory of supersymmetry is the rebirth of gauge field theory, which does not need application of renormalization.*

2. The genetic code of origin as the superposition

2.1 The illness of linear differential equations for description of energy conservation

First, I would like to discuss shortly literature information [9] to show that Newton's traditional differential equations, describing change of an event in abstract space and time, do not provide exact solution. It describes change of an event in abstract time by a smooth continuous vector field, which has local diffeomorphism, preserving only limited property of an event. Diffeomorphism is a function in smooth manifold, which describes differentiation where the original position is lost. The present mathematical knowledge does not provide any solution on how to eliminate the diffeomorphism problem of differentiation to save the initial property of a function. That is why all theories, describing symmetry, have to use renormalization, which drives them, similar to quantum uncertainty, to the local approximate symmetry.

Presently there is no mathematical solution on how to get nonlinear differential equation, which may describe change of dynamical local position from point to point on the discrete space–time, where space and time may change their dimensions relative to the energy-momentum flux to the space–time frame. This approach is very useful for a wide class of mathematical problems because it solves boundary value problems (which is in reality initial value problems) of differential equations.

Traditional differentiation describes change of interval of one variable in relation to other, for example, change of space in relation of time—**dS/dt.** We suggest a new mathematical theory for differentiation of a function without loss of the origin/local dynamical structure within multiple variables, such as space, time, and energy. Presentation of differentiation through coupling of intervals of change of multiple variables with their superposition origin produces deterministic outcome regardless of scale of interactions.

The theory, which we suggest, comprises the principle that any interaction, to hold symmetry, after change in the space–time frame, should look the same as its background superposition/dynamical local origin of energy-momentum content and space–time frame. The basic statement of such a concept is very simple: "particles may hold their "non-charged" state of rest only in discrete mode." Such an approach is the modification of Aristotle's concept [10] that "natural state of a body to be at rest" which does not present rest in discrete mode, therefore does not hold conservation of energy.

The nature of rest is well described by Nobel Laureate Anderson [11]. By Anderson opinion, a system at stationary state of rest could not stay long and stationary state can be only equal superposition and its inverse. By his opinion, only superposition and inverse mixture may describe the absence of dipole moment. Unfortunately, Anderson did not put his statement into mathematical formulation. However, Anderson's "equal mixture" is equivalent to the equal numbers of matter–antimatter, which, as we will show later, needs modification.

We found out that we could solve the gap in "Anderson's equal mixtures "and describe stable steady-state performance of matter if we will apply vector type of discrete exchange interactions between two symmetric states, which can bring the system to superposition in discrete mode and hold discrete CP invariance of strong interactions. The superposition displays the genetic particle, while displacement from the superposition appears as the antiparticle of the superposition. It is easy to show that the origin of this principle is the conjugation of discrete conservation of energy with the discrete space–time, which is in hold for any fields/particles regardless of scale and completeness.

2.2 The basic statement of symmetry

While symmetries are conjugated with the corresponding conservation laws, we will start our analysis from principles of energy conservation. Distribution of energy in a medium requires certain space locality and time duration. The portion of energy, consumed for displacement of space, appears as the potential ingredient, while the time portion of the total energy presents kinetic energy. We may present the potential and kinetic ingredients of energy in the form of conjugated space and time portions. The suggested approach is different from Lagrangian or Hamiltonian, because these functions present continuous conservation of certain abstract amounts.

The Lagrangian or Hamiltonian functions, as Feynman stated [12], describe an abstract mathematical principle, which involves a certain numerical quantity, which has to be conserved. These formulations present some abstract number, which does not change, and after the change, we should have the same number.

Due to the conjugation with the conservation of energy, an event symmetry after the change should look the same. Therefore, a mathematical formulation of symmetry should show that (a) we have the same number of energy after change of an event and (b) an event looks the same as origin. We will describe how we can get such a mathematical formulation.

The exchange interaction of superposition (initial neutral state) with its displacement may produce two outcomes: (a) the symmetry of particles is continuous, such as the outcome of change looks the same as origin continuously, but breaks down spontaneously; (b) the outcome of change looks the same in discrete mode, with invariant translation without violation of symmetry. Later we will show that the outcome of interactions after change may look the same if the fundamental laws describe conservation of energy only in discrete mode.

Mathematically this statement, in general, may have the following form:

$$\mathbf{F}'(\mathbf{s},\mathbf{t}) = \mathbf{F}(\mathbf{s},\mathbf{t}) \tag{1}$$

Eq. (1) after differentiation may display Maxwell equations in an alternative way. If a system after change looks the same in discrete mode in opposite phase, the equation of symmetry has a negative phase solution:

$$\mathbf{F}'(\mathbf{s},\mathbf{t}) = -\mathbf{F}(\mathbf{s},\mathbf{t}) \tag{2}$$

The positive and negative solutions of dynamic supersymmetry outcomes of a discrete event together will have a form:

$$\mathrm{F}'(\mathrm{s},\mathrm{t}) = \pm\mathrm{F}(\mathrm{s},\mathrm{t}) \tag{3}$$

We can assume that the positive solution of Eq. (3) presents the discrete symmetric function in space–time phase in the local gauge field, while the negative sign is an antisymmetric solution of the symmetric function of an event in an opposite energetic phase of the background gauge field. The positive and negative solutions of Eq. (3) appear as a discrete change of the symmetric function from one phase to another phase, which is a shift of energy conservation from space–time phase (holding by ordinary matter) to the energy phase. These phases as background/local gauge fields discretely transform to each other, leading conservation of energy and symmetry in discrete mode within opposite energy and space–time phases.

The classic physics Eq. (3) in some sense is similar with Schrödinger's wave function:

$$\frac{\mathbf{d}}{\mathbf{dt}}\boldsymbol{\psi} = -\mathbf{i}\mathbf{H}\boldsymbol{\psi} \tag{4}$$

The problem of Schrödinger's Eq. (4) is that it describes change of wave function only in one phase, which is time. The space phase representative is Hilbert space, which presents the original function but does not undergo any changes.

If classic physics could describe the symmetry and energy conservation law within the space–time frame with conjugation of space and time intervals with the dynamic local states of variables, there will be no need of application of Schrodinger's wave function (4), which uses probability approach. The wave function of quantum mechanics with the local states of space and time coordinates could have deterministic classic equation to describe the exact symmetry of Nature. The deterministic equation of background space–time symmetry after the change of an event in discrete mode may look the same:

$$\frac{\mathbf{dS}}{\mathbf{dt}} = \pm\frac{\boldsymbol{S1}}{\mathbf{t1}} \tag{5}$$

where the left side of the equation describes uniform change of space and time coordinates of an event, while the right side presents the original local space–time frame. The positive sign describes outcome of the ordinary matter phase, while the negative sign shows the outcome of antimatter. The statements of Eqs. (1)–(3) and (5), without Dirac's relativistic quantum mechanics, naturally predict existence of antiparticles to hold discrete conservation of energy within different states.

3. Development of a new mathematical theory for differentiation of change

Hoof't showed [8] that it is possible to eliminate the bad effect of small time intervals and small displacement in space by improvement of mathematical formulation of small-scale transformation, for example, by renormalization group. However, renormalization tool leads to the approximate symmetry and renormalized artificial outcome of a natural event. The other way, which he suggested, is to find a new, improved theory.

To find new theories, we need to eliminate two problems: nonindependence feature of uncertain space displacement and time intervals in combined space–time unit and linearity of the change. We cannot get any help from special relativity (SR) and Minkowski's space–time to eliminate independent features of space and time intervals because they do not involve local origin and connect opposite time interval with the three space intervals into a nonsymmetric four-momentum frame (3:1), which involves abstract intervals of neutral space and time variables without their local positions.

We cannot use principles of general relativity (GR) theory as well because general relativity does not provide boundary-mapped reversible dynamical law due to its continuous space–time frame. GR does not have a background, which is the reason that GR's geometric, continuous space–time structure at small-scale interactions cannot find origin and runs away to the infinity. Wheeler's suggestion [13] on "space tells mass how to move, mass tells to space-time how to curve" does not produce a complete concept in a sense that it produces uncertainty because GR's space–time cannot tell to mass the path and boundary to move and mass cannot tell space–time boundary where to stop.

First, we will look how the features of dynamics change if we gradually reduce time interval $\mathbf{\Delta t}$, moving from the high scale to the small-scale event, as was done by Hoof't [8]. However, we will analyze not an interval as Hoof't did, but a function $\mathbf{\Delta f/f_1}$, which as a mathematical operator may give information about change of a function in relation to its dynamical local origin. This function is a sufficient entity for the identification of change. The non-unitary function $\mathbf{\Delta f/f_1}$ shows quantum behavior and with the fractional feature (portion) produces the outcomes with the integer numbers $\mathbf{(f_2/f_1{-}1)}$. The mathematical operator in the form of $\mathbf{\Delta f/f_1}$ portion describes the fraction of the change in relation to its dynamical origin. Similarly, the operator $\Delta\mathbf{S/S_1}$ describes displacement of space with the applied force in relation to its origin, while the operator $\Delta\mathbf{t/t_1}$ describes the fluctuation of time about instant of action. The functions $\Delta\mathbf{S/S_1}$ and $\Delta\mathbf{t/t_1}$ describe the entanglement of the displacement with the initial superposition as the genetic code of the event. The relation of change around its origin $\mathbf{\Delta s/S_1}$ generates a spherical space, while relation of time interval to instant of time produces a round time structure. Therefore, there is no preferred inertial system and mathematical model, which may display an event better than its initial superposition state.

In planet-scale events, reduction of the distance twice, as was shown by Hoof't [8], does not affect significantly the linearity of change. The parameter $\Delta\mathbf{S/S_1}$ also describes a similar effect of the change to the linearity. However, if we reduce interval of time twice in a small-scale event, using $\Delta\mathbf{t/t_1}$ function, we will be able to describe catastrophic effect of the change to the linearity of the motion.

The relation of intervals of time and displacement to the origin creates entanglement of the final and initial states of coordinates. The origin of an event in this case "tells the body how to move and the final state of a motion gets the information where to stop." However, the entanglement of interval of change with the origin leads to the deterministic nature of the dynamical event within a certain boundary, and it is the only way for elimination of the infinity problem of small-scale interactions. The effect of initial/local coordinate of time and space of a body appears as an action of initial energy contents (such as inertial mass, inertial energy) of a body to the change of pathway. Presently all the physical laws use only independent intervals to describe the change of an event without relation of change to the initial local state. This is the main problem of physical laws, applying the renormalization group to remove uncertainty of initial position.

4. Energy-momentum: (a) charged antiparticle-particle pair and (b) neutral twin particles

Lagrange and Hamilton suggested conservation of energy in the form of linear differential equations as well. The main concern of these equations is that the position coordinates and velocity components are independent variables and derivatives of the Lagrangian with respect to the variables taken separately.

The specific feature of our approach is that energy, distributed within space and time portions, appears in the form of non-separable energy-momentum exchange entities. Energy in one phase appears as the consumed charged part in the space–time frame and in another phase appears as itself, comprising color ingredients of neutral photon-antiphoton pair. On this basis, we may present energy and momentum in two forms: (a) energy-momentum exists in the form of electrically charged matter–antimatter pairs, and (b) energy-momentum exists in the form of color charge pairs, where every part is an own particle of the other part. The condition of energy and momentum in forms (a) and (b) are completely different. However, the color charged bosonic pairs, which appear as "the neutral twin brothers" in the form of Majorana particles, are the superposition where it has a trend to move In space–time phase, energy appears as Dirac's particles. It seems obvious that, at superposition of color charge "neutral twin brothers," all the ingredients of energy-momentum, as internal products, will exist in the form of twin particles.

Now we may apply this mathematical tool for characterization of any type of change, particularly ingredients of space–time. The parameters $\mathbf{\Delta S/S_1}$ and $\mathbf{\Delta t/t_1}$ have no unity and are unit-less parameters, which makes easy to compare them as the equivalent entities. Using Wheeler's [13] statement that the equation of special relativity $\mathbf{E = mc^2}$ allows to transfer space and time equivalently to each other, we may show problems of such a statement. For this purpose, we may analyze the relationship between energy and mass portions without application of Lorentz transformations.

$$\gamma = \frac{\Delta \mathrm{E}}{\mathbf{E1}} : \frac{\Delta m}{\mathbf{m_1}} = \frac{\Delta \mathrm{E}}{\Delta \mathrm{m}} \cdot \frac{m1}{E1}; \frac{\mathbf{E1}}{\mathbf{m1}} = c2, \Delta \mathrm{E} = \gamma \Delta \mathrm{m}c2 \tag{6}$$

Eq. (6) describes change of energy-mass equivalence with the effect of initial condition (we may call rest mass and rest energy) in the form of "non-Lorentz transformation." By literature information [14], the exact value of Lorentz factor at velocity close to speed of light is **2.00**. If we use numeral value $\boldsymbol{\gamma} = \mathbf{2.00}$, as an exact Lorentz factor [14], at uniform speed of light $\mathbf{c^2} = \mathbf{1}$, Eq. (6) produces condition.

$$\Delta \mathrm{m} = \mathbf{1/2\Delta E} \tag{7}$$

for energy mass invariant translation. The energy and mass invariance (7) appears as the product of discrete exchange of energy-momentum relation and produces half-integer-integer spin interactions of mass and integer spin carrier particles.

5. Electromagnetic energy as the origin of space-time

5.1 Alternative model of space-time structure

Based on Planck's discrete energy radiation and empirical principle of energy conservation, we can formulate a nonempirical mathematical expression of energy.

The basic principle of energy conservation states that "Energy can only be transferred from one form to another." Transformation of energy from one form to another requires boundary within the space–time frame, carrying conservation of energy through space and time portions. While conservation laws associated with the time and space frame symmetries, we may consider that equally distributed space and time portions of energy hold simultaneous conservation of energy and momentum within symmetric frame. On this bass, the energy portions, equally distributed in space or time phases, both cover the half of the total available energy: **Es = Et = 1/2Ea.** This equation is the equivalent expression of Eq. (7). Similarly, the total energy comprises the mixture of energy portions, equally distributed within two parts of the space–time frame: 2**Es** = **Ea**.

Based on these simple equations, we may construct mathematical model of energy conservation, which has to combine energy-momentum conservations within the space–time frame. Conjugation of energy-momentum conservations within exchange interaction, which appears in discrete mode, generates principles of discrete symmetry. In this sense, special relativity's energy-mass relation $\mathbf{E} = \mathbf{mc}^2$ does not hold invariant discrete energy-mass exchange relation and cannot describe discrete symmetry of energy-mass relation, localized within the discrete space–time frame.

Based on such an approach, we may present space–time as a frame, which comprises cross product of space portion as materialization of energy and cross product of time portion, which at decay of space–time returns an energy to the origin:

$$\mathrm{Es}\frac{\mathbf{dS}}{\mathbf{S_1}} - (\mathbf{Ea} - \mathbf{Es})\frac{\mathbf{dt_1}}{\mathbf{t_1}} = \mathbf{0} \tag{8}$$

The first part of Eq. (8) presents the portion of consumed energy (**Es**) in space phase with the positive sign, while the second ingredient of the equation shows the remaining energy portions within the time ingredient of total energy with the negative sign. Model (8) gives the following equations:

$$\frac{\frac{\mathbf{dS}}{\mathbf{S_1}}}{\frac{\mathbf{dt}}{\mathbf{t_1}}} = \frac{\mathbf{E_a} - \mathbf{E_s}}{\mathbf{E_s}} \tag{9}$$

$$\frac{\mathbf{dS}}{\mathbf{dt}} = \frac{\mathbf{S_1}}{\mathbf{t_1}}\left(\frac{\mathbf{E_a}}{\mathbf{E_s}} - \mathbf{1}\right) \tag{10}$$

$$\lambda = \frac{\mathbf{E_a}}{\mathbf{E_s}} - \mathbf{1} \tag{11}$$

$$\text{at } \lambda = \mathbf{1}, \mathbf{Es} = \mathbf{1/2Ea} \tag{12}$$

$\mathbf{S_1}$ and $\mathbf{t_1}$ are the space and time variables, corresponding to the origin/dynamic local boundary, and **Ea–Es** and **Es** are the energy portions, distributed in space and time within energy-momentum exchange interaction at conditions corresponding to the background/dynamical local boundaries of $\mathbf{S_1}$ and $\mathbf{t_1}$. The background superposition as the gauge field holds the hidden initial space and time variables, which carry invariant translation of energy and corresponding symmetry from one form of energy to another and inverse. The local dynamical gauge position is the mathematical operator, which translates energy in the form of force from the local matter phase to energy phase. The **Ea** electromagnetic energy of model (9) is the symmetry generator of local gauge field, while **Es** appears as the local momentum ingredient of energy of the background gauge field.

The right side of Eq. (9) describes energy-momentum exchange interaction, relative to the original momentum of superposition, which generates shift of energy conservation from the space–time frame to the original energy phase. *The particle of space–time moves through the electric field, and electric field in reverse order propagates through the space–time field of matter.* Eq. (10) describes the cross product wave function where the local space–time wave $\mathbf{S_1/t_1}$ is carried by the flux of energy-momentum wave (**Ea/Es-1**) which changes wavelength and wave amplitude of space–time by S_1 and t_1. Conservation of energy-momentum is associated with the symmetry in time and space; therefore symmetry has to be the cross product of space–time and energy-momentum relation.

Model (10) has two important features on coupling of particle with the field. The eigenvector of model (9) connects force, field, and particle together (a) the mathematical operator describes change of symmetry generating electromagnetic field in relation to initial momentum of a particle, and (b) it presents change of space–time position of a "non-Aristotelian" particle in relation to the symmetry-generating field.

The superposition's genetic code in the form of dynamical space–time ($\mathbf{S_1/t_1}$) unit has discrete coupling with the electromagnetic field (10). Model (10) in general form describes relation of energy portions, distributed within space–time field, which generates discrete vector space as a product of discrete energy-momentum relation. The suggested approach is different from Sudarshan and Marshak's **V-A** theory of weak force [15], while without discrete eigenfunction, producing integer spin particles you cannot reverse a particle to the background gauge field. However, Hilbert space of quantum mechanics and **V-A** theory do not carry such a performance. *The other feature of Eq. (9) is reciprocal isomorphic discrete symmetry of space–time and energy-momentum exchange interaction, which became the outcome products of each other, forming the supersymmetric gauge equation.* Such an approach allows combining all the conservation laws within these symmetric interactions.

The background gauge field's force carrier **Ea** holds the symmetry of **Es** matter ingredients of eigenfunction **(Ea − Es)/Es** in the space–time frame of local gauge field. When the symmetry generator is turned off (**Ea = 0**), the **Es** through coupling of local and background particles return to the background gauge field in the form of neutral pairs of gauge field. Based on model (10), which combines space–time with the electromagnetism (energy-momentum conservation), the origin of space–time appears to be the background gauge field energy, which generates the basic unit of matter space–time frame and holds its conservation within conjugation of background/local gauge phases.

We found out that simplifying strong interactions to the linear exchange of photons or meson within continuous symmetry is the reason for appearance of problems of particle physics theories. Particularly, Yukawa's meson theory of strong interactions, describing linear exchange of mesons, and **V-A** theory of CP violation are examples of such theories.

5.2 The space and time particles of the space-time frame

Model (9) has a philosophical meaning: we do not present time as itself, which as an entity is different from space. We present a certain entity in time phase and this entity is the energy. That is why time has no independent existence from space energy portions and is not an abstract parameter, which may flow independently. The same philosophy is relative to the space as well. We did not present an event in abstract three-dimensional space or within four-momentum frame of SR; we describe the vector space, which changes dimension and direction in accordance with the flux of energy and momentum to this space. Such a space of space–time may change from three-dimensional frame to two- or round dimensional space–time. Based on the frequency of energy-momentum flux, space–time at the small scale moves to the round dimension, which is not possible by Hamiltonian's or Lagrangian's only time-dependent linear equations.

Replacement of intervals by combination of the origin with the displacement in the form of portions is the new algebraic expression of dimension, which as the mathematical operation carries *natural renormalization* of the change to the initial origin. Relation of the change to the initial origin generates $\mathbf{S_2/S_1{-}1}$ quantum operator, producing outcomes by the integer numbers.

Based on such a phenomenon, any particle of space–time field or antiparticle of energy field has no independent existence. The condition $\mathbf{E_a > 0}$ is the displacement from the superposition (field excitation) with the generation of three-dimensional space, which produces a local field and its particle (electrically charged mass). When an energy flux to space–time discontinued ($\mathbf{E_a = 0}$), the distinction between field and particle disappears, and the superposition and field merge. At $\mathbf{E_a = 0}$, the negative energy matrix produces **U (1)** symmetry group of gauge field, similar to Maxwell's theory of electromagnetism. Model (9) can be applied for any interaction of **(Ea − Es)** as a field and **Es** as a particle.

The other specific feature of our theory is that the relation of the change to its origin generates an original code as its own reference frame of an event. The state of origin of the space or time particle became their own antiparticle. When a particle does not change ($\mathbf{E_a = 0}$), its position in space, merging with its antiparticle, generates neutral particle of discrete rest. Conjugation and merging of two states (fields) became the main principle of discrete symmetry and conservation of energy within a certain boundary.

Our theory uses the background gauge field as the only possible reference frame, where all the interactions take their origin. The relation of an event or the particle's space–time frame to the background gauge field became the obvious concept, while the background gauge field is the source of interactions and mediates the space–time frame of a particle within the local gauge field. The background state is the source of symmetry generating electromagnetic force-gravitation exchange interaction in the local gauge field, which has to deliver energy back to its origin. *Gravitation appears as the short-range force, which holds discrete performance of electromagnetic force and generates stable existence of a nucleon in discrete mode.*

The inertial frame of reference in classic physics and special relativity is the same and states "the body with net zero force does not accelerate and such a body is at rest or moves at a constant velocity." Based on our theory, this statement is not completely true. When the net force flux to the space–time frame is zero **(Ea = 0)**, particles move to the reference background gauge field which holds discrete performance but not constant rest. We explain SR's time delay statement differently. By SR, "the clock of a moving body will tick slower than clock that is in rest in his inertial frame of reference." SR states that if the particle's speed approaches the speed of light, the massless particles that travel with speed of light is unaffected by passage of time.

First, the massless particles cannot have free travel due to the requirement of energy conservation within boundary. Based on model (10), time instant $\mathbf{t_1}$ is proportional to eigenvalue (12), and with the reduction of this value, the time instant and clock will tick slower than the background state.

The relation of change to initial origin eliminates unity which allows describing energy portions in space and time phases within any symmetric dimensions which may change from linearity of planet-scale event (string like dimension) to round dimension of baryon-scale interactions. The minimum portion of quanta, produced from the nonlinear energy-momentum exchange vector interaction, generates an elementary space–time frame. The condition **Es = 1/2Ea** of model (10) generates the invariant translations in a space–time frame:

$$\frac{\mathbf{dS}}{\mathbf{S_1}} - \frac{\mathbf{dt}}{\mathbf{t_1}} = \mathbf{0} \tag{13}$$

When symmetry generator electromagnetic energy is turned off (**Ea = 0**), we will get decay of space–time and shift of energy from local gauge space–time frame to the energy phase of the background gauge field:

$$\frac{\mathbf{dS}}{\mathbf{S_1}} + \frac{\mathbf{dt}}{\mathbf{t_1}} = \mathbf{0} \tag{14}$$

Eqs. (13) and (14) are the alternative presentations of Eq. (5). The portions of energy, carried by space and time identities $\mathbf{dS/S_1}$ and $\mathbf{dt/t_1}$, play a role of the quantum operator of annihilation or creation, through coupling with the energy-momentum exchange interactions.

The condition **Es = 1/2Ea** became the energy-momentum genetic code of particle-antiparticle interactions in the discrete space–time frame. The genetic code of Eq. (13) in the space–time frame generates a three-jet performance of ingredients of energy-mass exchange interaction.

The concept of supersymmetry, which we suggest in (9), describes the conjugated symmetry, which involves simultaneous symmetry of space–time frame and energy-momentum exchange interactions, carrying both in discrete mode.

5.3 The theory of spin as the product of discrete energy-momentum exchange interaction

In accordance with our concept, a change of particle's displacement around their superposition generates the conserved quantity called spin, which in quantum physics has identification, as the angular momentum. By quantum physics, the spin number for a point particle is the product of pseudo-vector position (relative to some unknown origin) and its momentum vector $\mathbf{r} \times \mathbf{p}$ [16].

In accordance with our theory, the spin is the conserved vector quantity, produced from conservation of energy within discrete energy-momentum exchange relation, which generates for ingredients of this interaction's spin numbers (12). The space and time portions of energy in exchange interaction (12) appear as interaction of fields, which produces the ingredients of this interaction in the form of fermions and bosons.

The quantum physics' presentation of spin, as a cross product of vector position with the momentum, does not produce quantity, which may carry energy-momentum conservation in a proper way. The quantum mechanic's specification of spin is a very abstract concept because the point particle is not a particle, which does not have a space–time frame of matter and therefore cannot produce half spin identity in the form of fermion. We suggest the identification of angular momentum as a product of the particle's space–time position vector and energy-momentum exchange interaction (10), which produces not the pseudo-vector but the local space vector. This vector generates a deterministic pathway of a particle's dynamics. In such a model, the dynamic local position became the deterministic position vector. Therefore, we may identify fermions and bosons only as the products of space–time frame. Due to these features, quantum mechanics cannot explain unusual feature of baryon frame where two identical quarks in proton or neutron frame do not obey the Pauli rules of quantum statistics.

The genetic code of supersymmetry **Es = 1/2Ea** explains this paradox. The antisymmetric wave function (13) holds the invariance of baryon performance through discrete symmetry, carried within background and local gauge fields. From supersymmetric genetic code.

Es = 1/2Ea (12) follows why quark ingredients should have **2/3** and **1/3** fractional charges. From three portions of energy (charges), only two portions describe one type of charge, and the other one portion describes another charge, holding the requirement of discrete **Es = 1/2Ea** symmetry.

6. The invariant translations within fermion-boson pairs

By Wilczek's [17] opinion of getting symmetry and maintaining the balance of conserved quantum numbers, the extra particles should exist by an equal number of antiparticles. However, our theory predicts that invariant translation of ingredients of energy-momentum exchange interaction should not involve equal numbers of particle-antiparticle pairs but has to follow the condition of Eq. (12) **Es = 1/2Ea**. We think that the concept of equal numbers appeared from wrong identification of a particle as a point-like particle, which cannot produce identity for fermion. In accordance with the invariant translation (12), from one charged fermion, we can produce only half-neutral boson. Therefore, based on invariant particle-antiparticle translation (12), to get a neutral bosonic particle, we have to double the number of particles to produce a neutral boson:

$$\mathbf{2}\,(\mathbf{Es} = \mathbf{1}/\mathbf{2Ea}) \rightarrow (\mathbf{2Es^t} = \mathbf{Ea^t}) \tag{15}$$

This operation is similar to quantum mechanic's doubling of wave function. However, in our case it is due to combining of energy portions, distributed in space and time phases to get full portion of energy at the origin in the form of boson. Elimination of dipole moment requires removal of charges in the space–time frame of matter, which requires decay of the space–time frame of ordinary matter and restoration of energy at the origin. However, the fractional charges of nucleons of baryon structure do not allow separation of quarks with elimination of charges. To eliminate this restriction, virtual particles with the fractional charges of baryon structure undergo coupling to pion families, which, as intermediate bosons, carry easy decay with production of neutral particles of the background gauge field.

The π-mesons generation through coupling of proton-antiproton or quark-antiquark pairs during decay of space–time frame was proven by experiments, carried out in Berkeley Center where it was observed that formation of neutral field, which could be accounted for neutral π-mesons, created by collisions of high-energy protons. In addition, it was shown that the neutral mesons decayed into two mesons with the lifetime of the order of $\mathbf{10^{-13}}$ **s** or less [18].

The produced π-mesons family became intermediate spin zero bosons due to the decay of the space–time frame of matter, while the spin number is the product of energy-momentum exchange interaction within the space–time frame of ordinary matter. On this basis, the coupling particles get the performance of the neutral particles of gauge field. The doubling of particles (14) at **Ea = 0** reverses the performance of the forces due to the transition of energy conservation from space–time phase to energy phase with the change of sign (13).

The shift of symmetry from space–time frame local field to the background gauge field symmetry leads to the disappearance of spin and generation of gauge field particles due to the coupling of initial and local momentum in the form (−**Es/Es**) of Eq. (10).

It is necessary to note a very important feature of translation of ingredients of Eq. (14) when the ingredients of this equation doubled. The half-integer fermions of this equation became integer carrier particles, while integer carrier particles became double integer carrier particles. Therefore, in energy phase the performance of forces holding space–time frame changes in the opposite order.

Based on model (10), depending on the energy flux to the space–time frame, the helicity of the ingredients of the space–time frame changes. In accordance with Eq. (14) when electromagnetic interactions turned off (**Ea** = **0**), the difference between space and time phase particles disappears, and all the particles behave as integer number particles of the background gauge field. In this case, the chirality and handiness of particles gets the same left-handed direction. When local symmetry generator force is not available (**Ea** = **0**), the momentum of matter space–time phase (10) transforms to the energy of the gauge field and gets a negative sign. Simply, "the ingredients of energy return to itself." In this case, Dirac's neutrinos transform to the neutral Majorana neutrinos having left helicity to the background gauge field where bosonic particles involve gamma rays, neutral fermions pair, and neutral neutrinos.

To understand the nature of particles and forces, we have to analyze the decay mechanism of produced pion families, where the W vector bosons were intermediate ingredients:

$$\pi^{+} \rightarrow \mu^{+} + \nu_{\mu} \tag{16}$$

$$\pi^{-} \rightarrow \mu^{-} + \nu_{\mu}^{-} \tag{17}$$

$$\pi^{0} \rightarrow 2\gamma \tag{18}$$

The produced ingredients of the decay form the balance equation:

$$\pi^{+}\pi^{-}\pi^{0} \rightarrow 2\gamma + \mu^{+}/\mu^{-} + \nu_{\mu/}\nu_{\mu}^{-} \tag{19}$$

The decay of the local gauge field to hold vector conservation produces a new vector, which comprises generation of intermediate W vector bosons from decay of π pions. The **2Es** = **Ea** code of the background gauge field requires equal numbers of electron and neutrino family pairs which is realized by the equal branching ratios of the decay of intermediate W vector bosons. Eq. (19) describes decay condition in average for muon family leptons.

The product stream composition generates composite of neutral particles, which exists in annihilation mode in the background gauge field to hold equation 2Es = Ea:

$$2\gamma \leftrightarrow -\mu^{+}/\mu^{-} + \nu_{\mu/}\nu_{\mu}^{-} \tag{20}$$

In the discrete energy conservation mode, the particles cannot hold annihilation process for a long time. Coupling of gamma rays with the neutral particles leads to the generation of charge and electromagnetic force of local gauge field:

$$2\gamma + \mu^{+}/\mu^{-} + \nu_{\mu/}\,\nu_{\mu}^{-} \leftrightarrow \mu^{-}/\nu_{\mu}^{-} + \mu^{+}/\nu_{\mu} + \mathbf{Ea}\;(\textbf{electromagnetic force}) \tag{21}$$

The intermediate step in the symmetric translation from fermions to gauge field is the transformation of proton-antiproton pair to neutron-antineutron Majorana-type particles, which decompose to kaon family mesons. Due to the existence of three fractional proton-antiprotons, comprising other flavors of quarks, the decomposition of neutron-antineutron pair produces three kaon-type mesons. The decay products of other unidentified two kaons, which we may call $\mathbf{Kaon_2}$ and $\mathbf{Kaon_3}$, can be described similarly by Eqs. (16)–(19).

In accordance with the equation $\mathbf{E_s = 1/2E_a}$, for transformation of fermions to bosons (transformation of mass back to energy), we have to double the numbers of particles through coupling of space and time phases, carrying energy portions. In reverse order, for generation of fermions from bosons (generation of mass from energy), we have to separate space and time phases (12) to produce charge and reduce spin numbers of particles.

While our theory involves meson families as intermediate particles within half-integer-integer particle transformations, we may compare our supersymmetry theory with the basic principles of Yukawa's meson theory. The background of Yukawa's theory is the spontaneous breaking of continuous symmetry [19] and involves interaction between scalar ϕ and a Dirac field ψ. Yukawa's vector in the form of pseudo-scalar field is the linear combination of nuclear force-electric dipole moment ($\boldsymbol{\varphi - \varphi_0}$) which is very similar to V-A theory [15]. Both theories cannot describe CP invariance of strong interactions, while the linear combination of vectors could not produce translation of interactions to the initial state.

In Yukawa model, meson is the force carrier, but in accordance with our theory, meson is the product of invariant translation from baryon frame and is the intermediate ingredient of the background gauge field where all the forces merged.

7. The Yang-Mills theory and the mass gap of Yang-Mills theory

The main feature of Yang-Mill theory [1, 20] is that to produce differentiable manifold it applies continuous elements of Lie group. Yang-Mills theory, using differentiable manifold of non-Abelian Lie group and continuous Lagrangian, tried to describe the behavior of elementary particles through the combination of electromagnetic and weak forces. The non-Abelian Lie group is opposite to discrete symmetry, while traditional differentiation is not applicable for discrete symmetry group. The Yang-Mills theory does not have mathematical formulation for proper matrix reduction to get the nonzero mass particles of the local gauge field.

The Yang-Mills theory does not explain why the weak force has continuous energy spectrum. Pauli [21] suggested the production of massless neutrino together with the electron to explain continuous spectrum, but this explanation was not valid because quantum field theories have no mechanism for translation of space–time fermions to the gauge field bosons, showing continuous energy spectrum. Yang and Mill had no choice and selected the only possible way—application of non-Abelian Lie group, having Lagrangian manifold. The other problem of Yang-Mills theory is the application of energy-momentum four-vector (R^4), which leads to the V-A-type energy-momentum spectrum that produces a gap in energy between zero and some positive number.

However, the Yang-Mills theory is the only correct concept among all particle physics theories, used to describe strong interactions. The very important feature of the Yang-Mills theory is that it suggests simultaneous production of massless photons in addition to three massive bosons. Unfortunately, this excellent suggestion has no proper mathematics, which could describe invariant translation of gauge bosons to fermions of strong interactions.

8. Dirac's relativistic quantum theory and problems of Dirac's "electron sea"

Dirac [22] applied the relativistic theory to Schrodinger's equation to get relativistic wave function of electron motion. The problem of Dirac equation was negative energy solution. To solve this problem, Dirac assumed interaction of electron with the electromagnetic field where the electron was placed in a positive-energy eigenstate to get decay into negative-energy eigenstates. However, such an approach had a problem that the real electron would disappear by emitting energy in the form of photons.

Based on our theory on discrete performance of nucleon ingredients, it is easy to show that generation of photons from fractional electron charges at coupling mode predicts existence of its antiparticle-positron. Model (9) presents electromagnetic interaction of space–time particle (**Es**), particularly electron, with the electromagnetic field (**Ea**), through deterministic energy-momentum exchange interaction without the application of relativistic quantum approach.

In the absence of electromagnetic field (**Es** = **0**), this interaction moves to the background energy field through merging of photon's fractional electric charges to particle-antiparticle pair **e/e** and **ν/ν** with the generation of a neutral current instead of Dirac's electron "sea."

For formulation of gauge field theory, instead of Dirac's relativistic approach, we used classic principles: (a) formulation has to hold symmetric space and time derivatives, in relation to origin, and (b) energy-momentum exchange relation has to present the momentum and energy as the space and time parts of a space–time vector instead of a four-momentum frame of Dirac's relativistic theory.

The basic principles of our supersymmetric theory replaced Dirac conditions through (a) and (b). The problem of Dirac's approach is that the space and time derivatives enter to the equation with the second order, which led to the loss of the original function and its first-order derivative. That is why Dirac's equation could not find the local position of an electron in motion in a deterministic way and used probability density. The second problem of Dirac equation is that he introduced to his equation relativistic energy-momentum relation in the form of linear space–time vector, similar to Sudarshan's V-A vector that could not produce integer spin carrier neutral particle field from half-integer spin carrying fermion particle. This was the reason for the theory to produce "electron sea." In accordance with the supersymmetric theory, fermion-showing performance as a particle in local space–time gauge phase became a field of neutral bosons of background gauge phase of energy. This is the supersymmetric feature of nature.

In gauge energy phase, electron and positron coupling to **e/e** generates, together with the Majorana **ν /ν** neutrinos, a vacuum "sea" of neutral current. Therefore, the "Dirac sea of electrons" in reality is the gauge field of neutral particles. The energy of the field is finite and has a boundary within the space–time frame, existing through discrete shift between space–time and energy phases of energy conservation.

9. Performance of Dirac's and Majorana neutrinos in model

Particle physics does not provide any information why Dirac's neutrinos have to transform to Majorana neutrinos. The exchange interaction (**Ea/Es** – **Es/Es**) of Eq. (10) determines the nature of neutrinos. When interaction with electromagnetic energy is off (**Ea** = **0**), the difference between **Es** in the denominator and nominator of Eq. (10) disappears. The **Es** of nominator presents local momentum, while the **Es** in the denominator describes initial momentum as genetic code of superposition. At (**Ea** = **0**), the momentum ingredients of expression **Es/Es** became equivalent and cancel each other with the disappearance of charges of quarks **(e–/ν–)/(e+/ν)** → **(e/e)/(ν/ν)** and generation of Majorana neutrinos (**ν/ν**) and neutral **e/e** fermion pairs of gauge field in the form of bosons. In the energy phase, the gauge field Majorana particles became boson particles, and the chirality and the handiness get the same direction. *Transformation of fermions of local gauge field to the* *background gauge bosons as the intermediate particles is the necessary step for invariant translations of strong interactions.*

The mixture of neutral electron-positron pairs and Majorana neutrinos generates the spin 2 neutral particles of graviton of spin zero gauge field, carrying gravitation force to the background vacuum with the velocity faster than electromagnetic force in any medium. Therefore, gravitation force appears in reverse order from electromagnetic energy through coupling of entangled space and time portions of energy to restore it at the background vacuum (**Ea = 0**). It is not "spooky action at a distance" [23] but coupling of entangled non-separable portions of energy, existing in different forms.

The mass of Majorana neutrinos (Majorana mass) in the background gauge field is very low, but they became massive as Dirac neutrinos of baryon structure due to the entanglement with the charged electron family particles. Due to the decay of space–time phase at (**Ea = 0**), the particles of gauge field has the continuum spectrum. The **e/e** and **ν/ν** pars have no independent existence, but with the gamma rays, they form dark matter and energy content with ratio 33 and 66%, holding **Ea = 2Es** gauge field frame of energy conservation.

At vacuum expectation value takes place discrete shift of gauge field energy back to the local gauge field of space–time. Majorana neutrinos became again Dirac neutrinos with the generation of charge and massive particles of exchange interaction **(Ea−Es)/Es**.

10. The problems of Weyl spinors and quantum field theory

Quantum field theory suggests existence of massless half spin fermions and provides relativistic Weyl equation for description of massless half spin fermions. Due to the connection of Weyl's spinor to Dirac's theory of half spin electron, Weyl spinors describes Dirac fermions in the form of two ½ spin massless fermions. Quantum field theory does not explain physical nature of Weyl's massless spin ½ particles, while spin ½ fermions in Dirac structure are massive particles. Dirac's theory describes energy-momentum relation as a continuous function that is why physical nature of predicted massless ½ spin fermions remains open. In accordance with our theory, neutrinos, existing in pair with the electron and positron, as Dirac fermions in quark's structure, in energy phase transform to integer spin "twin" Majorana particles. In mathematics, usually such an inversion has to meet requirements of spinors.

The spinor is the mathematical operation [24], which produces vector space by addition of vectors together or multiplication by numbers, called scalar. The vector addition or scalar multiple operation must satisfy requirements, called axioms. The real vector space presents a physical quantity such as force and multiplication of a force by a real multiplier which produces another force vector.

Spinor, as a vector, exhibits inversion, when a physical system constantly rotates through a full turn (360°). In the following chapters, we will explain scalar multiplication in strong interactions and will provide mathematical framework, carrying the inversion of particles from half spin to integer spin neutral particles with the simultaneous change in the nature of existing force.

11. Development of new gauge field as the frame for discrete conservation of energy

11.1 General principles

The idea of a gauge theory appeared from Weyl spinors, but Weyl's theory, as we mentioned, could not produce the gauge scalar field due to the helicity problem of neutrinos, and the produced particles remained spin ½ fermions. The neutrinos helicity was the main problem of all of the field theories which did not allow to describe fundamental laws of nature in a proper way. The question why nature has no right-handed neutrino produced an opinion [25] that "God decided that Nature should be left handed." Due to this problem, particle physics theories suggest that nature respects parity with regard to all the fundamental forces with the exception of the weak interaction, which involves neutrinos.

Based on our model (9), the generation of free neutrinos and antineutrinos takes place at cutoff electromagnetic interactions (**Ea** = **0**), which reverses the momentum to the background state. The problem is that Weyl's theory determines the spin only relative to the positive momentum vector, and individual ½ spin carrier massless Weyl spinors violate conservation of parity. For this reason, Pauli specified Weyl spinors "unphysical" [21]. Weyl's theory could not combine neutrinos to hold parity conservation in the form pair of virtual Majorana particles.

Weyl's theory cannot explain why production of his spinors in the unitary transformations takes place only in the presence of half angle. Hamilton rotation about some axis, in a similar way, connects half angle and the Pauli matrixes. The presence of half angle in both cases was unavoidable [25].

Model (10) shows that at **Es** = **1/2Ea**, the invariant translation within complex space–time coordinates are connected within tangent 45 which describes space–time symmetry $\mathbf{t_1\,\Delta S = S_1\,\Delta t}$ in the form of coordinates **y** = **x** symmetry. This is the half-angle mystery of Weyl and Hamilton translations. The two-dimensional space–time frame in association with the **Es** = **1/2 Ea** discrete energy-momentum symmetry carries this translation.

Hamiltonian and momentum are the adjoint elements of the Lie Algebra group that generate linear transition in space and time. Model (10) presents the nonlinear energy-momentum exchange relation as the adjoint elements of three-dimensional **SO (3)** group and shows that Hamiltonian linear transformation alone cannot do invariant translation. Due to the involvement of symmetry generator force **Ea**, the translation has to be with the change of dimension. The invariant translation requires conjugation of invariant space–time frame **SU (2)** with the three-dimensional energy-mass exchange transformation through **SO (3)** group where change of space–time dimension is the driving force of translations.

Under unitary transformations, one rotation (360°) does not bring the state of a body to the origin. One rotation brings **SU (2)** x **SO (3)** local gauge symmetry to **U (1)** matrix with simultaneous transformation of a three-dimensional particle frame to linear gauge field. Therefore, full translation of opposite phases holds condition: **SU (2) x SO (3)/U (1).** Doubling the spin numbers of quark ingredients through coupling of space and time portions of energy **Es** = **1/2Ea** to **2Es** = **Ea** produces unstable pions which produce non-charged boson-like ingredients of the background gauge field **U (1).**

The second transformation with the reversing of **U (1)** symmetry group brings the linear gauge field back to the space–time (**SU (2)** x **SO (3)**) frame of three particles of baryon structure. Coupling of neutral **e/e**, **ν/ν**, and **γ/γ** ingredients in such a translation generates quarks of baryon structure. The invariance between bosons and fermions in the form of strong interactions is possible only in discrete mode with the change of space–time dimensions.

With the reversed momentum line (**Ea** = **0**), the antineutrino changes its helicity and becomes the left-handed particle, which leads to the coupling of two neutrinos in the form of bosonic twin particles. The right-handed neutrino would block generation of **U (1)** field and its translation back to **SO (3)** matrix, that is why nature does not allow its existence. Weyl's theory due to the wrong helicity misses these translations.

The other problem of quantum mechanics is that it eliminates participation of neutrinos in strong interactions due to the absence of charges. However, our theory shows that neutrinos are the necessary ingredients for generation of strong force, while coupling with the neutral **e/e** pairs generates formation of charges through their reproduced right–left helicity in **SO (3)** group.

Without conjugation of **SU (2)** and **SO (3)** symmetry groups, the description of chiral symmetry is not possible. At **Ea = 0**, Dirac particles transform to Majorana neutral massless particles which eliminates the difference between handiness and chirality.

11.2 Mathematical framework of discrete gauge field

Gauge, in common sense [26] is a measurement of a relative position of a system with reference to another abstract system to determine boundary of measurement. The gauge theory has no mathematical framework relative to the proper reference frame for measurement of change. In this aspect, the gauge theory has the same reference frame problem of classic physics.

Based on literature [26], the gauge symmetry has specification, as "is a lack of change when some field being applied." The meaning of this statement is that the measurable quantity after the change looks the same. Linearly differentiable Lagrangian of non-Abelian algebra due to the absence of the space–time frame cannot provide a mathematical formulation on how the gauge field after the change may look the same.

The theory, which we apply, provides a mathematical framework to the gauge theory to measure a quantity, relative to its initial superposition state. We suggest that change of energy and momentum in discrete mode generates the dynamical operator of gauge field, which describes the measurement in relation to the initial origin with the integer numbers of energy portions.

When the superposition of a gauge field after displacement within space, time, and energy looks the same, the invariant transformation produces invariance for all the inner ingredients of the change $\mathbf{\Delta S = S_1}$, $\mathbf{\Delta t = t_1}$, and **Ea –Es = Es** with the realization of condition **1 = 1** (9).

The genetic code of exchange interactions **Es = 1/2Ea** keeps the discrete symmetry of force carrier and electrically charged ingredients of space–time at different spin numbers (12). In the energy phase (13) of gauge field, the genetic code **Es = 1/2Ea** undergoes multiplication by scalar **2** to **Ea = 2Es** which holds the discrete symmetry within color charge ingredients of gauge field, leaving spin untouched. When the electromagnetic interactions is off (**Ea = 0**), coupling of space and time portions of energy generates transformation of half spin matter fermions to integer bosons ν/ν + **e/e** + **ɣ/ɣ** family of background vacuum:

$$\frac{\mathbf{dS}}{\mathbf{S_1}} = \frac{\mathbf{dt}}{\mathbf{t_1}}\left(-\frac{\mathbf{E_s}}{\mathbf{E_s}}\right) \qquad (22)$$

When electromagnetic energy is off (Ea = 0), merging of space and time portions of energy generates left side helicity for all the non-charged ingredients of gauge field of background vacuum. At this condition, all the half-integer fermions merging with their own antiparticles form neutral integer spin carrying bosonic particles. Eq. (22) is the equation of vacuum, where space and time portions of energy merging generates a one-dimensional space–time frame of vacuum.

The integer spin carrying particles, produced at zero electromagnetic interactions within baryon's space–time frame, can hold invariance of gauge field only in discrete mode:

$$(\mathbf{e} - /\boldsymbol{\nu}-)/(\mathbf{e} + /\boldsymbol{\nu}) \leftrightarrow (\mathbf{e}/\mathbf{e})/(\boldsymbol{\nu}/\boldsymbol{\nu}) + \mathbf{p}/\mathbf{p} \tag{23}$$

The symmetries of space–time (**Es = 1/2Ea**) and energy (2**Es** = **Ea**) phases do not have independent existence and only in conjugation carry discrete conservation of energy.

We assume that **2Es** = **Ea** frame of the background gauge field involves a combination of elastic (Thomson effect) and inelastic scattering (Compton effect) where inelastic scattering gradually transforms to elastic scattering. At background vacuum expectation value takes place translation of the background gauge field energy to space–time frame of local gauge field by elastic scattering, which involves absorbing of gamma rays by the virtual matter bosons. This process shifts the continuous spectrum of longitudinal waves of the background field of bosons to the discrete spectrum of transference waves of charged matter particles. This is the process, which eliminates generation of ultraviolent divergences.

According to quantum mechanics, vacuum energy without renormalization mathematically is infinite. However, this statement is true only if the background's gauge field energy has no shift to the local gauge field of matter's space–time frame. During shift of the background gauge field's energy to the local space–time field, Majorana neutrinos transform to Dirac neutrinos with the transformation of color charges to the electric charges of quarks.

11.3 Mechanism of conjugation of background and local gauge fields

In accordance with our theory, if field does not change, it cannot hold energy conservation and symmetry within reversible dynamic translations. Energy can exist only through propagation in space–time frame, and in reverse order, space–time is the matter product of energy distribution. On this basis, conjugated existence of background energy and local gauge matter fields is the necessary condition for conservation of energy.

The energy-momentum exchange relation of the model (12) in the form of eigenvector generates exchange of particle with the field. The energy-momentum exchange relation of eigenvector (12) describes the relation of two fields, such as electromagnetic-gravitation fields, which carry invariant translation to each other. Electromagnetic force in the form of **Ea** can be a vector field and at the same time photon particle. At **Ea** = **0**, the electromagnetic force disappears as a field/particle and transforms to gauge field of boson ingredients.

Model (10) describes local gauge field $\mathbf{S_1/t_1}$, which carries energy at each point of space–time. Local gauge field carries electromagnetic force in space–time frame in the form of energy-momentum content and strength of electromagnetic field determined by its coupling with the local space–time field. Model (10) combines all types of interactions and translates them to each other through energy-momentum exchange interaction. In this case, background/local gauge field of articles appear as the "two worlds of particles."

The vector space, as specified in mathematics, moves through plane wave, which is field. By the requirement of vector space [24], the field where the vector has to move requires existence of two equivalent field functions that determine the field value. These functions involve two parameters, which are time and displacement along the direction. In accordance with our model (13), the symmetry of energy portions, distributed evenly within space and time phases, generates two field functions of negative displacement by the conjugation of space and time variables:

$$\mathbf{s_1 \Delta t} = -\mathbf{t_1 \Delta s} \tag{24}$$

Therefore, the ingredients of Eq. (13) generate two equivalent functions:

$$\mathbf{F_t(s_{1,} \Delta t)} = -\mathbf{F_s(t_{1,}\ \Delta s)} \tag{25}$$

The function **Fs** describes displacement in space, while the function $\mathbf{F_t}$ describes duration of change. When the values of field function are vectors, the plane wave is longitudinal. The space and time portions of energy in the background gauge field form a one-dimensional space–time frame that is why the plane wave in this case is longitudinal. Multiplication of equation **(Es = 1/2Ea)** by scalar **2** leads to the formation of neutral particles **e/e**, **ν/ν**, and **ṿ/ṿ** of the background gauge field **2Es = Ea** in the form of spinors, similar to Hamilton quaternions spinors [25]. The ingredients of Hamilton's equation **(I; J; K)** are imaginary quantities, while the products of our model are virtual particles:

$$\mathbf{I^2 = J^2 = K^2 = ijk = -1\ (Hamilton)} \tag{26}$$

$$\mathbf{e/e = \nu/\nu = ṿ/ṿ = (e\nu ṿ) = -1} \tag{27}$$

$$\mathbf{(Es = 1/2Ea)\ (Space\text{–}time\ phase) \rightarrow Inversion\ to\ energy\ phase\ (2\ Es = Ea)} \tag{28}$$

The equation **(2Es = Ea)** produces a new vector where the scalar is the real number. The inversion transforms integer spin electromagnetic force to the other force being integer **2** spin carrying force. The new force is the gravitation, which with continuous longitudinal wave moves to the background through conjugation of ingredient **(eνṿ)** of produced neutral spinors.

It is necessary to note that generation of inversion vector space for transformation of energy conservation from space–time phase to energy phase meets all the requirements, required for scalar multiplication procedure, given in the form of axioms [24]. One of the requirements of axioms is the condition **x + (−x) = 0**, which is in hold within discrete annihilation of space and time variables (14). Due to the conservation of energy within discrete energy and space–time phases, an event comes to the origin after two full rotations (720°). The generated field has an algebra of zero-dimensional geometric spinor with one-directional helicity to the background origin.

Model (9) describes conservation of energy through non-unitary space–time variables, which unifies fields, particles, and forces within non-unitary energy portions allowing transformation of all the identities to each other. The background plasma-like gauge field can hold the discrete symmetry only through coupling with the local gauge field. The local gauge field of virtual matter, which is the combination of electric and magnetic fields, holds interaction of electrically charged space and handiness carrying time particles within the discrete space–time frame. The driving force for generation of local gauge field's space–time frame of matter is the discrete conservation of energy-momentum pairs.

Due to the absence of space–time frame, the neutral particles of the background gauge field have only color interactions with the feature of "neutral crystals" of time and space portions, bubbling in gauge field condensate with the small wavelength. Recently Wilczek [27] described the similar idea in more details.

11.4 Invariant translation of symmetries within background/local gauge fields

Background gauge/local gauge fields exist in the form of field-anti-field pair. Due to the energy-momentum non-commutation, the local gauge field is the non-Abelian, while background gauge is Abelian field. Generation of non-Abelian local gauge field from background Abelian's gauge field is not spontaneously symmetry breaking. The local gauge field at **Ea = 0** of model (9) merges with the background gauge field, as particle-antiparticle pair. In this case, the difference between particle and field disappears. The background and local gauge fields are connected through **SO (3)** rotational matrix which carries invariant translation of particle to field. Quantum mechanics mediates physical quantity by the square of the wave function, but **SO (3)** group of model (9) mediates physical quantity of the background gauge field by coupling of space–time portions of energy, carried in the form of matter–antimatter pair.

Model (9) suggests that CP symmetry of strong interactions is in hold only through cross product of **SU (2) x SO (3)** symmetry groups within two transformations: one is charge cancelation translation, and the second is parity transformation.

Yang-Mills [24, 28] attempted to apply gauge theory to the strong interactions through elevating of global symmetry to local gauge symmetry, but this attempt produced symmetry breaking. Without conjugation of **SU (2)** matrix of space–time frame and energy-momentum exchange interactions **SO (3)** (9), the background gauge field **U (1)** cannot carry invariant translation of mass to the opposite phase of local gauge field of strong interactions.

By Glashow's opinion [2] electromagnetism is mediated not only by photons; it arises from the requirement of local gauge invariance. However, based on our theory, this statement is true only partly because the role of local gauge field is reversible and symmetric. The local gauge field is needed for generation of electromagnetic interaction and cancelation it takes place by gravitation for discrete conservation of energy within space–time of baryon frame. The **SO (3)** symmetry group of gauge field translates electromagnetic force to the gravitation force. Therefore, without gravitation force, it is impossible to get invariant performance of strong interactions. The Standard Model, as Kibble showed [29], did not find place for gravity, and that is why it cannot not explain why the elementary particles come in three families with very similar structure but wildly differing masses.

In our theory electromagnetism and gravitation are unified within **SU (2) x SO (3)** symmetry of local gauge field which involves unification of charges as the internal products of baryon's space–time frame. The genetic code of baryon particles **Es = 1/2 Ea** holds all the internal conservation laws: baryon conservation, isospin conservation, hypercharge conservation, and boson-fermion spin invariant translation. It is known that the hypercharge of **SU (3)** symmetry is one of two quantum numbers of the hadrons and alongside with isospin I_3 follows the formula: $\mathbf{Q = J_3 + 1/2Y}$. For multiples of particles, the hypercharge gets formulation **J** = 2Q.

According to model (9), at local gauge field, the hypercharge current coupling is the condition **Es = 1/2 Ea** which describes local space–time symmetry at J_3 = **0**. At the background gauge field, the hypercharge conservation **Ea = 2Es** describes multiple bosons, similar to **J = 2Q.** Based on Eqs. (8)–(12), the discrete conservation of energy at the background gauge field produces condition:

$$\Delta\mathrm{F} = \frac{\mathbf{\Delta S}}{\boldsymbol{\Delta t}} + \frac{\mathbf{S1}}{\mathbf{t_1}};\ \frac{\mathbf{\Delta S}}{\boldsymbol{\Delta t}} = -\frac{\mathbf{S1}}{\mathbf{t_1}}, \Delta\mathrm{F} = \mathbf{0} \qquad (29)$$

The equation **ΔF = 0** describes discrete nonvanishing energy state of spin zero boson's condensate of the background gauge field.

12. Principles of isomorphism of SU (2) and SO (3) symmetry groups

We developed a new algebra for the isomorphism of **SU (2)** matrix to **SO (3)** group which holds 3D rotation about three-dimensional R^3 Euclidean space, to preserve the origin in discrete mode. The **SO (3)** three-dimensional matrix describes the three-jet performance of elementary particles.

The left side of model (9) is **SU (2)** matrix of space–time, but the right side describes three-dimensional energy-momentum exchange interaction within **SO (3)** group. The inseparable **SU (2)** and **SO (3)** matrixes make inseparable position and momentum. Therefore, the non-separation phenomenon of position and momentum, called uncertainty of quantum mechanics, is the necessary condition to hold discrete invariant translation of symmetries.

The new space and time geometry, which we suggested, is the presentation of new Hilbert space, which is equivalent to Euclidean space where the dimension of a Euclidean-type space may change in accordance with the associated vector space. The **SU (2)** and **U (1)** symmetry groups of standard model do not exist in the same phase, and **U (1)** x **SU (2)** is not the cross products due to the existence of these groups in opposite phases of discrete conservation of energy. Therefore, even the extension of **SU (2) x U (1)** matrixes [2] of standard model to symmetry group **SU (3) x SU (2) x U (1)** [30] cannot describe strong interactions.

For restoration of origin, the eigenvalue of rotation has to have signs ± 1, and model (9) provides this condition. The eigenvector with eigenvalue **+1** describes extension of baryon space–time frame, while (**−1**) in the form of reflection returns the space–time to the origin. Conservation of energy at the origin holds conservation all of the inner products of translation.

The **SU (2)** and **SO (3)** are not subgroups of **U (1)**, as common algebra states; the **SU (2) x SO (3)** and **U (1)** are the products of each other in opposite phases of discrete symmetry. The special orthogonal **SO (3)** rotation symmetry group describes rotation about the origin of the three-dimensional Euclidean space. Orthogonal matrix is the square matrix; a matrix is orthogonal if its transpose is equal to its inverse within equations **Es = 1/2Ea** and **2Es = Ea** which are equal to each other. First is the matrix Q, and second is the inverse matrix.

At **Es = 1/2Ea**, the **SU (2)** symmetry has isomorphic relation to **SO (3)** symmetry, but at **2Es = Ea**, the **SU (2)** symmetry undergoes surjective homomorphism to **SO (3)** symmetry. The surjective homomorphism of the Lie group describes [30] two algebraic structures of the same type, which generates coupling of particle and antiparticle to the same structure. The isomorphism requires symmetry in opposite phases, but Lie algebra does not explain why isomorphic symmetries exist in opposite phases. The surjective homomorphism requires that the ingredients of the homomorphism should have one element, which should be the same for these ingredients. The same ingredient is the mass of particle and antiparticle, which makes them coupling by surjective homomorphism. The particle-antiparticle pair forms a domain-codomain pair where antiparticle codomain is the mathematical image of superposition origin and completely covers the domain function.

At **Ea = 0**, the **SU (2)** matrix gets smooth **2:1** surjective homomorphism to **SO (3)** group matrix which generates **U (1)** symmetry of the background gauge field.

13. Mathematical formulations of SU (2) x SO (3) matrixes

Multiplication of energy-momentum spin relation **Es = 1/2Ea** to **2Es = Ea** eliminates the difference in energy portions, distributed in space and time phases. Using these equations and model (9), we can get the following equations:

$$\frac{\frac{\Delta S}{S_1}}{\frac{\Delta t}{t_1}} = \frac{2\frac{\Delta s}{s_1} - \frac{\Delta t}{t_1}}{\frac{\Delta t}{t_1}} \tag{30}$$

From this formula, we will get:

$$\left(\frac{\Delta S}{S_1}\right)^2 - 2\left(\frac{\Delta t_1}{t_1}\right)\left(\frac{\Delta s_1}{s_1}\right) + \left(\frac{\Delta t}{t_1}\right)^2 = 0 \tag{31}$$

$$\left[\left(\frac{\Delta S}{S_1}\right) - \left(\frac{\Delta t}{t_1}\right)\right]^2 = \mathbf{0} \tag{32}$$

$$\frac{\Delta S}{S_1} - \frac{\Delta t}{t_1} = \mathbf{0}\ (a), \frac{\Delta S}{S_1} - \frac{\Delta t}{t_1} = \mathbf{0}\ (b) \tag{33}$$

The equations (32) and (33) describe the combination of space–time and energy-momentum symmetries in the form **SU (2) x SO (3)** product, which holds conservation of energy within invariant translations.

In mathematics isomorphism is a mapping between two structures of the same type that can be reversed. Model (10) describes isomorphism of SU (2) and SO (3) matrixes not only from the point of view of reverse mapping structures; it shows that due to the reciprocal transformation of space–time and energy-momentum identities, these symmetry groups are not separable from each other.

The rotational symmetry group ***SO (3)*** *cannot carry translation if the model does not provide the state of origin.* Without initial position of space and time, you cannot build a gauge field theory where the antiparticle cannot find its twin brother in the background gauge field. The energy-momentum exchange eigenvector (12) through angular momentum generates the rotational **SO (3)** symmetry, while the **SU (3)** group of standard model describes only continuous symmetry. The **SO (3)** generates rotation about the origin in Euclidean space. Only this symmetry group with matrix multiplication may produce elementary particles.

The quadratic Eq. (31) with two variables, which is generalized to vector space, is an algebraic expression of quadratic polynomial **P(x, y) = 0** equation. Such a polynomial fundamental equation takes place in conic sectors, having the expression **f(x, y) = 0.**

At **Ea = 0**, the space and time variables became asymptotically equivalent. The asymptotic limit for these variables having binary relation **f (Δs/S1), f (Δt/t1)** can be described as follows:

$$\frac{\text{Lim}\ \ f\left(\frac{\Delta t}{t_1}\right)}{\frac{\Delta s}{s_1} \to 1 f\left(\frac{\Delta s}{s1}\right)} = 1 \tag{34}$$

14. Translation of space dimensions. Transmutation of dimension-based physical laws to frequency

We replaced velocity in linear equation of classic electromagnetic field by the frequency to present electromagnetic field, conserved as the cross product of energy-momentum exchange and local gauge field position of space–time. The electromagnetic field **Ea** of model (9) involves electromagnetic fields, and its relation with the magnetic field **Es** produces the three field symmetry **Es = 1/2Ea.**

Model (10) shows that generation of new space vector takes place when dimension of space–time changes. The energy flux, coupled with the space–time (10), determines the space–time structure and dimension of space and time variables. The third dimension of space and generation of mass takes place in discrete mode at positive value of function $\mathbf{Ea} - \mathbf{Es} > \mathbf{0}$.

Based on model (10), the wave amplitude of space–time is the composite product of instant of time and displacement in space, while the wavelength is the composite product of change of time and local space. The time instant $\mathbf{t_1}$ appears as the genetic code of wave amplitude, while the local space $\mathbf{S_1}$ appears as the genetic code of wavelength in the form of superposition.

Conservation of energy in the form of space and time portions requires transmutation of dimension-based physical laws to unit-less dynamics of frequency, which changes through integer numbers (10). While the energy-momentum content of photon is constant, the phenomenon called mass appears as a unit of change of frequency of energy distribution in space–time frame. Change of the frequency leads to the change of the space length and duration of the interaction, keeping the same physical law regardless of scale. At $\mathbf{\Delta S/S1 = 0}$, we get $\mathbf{Ea = Es}$ which shows that when particles move to short or zero distance, the difference between energy and momentum disappears. Thus, mass appears as the space phase equivalent of energy.

To hold conservation of finite amount, energy generates space–time phase through which it moves from one form to another. The **SO (3)** group generates a two-dimensional non-unitary isomorphic space–time symmetry of **SU (2)** matrix which holds the three-dimensional discrete performance of baryon structure through discrete invariant in-out of energy (in the form of so called gluons) to this frame.

Therefore, the discrete in-out external energy (gamma rays, transformed to electromagnetic force) generates additional in-out space dimension in baryon structure of local gauge field. When neutral particles are translated to the background gauge field, the **SO (3)** group eliminates external dimension in baryon structure and returns energy back to vacuum.

Such a performance of **SO (3)** symmetry group is missed in the standard model, and the known symmetry groups of strong interactions do not involve this symmetry. Combination of **SU (2)** non-unitary group with the **SO (3)** matrix generates new principles of fundamental laws which hold invariant translations of all of the natural symmetries through the background gauge field **U (1)**. However, the background gauge field cannot hold its state in continuous symmetry. We may describe the uniform state of a particle in gauge field as $\mathbf{\Delta S/\Delta t = 0}$, which has isomorphism with the matter–antimatter symmetry at conditions where there is no change in space–time:

$$\left(\frac{\mathbf{Ea} - \mathbf{E_s}}{\mathbf{E_s}}\right) = \mathbf{0}, \mathrm{Ea} = \mathbf{Es} \tag{35}$$

Such a state of a particle generates timeless matter–antimatter annihilation, which violates conservation of energy and leads to the ultraviolence divergences. According to the condition (35), it is very difficult to suggest any valid mechanism without renormalization to eliminate ultraviolence divergences with the equal numbers of matter–antimatter.

Wilzcek [17] suggested that instead of number of virtual particles, we have to speak of the numbers of internal loops in Feynman graphs. However, instead of Feynman diagram, we suggest the energy-momentum loop. Wilczek showed that proton mass in Planck unit arises from the basic unit of color coupling strength, which is of order ½ at the Plank scale. We showed that the color coupling code ½ arises from energy-momentum exchange interaction.

15. The new theory of photon

15.1 Dual performance of photons

Quantum mechanics suggests that photons are electrically neutral and do not couple to other photons. Based on our theory, a photon in the background gauge field behaves as color charge "twin pairs" while in the local gauge field became an electrically charged virtual particle-antiparticle pair. In the local gauge field of matter space–time frame, the interaction of photon with the quarks takes place through generation of fractional charge ingredients of photon and their cross coupling with the quark charges which produces Majorana bosons of the background gauge field. Neutrinos in the local gauge field separate colors of gamma photons for generations of quarks for three fractional protons. You cannot see quarks because photon-photon cross coupling eliminates quarks, translating them to the gauge field.

Photon-antiphoton in the form of Majorana particles do not have independent existence and for conservation of energy have to generate the discrete space–time frame of baryon's matter, holding strong interactions. When an electromagnetic field is on, it keeps conservation laws in baryon structure but, when it is off, transforms conservation laws to the background energy phase of gauge field.

Photon in the gauge field is not a single boson, but it is the composite frame of neutral bosons. Invariant translation of fermions to bosons requires cutoff electromagnetic force (**Ea = 0**) where cross coupling of **Es = 1/2 Ea** shape particles to particles of the background gauge field **2Es = Ea** takes place. The color charge mass of photons of the background gauge field appears in the local gauge field in the form of space mass of fractional electric charges of baryon space–time frame.

In accordance with model (9), quanta are the energy-momentum carrying elements, and only the energy-momentum content determines the existence of photons in the form of finite amount of quanta. The portions of energy, carried by space–time portions of energy, appear with the integer numbers (10).

Without clear understanding of half spin phenomenon and the Pauli exclusion principle, we cannot describe photon as a boson. The exclusion principle states that two identical half spin carrying fermions cannot occupy the same quantum state. Pauli's "quantum state" is an abstract point-like state of a particle which does not involve the space–time frame, and his rule does not explain the fact of the existence of two same quarks in baryon's space–time frame. In this sense, Wilczek [16] also raised the question that two identical quark fermions did not appear to obey the normal rules of quantum statistics. It is difficult to understand the pattern of observed baryons using antisymmetric wave functions, as it requires symmetric wave functions.

The formula **Es = 1/2Ea** explains that space and time portions of energy in the form of particle and antiparticle discretely share the space and only half of the available energy belongs to matter's space portion. On this basis, we modified the exclusion principle to the statement that matter fermion with half spin can present only half portion of the available energy in the form of space. With multiplication of fermion space to scalar, we can produce two fermions that can occupy similar space at different times, holding **2Es = Ea** condition. The vector space of matrix **SO (3)** does multiplication of the code **Es = 1/2Ea** by two to produce two quarks, existing in opposite phases with one force carrier quark **2Es = Ea**: decay of proton's **(+2/3)** quark produces in the opposite phase two other **(−1/3)** quarks in the opposite phase. The antiquarks, following the same rule, keep the existence of baryon structure. The Pauli exclusion principle cannot predict such a translation of half spin fermions to integer spin particles, while the standard model has no cross **SU (2) x SO (3)** matrixes to carry this translation.

In a similar way we can explain why light cannot be at the same time matter and antimatter, which is the necessary condition to carry a finite amount of energy. Distribution of photon energy within space and time phase colors in the space–time frame generates fractional charges in time phase in the form of positron **(+2/3, +2/3, −1/3)** and electron **(−2/3, −2/3, +1/3)** of space phase to hold the genetic code **Es = 1/2Ea**. The quarks appear as the ingredients of photon's fractional charges within the space–time frame which carry a virtual baryon structure and the ingredients of nucleons. You cannot cut and separate fractional charges of entangled quarks into two separate species. That is the reason why the ingredients of quark-antiquark pair do not have independent existence, which is specified as *the confinement problem of quantum physics*.

Quarks in the proton-neutron frame exist in the form of fractional charges; that is why we cannot see a fractional proton or fractional neutron, but we can see a pion, which appears from doubling of photon's fractional charges. This mechanism explains the phenomenon that when the quarks of nucleon are poked by high-energy photons, the quarks show behavior as they were free particles [17]. Cross coupling of photon's quark ingredients with the second photon leads to the scaling of the genetic code **Es = 1/2 Ea** to **2Es = Ea** which generates free neutral particles. The cross coupling of fractional charges of a photon in baryon frame through **SO (3)** matrix leads to the formation of a pion—the lightest particle to produce the background gauge field boson which plays a role of a Goldstone boson. In the local gauge field, the photons, as neutrinos, became Dirac particles, while in the background gauge field, they are Majorana pairs.

The condition **Es = 1/2Ea** is the threshold energy to hold a photon within fractional electric charges of baryon frame. Photon seems to be not a fundamental unit and conserved in space–time in the form of fractional quark unit. In such a mechanism, the threshold energy is not the bound energy of electron in metal, but it is the energy required to hold **2Es = Ea** transformation of fractional electric charges which is necessary to produce integer charges. The integer electric charge is the combination of fractional charges, produced by coupling of condition **Es = 1/2Ea**:

$$\mathbf{1} = \mathbf{3/3} = (\mathbf{2/3}) + (\mathbf{2/3}) - (\mathbf{1/3});\ -\mathbf{1} = -\mathbf{3/3} = (-\mathbf{2/3}) + (-\mathbf{2/3}) - (+\mathbf{1/3}) \tag{36}$$

That is why photoemission is not a one-step process, described by Einstein's linear equation, which does not cover these steps. Generation of integer electrons depends on the energy-momentum genetic code **(Ea/Es−1)**, which determines a threefold frequency.

Model (10) suggests that Planck's emission of photons takes place only through merging of fractional charges. At **ΔS = Δt**, we can get the equation for photon radiation:

$$\frac{\mathbf{E_s}}{\mathbf{E_a}} = \frac{\frac{\mathbf{S1}}{\mathbf{t_1}}}{\frac{\mathbf{s1}}{\mathbf{t_1}} - \mathbf{1}} \tag{37}$$

Radiation takes place uniformly through the reduction of frequency by integer numbers, which describe numbers of energy portions in relation to total energy. The **Ea** in Eq. (37) presents the total numbers of elementary quanta. At Planck scale with the uniform distribution of energy in space and time phase, using condition (37), we can get the Planck formulation **Es = hν** where **h** presents the vacuum expectation value of background energy **(Ea)**.

15.2 The three fractional proton families of baryon frame

Presently there is no quantum field theory, which may include space–time as the main ingredient of strong interactions. By Weinberg's opinion [1], isospin conservation, which governs strong interactions, has nothing to do with space and time. However, without space–time, it is impossible to produce the theory of strong interactions because space–time is the matrix for flux of energy to baryon frame. Discrete conservation of energy, carried in the space–time by minimum elementary grain of space–time frame of matter (baryon frame), is the same phenomenon called strong nuclear interactions. The strong interactions arise from conservation of energy within the space–time, which has to hold basic elementary baryon's space–time unit.

If displacement in the space–time frame of baryon frame has a trend for contraction (**ΔS = 0**), the space–time frame of baryon frame disappears, and the ingredients of baryon structure became free particles (**Ea = Es**) which appear as the "asymptotic freedom phenomenon" of gauge field. By quantum physics, energy is borrowed for the generation of particles-antiparticles, but the energy, borrowed from the background gauge field, in reality is required for discrete performance of baryon's frame. According to our theory, the integer proton-neutron pair may exist only within three fractional families, with involvement of other quark flavors, existing through internal color charge interactions between them with untouched spin relations. The condition **2Es = Ea** produces all types of symmetry ($\mathbf{n}, \mathbf{\iota}, \mathbf{m_\iota}$) within three fractional proton-neutron families, but the ingredients of this symmetry have a difference only in color mass ($\mathbf{m_s}$). The proton mass does not come from quarks, but it is comprised of the energy which keeps invariant interactions of three fractional proton-neutron families. To hold color-based interactions between quark flavors of fractional protons, quarks have different colors.

Quantum mechanics suggests that isospin, which identifies proton and neutron as the different states of same particle due to the small mass difference, is an approximate symmetry. In accordance with our theory, the extra mass of neutron in comparison with proton arises from coupling of proton-antiproton pairs, which adds mass of color interactions within fractional charges. On this basis, proton and neutron are not the different states of the same particle. The neutron-antineutron pair is the different state of proton-antiproton pair.

According to the Yang-Mills theory [20], when electromagnetic interaction is neglected, the isotopic spin has no physical significance, and all physical processes would be invariant under isotopic gauge transformations. It was shown that when electromagnetic field is not involved, all interactions are invariances at all space–time points. But these statements could be true only partly because when electromagnetic field is not involved (**Ea = 0**), all transformations move to the background gauge field where space–time forms the frame of integer spin carrying particles.

In accordance with the **SO (3)** symmetry, the local **Es = 1/2Ea** and the background gauge field **2Es = Ea** require existence of **uud –ddu** proton-neutron relation within two rotations. Such an existence of quarks determines similar existence of other two fractional proton-neutron families, which occupy top-down location of **uud-udd** in an alternative mode. In this case, the color charges of quarks cancel each other. This principle explains problems, raised by Wilczek [16] who showed that it is difficult to get quark-antiquark color cancelation which needs energy. In accordance with our theory, rotation of fractional proton families' realizes charge cancelation during locating them alternatively at top-down positions.

The ingredients of background gauge field appear in the form of dark matter/dark energy, the composition of which is the same as dark energy/dark matter composition of universe. The portion of every boson in gauge field is 33%, which explains the predicted dark matter composition.

15.3 What is the Planck scale? Where did it come from?

It is well known that the Plank scale is the magnitude of space, time, and energy below which the prediction of quantum theories is no longer valid and quantum effects of gravity are expected to dominate. Planck units are derived by normalization of the numerical values of certain fundamental constants to 1: $\mathbf{c} = \hbar = h = \mathcal{E}\mathbf{o} = \mathbf{k} = \mathbf{1}$.

Planck did normalization of different constants regardless of their dimensions. However, as we showed through the example of energy-mass transformation of SR, such a normalization can be done if physical quantities were expressed with dimensionless units, which give numbers. The relation of changes to their initial value, which we applied, gives proper normalization, which is a dimensionless non-unitary operator. Model (9) describes the Planck scale as the boundary position of a particle in space–time where the change of space presents wavelength, while amplitude is the initial space locality. The conditions of model (9) $\mathbf{\Delta S = S1}$, $\mathbf{\Delta t = t_1}$ are the boundary condition for existence of the space–time which may present the Planck space and time. Model (9) describes normalization of all the dimensionless parameters to 1. The space–time triangle wave with the equal wavelength and amplitude is the Planck scale of space–time. When $\mathbf{\Delta S < S_1}$, there is the no space–time frame and strong interactions of baryon frame. This is the phenomenon called vanishing of the effective coupling at short distances.

At high-energy region, close to $\mathbf{\Delta S = 0}$, there is no consumption of energy for displacement of space which presents quarks as a point-like particle. The point-like interaction out of space–time "is equivalent to no interaction," because at point-like particles, there is no conservation of energy and there is no particle. In the similar way light cannot be identified as a point-like particle because light without emission from the space–time frame cannot exist. Without space–time with local position, energy is not conserved, and baryon structure does not exist. Therefore, without mathematics of space–time frame, we cannot explain strong interactions and "asymptotic freedom of baryon quarks at short distances."

When the local momentum merges with the initial momentum, the local position also merges with the initial position. Therefore, due to the non-separable conservation of energy-momentum in the space–time frame, momentum and position are not separable identities. The time-energy relation of the uncertainty principle involves interval of external time, which flow independently of measurement. However, in the concept of production of space–time position from energy conservation, the outcomes of the uncertainty principle probably will be different. The condition $\mathbf{Es \geq 1/2Ea}$ of the model describes the limit above which **Ea** may present the Planck scale, where the space–time and local position do not exist.

Following to the genetic code $\mathbf{Es = 1/2Ea}$, existence of position and momentum in different phases generates non-commutation of these identities.

16. Gluons

The standard model does not provide any information on gluon's origin. Based on our theory, only the origin of a particle can give information on how it will behave. This is the requirement of the causality that "past determines future."

Invariant translation from the local gauge field to the background field shows that gamma rays are the products of transformation of electromagnetic energy during decay of the space–time phase of baryon frame and within $2\mathbf{E_s} = \mathbf{E_a}$ invariance translation became vector boson of the background gauge field. The **e/e** and **ν/ν** pairs, produced simultaneously, do not have independent existence, and with gamma rays they form three-jet particles to hold $2\mathbf{E_s} = \mathbf{E_a}$ symmetry of the background gauge field.

When the flux of electromagnetic interactions to baryon structure is neglected **(Ea = 0)**, electrically charged interactions disappear, but color interactions without change are translated to the background gauge field. The color interactions in baryonic frame do not touch spin interactions of baryonic quarks but hold interactions within three fractional proton-neutron families. Therefore, the color of light photon as a variable is needed to generate translations between electrically charged spin **Es = 1/2 Ea** and color **2Es = Ea** interactions.

The color charge of quarks is required to carry interaction between fractional protons. The correct mass of proton can be calculated only from color-based interactions within fractional protons, and the theories based on a common proton-electron structure cannot produce correct proton mass. The interaction between fractional protons is spin invariant and determined by the color interactions. The six of eight gluons participate within the three fractional proton-neutron families, two between proton-antiproton-neutron-anti-neutron interactions.

The ingredients of exchange interaction in baryon's space–time frame carry three symmetric interactions. The first is the symmetric energy-momentum interaction, regulated by conservation of spin numbers **(Es = 1/2Ea)**, which takes place between quarks. The second symmetric interactions take place internally within ingredients of baryon frame, (a) the internal color-based symmetric flavor interactions within quarks (called gluons color charge interactions), which combine two symmetric internal interactions **a** = **b1**+ **b2**: (a1) the internal mass-based symmetric interactions within neutrinos and (b2) the internal mass-based symmetric interactions within electron families.

At **Ea** = **0** takes place invariant translation of color- and mass-based internal symmetric interactions to the background gauge field to hold the symmetric internal interactions within neutral electron and neutrino families. In reverse translation of energy conservation from background energy phase to local space–time phase, the color charge transforms to electric charges of quark-antiquark families. The energy inserted to quark families of baryon frame is the gluon of gamma photons from the gauge energy phase. Due to the discrete insertion of gamma photons to quark frame of baryon structure, the mass of individual quarks is very less than the proton-neutron masses.

17. New principles of quantum chromodynamics theory

The QCD theory is a non-Abelian gauge theory (Yang-Mills theory) and based on approximate **SU (3)** symmetry. Gell-Mann suggested [31] that quarks do not have space–time frame. Such an approach was the main reason for the appearance of approximate SU (3) symmetry because point-like behavior of quarks cannot carry conservation of energy. The other problem of the Gell-Mann approach was due to the application of Lagrangian continuous field, which produces approximation for perturbative theories. In addition, the theory used nonsymmetric four-momentum frame of special relativity. In accordance with our theory, without the discrete space–time symmetry, all field theories will produce approximate symmetry.

Our theory shows that quarks are not Gell-Mann's mathematical construct; they are ingredients of photon's fractional charges, distributed within the space–time frame of baryon frame. Han [32] desired to construct models in which the quarks had integer value electric charges but was not able to deliver a theory.

The other problem of the field theories is that, as Gross [28] perfectly suggested, quantum field theories do not know which field to use and cannot explain why all the hadrons, baryons, and mesons appeared to be equally fundamental. The field theories do not clarify properly the nature of gauge field and unify the four forces for description of fundamental interactions.

Model (10) combines all fundamental interactions within only electromagnetic and gravitation forces, strength of which changes with the integer numbers. The background gauge symmetry is the vacuum, which has no independent existence; that is why it mediates local gauge field. Such features of the background gauge field eliminate renormalization procedure, which is widely applied in quantum field theories. At maximum boundary energy (vacuum expectation value **Ea**), energy does not runaway to ultraviolent divergence due to translation of energy for separation of space **e/e** and time **ν/ν** spin one neutral pairs. In this case, separation of **U (1)** matrix into two symmetries takes place with the generation of space–time **SU (2)** and energy-momentum **SO (3)** matrixes. It is reduction from **2Es = Ea** background symmetry to the local gauge field symmetry **Es = 1/2 Ea** with generations of ½ spin carrying fermions and integer spin carrying photons of electromagnetic force. This invariance translation generates electromagnetic force with the positive sign **(Ea−Es)/Es.**

QCD is based partly on Poincare symmetry [33] that involves: (a) Abelian Lie group, (b) rotation in space to the non-Abelian Lie group, and (c) transformations connecting two uniformly moving bodies. However, having the excellent statements of (a) and (b), Poincare symmetry is not free from the problems due to the application of Minkowski's four-momentum space–time isometries that produces a semi-direct product of the translations. However, the statement (c) does not hold conservation of energy because it ignores boundary of motion.

The Wikipedia discussion [33] on Poincare symmetry shows that it might be possible to extend the Poincare algebra to produce super-Poincare algebra that may lead to the supersymmetry between spatial and fermionic directions. However, Poincare symmetry due to the absence of initial position cannot deliver conservation of energy at origin.

Nambu [34] suggested that the nucleon mass arises largely as self-energy of some primary fermion field, similar to the appearance of energy gap in the theory of superconductivity. According to his opinion, the nucleon mass is a manifestation of some unknown primary interaction between originally massless fermions. In addition, the pion is not the primary agent of strong interactions, and the nature of primary interaction is not clear.

In accordance with our theory, Nambu's coupling is the discrete cutoff electromagnetic energy **(Ea = 0)** to baryon space–time frame, turning fermions to bosons of the background gauge field, which performs as the superconductive medium due to the absence of fermionic space–time frame of "free" boson particles of condensate.

Our theory explains the ratio of spin constituents on the basis of ratio of transference $\boldsymbol{\sigma_T}$ and longitudinal waves $\boldsymbol{\sigma_l}$ of virtual photon ($\mathbf{R} = \boldsymbol{\sigma_T/\sigma_l}$) discussed by Gross [28]. At **Es = 1/2Ea** we get transference waves $\boldsymbol{\sigma_l} = \mathbf{0}$, while at **2Es = Ea** it transforms to longitudinal wave of virtual bosons $\boldsymbol{\sigma_T} = \mathbf{0}$. If the constituent has spin zero, the $\boldsymbol{\sigma_T}$ became zero $\boldsymbol{\sigma_T} = \mathbf{0}$, but if spin is ½ the $\boldsymbol{\sigma_l}$ became zero.

18. The triplet model of hadron particles and problems of quantum mechanics

It is necessary to note that the three particles performance of nucleons was the mystery of strong interactions.

Heisenberg [35] suggested that the proton and neutron are different states of the same particle, which should produce integer spin for the nucleon because of the addition of the angular momentum of the constituents. He called this rule addition law and suggested that full spin of the nucleon is always integer if the mass number is even; the full spin is half-integer if the mass number is odd.

Sakata suggested [36] that the even-odd rule and addition law can be applied for other particles as well. He suggested the model of hadrons, which comprised triplet of proton, neutron, and lambda, but later the quark model was suggested where triplet of **uds** quarks replaced **pnλ**. Sakata's model could not explain why hadrons should follow triplet performance of particles, and Sakata suggested that three **pnλ** particles are composite states of some hypothetical object called B matter.

The mystery of triplet particles generated significant concern for particle physics theories when in 1964 unusual decay spectrum of kaon was reported [37]. Decay of neutral kaon produced mixture of $\pi - \pi + \nu$, which by the author's opinion "no physical process would accomplish this decay and any alternative explanation of the effect requires highly nonphysical behavior of three body decay of neutral kaon." The author suggested that the presence of two-pion mode implies that the neutral kaon meson is not pure eigenstate. Such a decay process leads to the new direction of studies of particle physics, called spontaneous symmetry breaking.

The eigenvalue (12) of model (9) shows that the triplet performance of hadron holds condition 2**Es** = **Ea** and has pure eigenstate to hold symmetry. This eigenstate requires existence of symmetry of integer-half-integer particles with the condition $\pi - \pi + \nu$, which meets the requirement of eigenstate (12). Therefore, there is no symmetry breaking in kaon decay to $\pi - \pi + \nu$, and the force called weak interaction is the gravitation force which holds the existence of the nucleon in discrete symmetry within **Es** = **1/2Ea** and 2**Es** = **Ea** invariant energy translations.

Translation of local gauge symmetry **Es** = **1/2Ea** to background symmetry **2Es** = **Ea**, due to the existence of quark flavors in three families of fractional protons, requires counterpart mixing of quark flavors with generation of kaons. Flavor mixing appears through mixing of **SU (2)** and **SO (3)** matrixes. *Due to the existence of three fractional proton-neutron pairs, the formation of three kaons is the necessary condition to hold discrete symmetry.*

Kobayashi [38] showed that CP violation would occur if irreducible complex number appears in the element of mixing. By terminology, the irreducible polynomial has a meaning that it cannot be factored into the product of two nonconstant polynomials. The symmetric reduction of condition 2**Es** = **Ea** to **Es** = **1/2Ea** meets this requirement. The other condition for CP violation, as Kobayashi mentioned, is that the complex number remains in the polynomial equation, which cannot be removed by the phase adjoint of the particle state. The polynomial Eq. (32), describing mixing of **SU (2) x SO (3)** matrixes meets this requirement as well. By Kobayashi's opinion, flavor mixing arises between gauge symmetry and particle states. Kobayashi's statement is partly equivalent to our approach only while flavor mixing is the requirement of invariant translation within the background local gauge fields.

The standard model suggests that the CP violation is due to the essential difference between particles and antiparticles. Based on our theory, particles and antiparticles exist in different phases and are connected through symmetry mediator electromagnetic energy; when it is off, the antiparticle as the displacement merges with its superposition twin particle.

Formation of integer spin within proton-neutron pair in the nucleon through the addition law of proton and neutron is not possible because the proton and neutron do not exist in the same phase. The integer spin at an even mass number is described by symmetry **2Es = Ea**, while half spin at an odd mass number is expressed by the condition **Es = 1/2Ea.**

Transformation from kaons to bosons of gauge field involves the following steps: coupling of fractional proton-antiproton pairs to kaon → coupling to neutral kaon → decay to pions → coupling to neutral pions → decay of neutral pions to quark ingredients with participation of intermediate W bosons through discrete translation of **Es = 1/2Ea** and **2Es = Ea** symmetries to each other.

19. The principles of isospin symmetry

Kibble [29] showed that the proton and neutron are not identical which the reason for generation of approximate symmetry. The proton has an electric charge, but the neutron does not. On this basis, the isospin symmetry, which describes proton-neutron symmetry by **SU (2)** group, was accepted as an approximate symmetry.

The standard model suggests that while isospin is an approximate symmetry, it must be broken in some way [29]. However, the addition of symmetry breaking terms generates non-renormalizable theories, producing infinite results. Therefore, the reason why the symmetry must be broken remained a mystery of particle physics. On this basis, the standard model, avoiding the need to add explicit symmetry breaking terms, suggested spontaneous symmetry breaking [29]. However, the spontaneous symmetry breaking theory of the standard model, producing massive bosons, did not explain the main problem of isospin that generates asymmetry: why the neutron has more mass than the proton or the proton has less mass than the neutron.

In accordance with quantum mechanics, the physical situation is unchanged if the electron wave function is multiplied by a phase factor [29]. This transformation involves a constant (α) and an imaginary number. The problem of such a transformation is that the constant (α) describes space–time in exponential function without involvement of space–time variables and their boundary. The other problem of this transformation is that if the electron wave function is multiplied by the phase factor, the physical situation changes and produces different phase symmetries.

In Kibble's analysis there is one excellent statement that spontaneous symmetry breaking occurs when ground state or vacuum does not share the underlying symmetry of the theory. As we showed, the background gauge symmetry does not exist independently and exists only in conjugation with the local gauge field, which appears as invariant translation of energy from the background vacuum.

Therefore, the isospin symmetry is not related to proton-neutron translation. As we showed in this chapter, isospin between the neutron and proton exist within the translation of proton-antiproton pair to neutron-antineutron pair with rotation of local gauge field to the background gauge field. The $\mathbf{\Delta S/S_1}$ and $\mathbf{\Delta t/t_1}$ operators of model (9) within **2 x 2** non-unitary matrix have the exact **SU (2)** symmetry and carry this translation. This symmetry within the **Es = 1/2Ea** genetic code of the energy-momentum isospin symmetry generates a supersymmetry within the three particles' performance of baryon frame which exist in discrete mode in conjugation with the background vacuum.

20. Conclusion

We developed a new supersymmetric gauge field theory of photon, which describes fundamental laws of physics through invariant translation of discrete symmetries of nature. Simply, we developed a new gauge theory of photon, which describes all the fundamental laws through conjugation of the discrete space–time **SU (2)** frame and energy-momentum **SO (3)** symmetry group. At background gauge supersymmetry **Ea = 2Es**, all the forces and interactions are symmetrically entangled. Based on the theory, *gravitation appears as the short-range force, which holds discrete performance of electromagnetic field for the existence of the nucleon in discrete mode.* Nature outlined this rule to avoid approximate symmetry in its fundamental laws.

Author details

Aghaddin Mamedov
SABIC Technology Center, Sugar Land, TX, USA

*Address all correspondence to: amamedov@sabic.com; aghaddinm@gmail.com

References

[1] Weinberg S. Conceptual foundations of the unified theory of weak and electromagnetic interactions. Nobel Lecture. 1979:543-559

[2] Glashow LS. Towards a unified theory—Threads in a tapestry. Nobel Lecture. 1979:494-504

[3] Feynman RP. The development of the space-time view of quantum electrodynamics. Nobel Lecture. 1965

[4] Feynman RP. The Feynman Lectures on Physics. Vol. 1. California Institute of Technology, California: Addison Wesley; 1970

[5] Mamedov AK. The Concept of Mass based on Accelerated Conservation of Energy within Asymmetric Space-time Phases. Rijeka, Croatia: InTech; 2018

[6] Mamedov AK. Unification of Quantum Mechanics and Relativity Based on Discrete Conservation of Energy. In: The Selected Topics of Quantum Mechanics. Rijeka, Croatia: InTech; 2014

[7] Mamedov A. The Hot Disputes Related to the Generation of a Unified Theory Combining The Outcomes of ER and EPR Papers. Rijeka, Croatia: InTech; 2019

[8] Hooft G't. Confrontation with infinity. Nobel Lecture. 1999: 359-370

[9] Banyaga A. Wikipedia.org/wiki/ local_ Diffeomorphism. The structure of classical diffeomorphism groups, Mathematics and its Applications. Vol. 400. Dordrecht: Kluwer Academic; 1997. p. 200

[10] Hawking S. The Brief History of Time. USA: Bantam Dell Publishing Group; 1988

[11] Anderson PV. More is different. Science. 1972;**177**(4047):393-396

[12] Gottlieb MA. Conservation of Energy. R. Feynman Lectures (1963, 2006, 2013 editions). California: Copyright California Institute of Technology

[13] Taylor EF, Wheeler JA. Space-Time. New York: W.H.Freeman and Co.; 1992

[14] Forshaw JR, Smith AG. W. Lorentz factor—Wikipedia. Dynamics and Relativity. Wiley, UK: Manchester Physics Series; 2009

[15] Okubo S, Marshak RE, Sudarshan ECG. V−A theory and the decay of the Λ hyperon. Physics Review. 1959;**113**:944

[16] Wikiped.org/wiki Angular_ mometum

[17] Wilczek F. Asymptotic freedom: From paradox to paradigm. Nobel Lecture. 2004:100-124

[18] York HF, Moyer BJ, Bjorklund R. High energy photons from proton-nucleon collisions. Physics Review. 1949;**76**:187

[19] Yukawa H. Meson theory in its developments. Nobel Lecture. 1949: 128-134

[20] Yang CN, Mills R. Conservation of isotopic spin and isotopic gauge invariance. Physical Review. 1954; **96**(1):191-195

[21] Pauli W. In: Meyenn KV, editor. Wissenschaftlicher Briefwechsel, Vol. IV, Part II. Germany: Springer-Verlag; 1999

[22] Dirac PAM. Principles of Quantum Mechanics. International Series of

Monographs on Physics (4th ed.). Ely House, London: Oxford University Press; 1958. p. 255

[23] Press Release by Popkin G in AAAS. Einstein's 'spooky action at a distance' spotted in objects almost big enough to see. 25 April 2018

[24] Brown WA. Wikipedia.org/wiki/vector_ space. Matrices and vector spaces, New York: M. Dekker; 1991

[25] Straub WO. Weyl Spinor and Dirac's Electron Equation. Pasadena, California; 2005. Available from: www.weylmann.com

[26] Cheng T-P, Li L-F. Gauge Theory of Elementary Particle Physics. Oxford University Press; 1983

[27] Wilczek F. A physicist crystallized in time. Press release in AAS. Posted by Gabriel Porkin; 25 April 2018

[28] Gross DJ. The discovery of asymptotic freedom and the emergence of QCD. Nobel Lecture. 2004:59-82

[29] Kibble TWB. History of electroweak symmetry breaking. Journal of Physics: Conference Series. 2015;**626**:012001

[30] Adams JF. Wikipedia.org./wiki/Lie_ group. Lectures on Lie Groups, Chicago Lectures in Mathematics. Chicago: University of Chicago Press; 1969

[31] Gell-Mann M, Neeman Y. The Eightfold Way. New York: W.A. Benjamin Inc.; 1964

[32] Han MY, Nambu Y. Three-triplet model with double SU(3) symmetry. Physics Review. 1965;**139B**

[33] Shifman M. WQCD Vacuum_ Wikipedia. Lessons for QCD from supersymmetry Nuclear Physics B - Proceedings Supplements. Vol. 108. 2002. p. 29

[34] Nambu N, Jona-Lasino G. Dynamical model of elementary particle based on an analogy with superconductivity. Physics Review. 1961;**122**(1):345

[35] Heisenberg W. Über den Bau der Atomkerne. Zeitschrift für Physik. 1932;**77**:1

[36] Sakata S. On a composite model for the new particles. Progress in Theoretical Physics. 1956;**16**:686

[37] Christenson JH, Cronin JW, Fitch VL, Turlay R. Evidence for the 2π decay of the K20 Meson. Physical Review Letters. 1964;**13**:-138

[38] Kobayashi M, Maskawa T. CP violation in the renormalizable theory of weak interactions. Progress in Theoretical Physics. 1973;**49**:652

Chapter 12

Realization of the Quantum Confinement

Eugen M. Sheregii

Abstract

In this chapter the three main technologies are described, which allows for the implementation of quantum structures (QS)—quantum wells (QWs) and hetero-structures. These are liquid phase epitaxy (LPE), molecular beam epitaxy (MBE), and metal-organic chemical vapor deposition (MOCVD). The most important properties, including the quantum Hall effect (QHE), of two-dimensional electron gas (2DEG) arising in a heterojunction on the boundary of two phases—the so-called interface—are also presented. The 2DEG properties in different kinds of QW are described. Double quantum wells as interesting example of quantum structure is considered also including such a spectacular quantum-mechanical phenomenon as splitting into symmetrical and anti-symmetrical states.

Keywords: hetero-structure, liquid phase epitaxy, molecular beam epitaxy, interface, single quantum well, quantum well, electron transport in quantum structures, two-dimensional electron gas (2DEG), quantum Hall effect, Shubnikov-de Haas effect, high electron mobility transistors

1. Introduction

The entrapping of electrons in an infinite quantum well (QW) is one of the basic issues of quantum mechanics showing its difference from classical mechanics. In fact, nature has given us a natural quantum well—atom. Coulomb's potential of the atomic nucleus creates the edges of this well (see **Figure 1**)—a very narrow well, about 1 Å (10^{-10} m) width. As it was shown in previous chapters, in such a narrow well the electron can occupy only certain energy states—discrete and not continuous energy values—as it is in the macro-world. This was indicated by the linear emission spectra of atoms discovered at the end of the nineteenth century. Their interpretation forced Niels Bohr to introduce discrete electronic states, so alien to classical physics, into the historically first quantum atom theory, which is familiar from the course of high school physics.

However, this rectangular quantum well, which is the subject of students' exercises in quantum mechanics course, until the early 1980s was a theoretical issue. As will be shown in the next paragraphs, the progress of semiconductor technology, particularly the development of the *molecular beam epitaxy* (MBE), allowed the production of hetero-structures with *very sharp interface* (transition between two material phases) and later also quantum wells with width less than 100 angstroms and with finite potential edges, which changed the situation cardinally: the issue of two-dimensional electron gas (2DEG) appeared and the quantum Hall effect (QHE)

IntechOpen

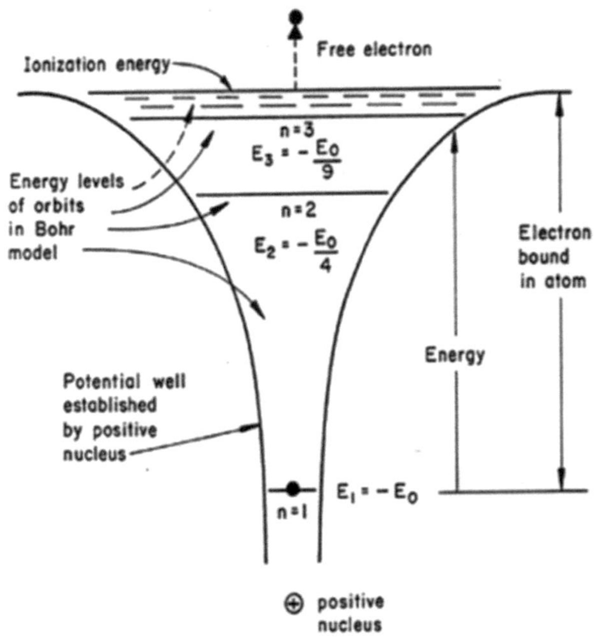

Figure 1.
Natural quantum well of the hydrogen atom created by the potential of the atomic nucleus E = −e/r.

was discovered—a qualitatively new phenomenon. In this manner, quantum wells are soluble models and have provided tests for quantum theory. Also, the applications appeared very quickly—already in the 80-ch formed hetero-lasers—the first solid-state lasers, and transistors on hetero-structures sizes that do not exceed of 100 nm, which led to the production of modern micro-processors with a packing density of 20,000 transistors in a spatial centimeter.

In this way quantum mechanics contributed to the emergence of the third industrial revolution—electronics and radio-communications—as well as the promised fourth one, computerization (without microprocessors would be impossible) and global communication network, the Internet, which without semiconductor lasers would not have been created either. To achieve this duty, it was necessary to develop appropriate technologies. The first was *liquid phase epitaxy* (LPE).

2. LPE technology and the hetero-structure production

LPE is the deposition from a liquid phase (a solution or melt) of a thin monocrystalline layer which is isostructural to the crystal of the substrate [1]. For the production of hetero-structures, the LPE was first used by Zhores Alferov with colleagues at the Ioffe Physical and Technical Institute in St. Petersburg [2]. They produced the hetero-structures based on the GaAs/AlGaAs n-p heterojunctions (unlike the usual p-n junction, which can be called a homo-junction) [3]. The zone scheme of such p-n heterojunction is presented in **Figure 2**. This diagram clearly shows that a quantum well forms at the interface in the conduction band from the side of GaAs, i.e., a semiconductor with a smaller energy gap (about 1.4 eV). In the case of the solid solution $Al_{0.3}Ga_{0.7}As$ that is 1.9 eV, QW is created by the discontinuity in the conduction band profile as a function of distance *x*. The discontinuity takes place in the case of the valence band too, and it manifests itself as a leap called

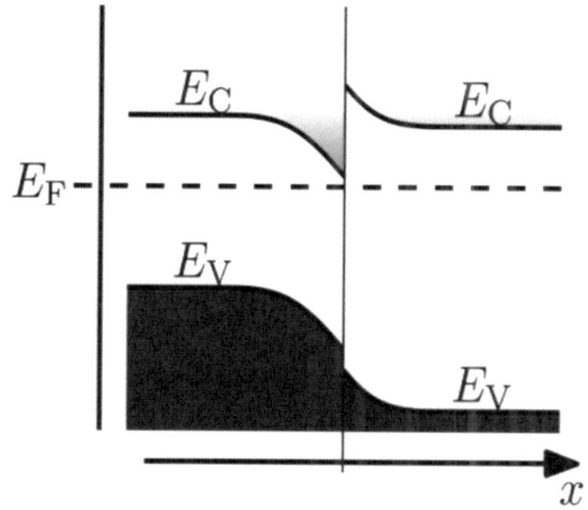

Figure 2.
Energy band diagram for the GaAs/AlGaAs heterojunction point QW, which would mean that the well is filled with electrons.

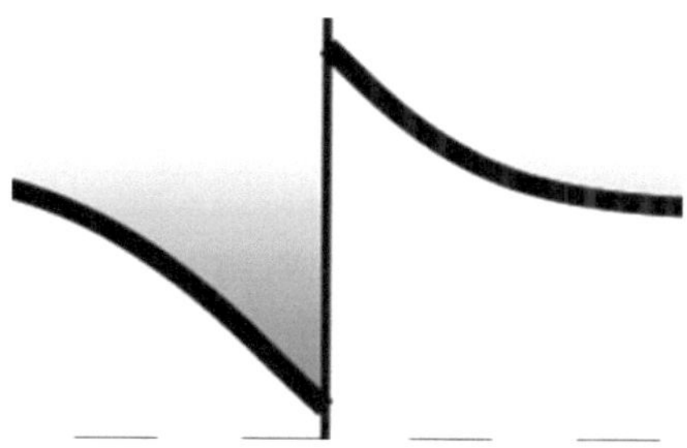

Figure 3.
Triangular QW formed in the conduction band at the heterojunction GaAs/AlGaAs shown in ***Figure 2****.*

offset. The QW exists only for electrons, and if the electron concentration increases, the Fermi level (FL) moves upwardly, and the conduction band will intersect at the QW, shown enlarged separately in **Figure 3**. You can see that it is a QW with a triangular shape. It should be noted that the shape of QW takes place also in the metal-oxide semiconductor (MOS) structures. Energy states for electrons in the real QW for the $Al_{0.3}Ga_{0.7}As$/GaAs junction were calculated by Zawadzki and Pfeffer [4]: the well depth is about 500 meV, and the resonance states occur at 200 meV and 360 meV from the bottom of the well. According to these calculations, the width of the well is about 200 Å or 0.02 μm. This fact explains why heterostructures with QW visible in the experiment could not have been obtained earlier using known crystal growth technologies and obtaining the p-n junctions by diffusion method. These methods did not allow for such a required change in the composition over several crystal lattice parameters. LPE methods have the advantage that with relative simplicity, a liquid AlGaAs solution with the necessary composition is poured onto a previously prepared (well-polished and heated to a temperature of about 600°C) GaAs substrate and the substrate will not melt during crystallization. Also, the diffusion of atoms is too slow for them to penetrate into the solid phase. Thanks to this, the required sharpness of the transition (junction) is preserved. However, the thickness of the AlGaAs layer should not exceed 1–2 μm. The last limitation is related to the upper layer stresses, resulting from incompatibility of the crystal lattice parameters of the substrate and the applied layer—so-called the lattice mismatch—minimal in the case of the GaAs and the AlGaAs solid solution with 30% AlAs as it is only 0.01 Å (4.65 Å for GaAs and 4.66 Å for $Al_{0.3}Ga_{0.7}As$) [4]. The first semiconductor lasers were produced thanks to LPE technology developed for the GaAs/AlGaAs heterojunction in the early 1970s [3]. However, more advanced technology was needed to improve the production of the heterojunction lasers.

3. MBE technology and the production of the solid-state QW

3.1 Description of the MBE technology

The width of the quantum well in the case of GaAs/$Al_{0.3}Ga_{0.7}As$ heterojunction is on the order of 150–200 Å, and the technology capabilities of LPE technology are on the border of these requirements to keep the production of devices based on them. For this reason, in the 1970s, a fundamentally new technology was developed that allowed a significant leap in the development of the semiconductor devices as well as the solid-state physics, generally. The MBE technology is based on the method of the crystal growth from the gas phase, but the use of computers made it possible to achieve precision previously unattainable [5]. First of all, it concerns the composition control (the composition control is so closely that practically every atom deposited on the substrate is calculated) but it is also the substrate temperature is much lower (450°C) than in the LPE method what is important because it reduces the diffusion intensity of atoms and has significantly improved the quality of the interface. But on the other hand, it is an expensive technology because it requires a high vacuum—10^{-11} Torr—which must be sustained continuously over several years. On the other hand, this extraordinary high vacuum allows the use of mass spectroscopy in the reactor for precise control of the composition and existing impurities. It should be recalled that the intrinsic properties of semiconductor materials were achieved only after chemists learned to clean the input materials from impurities at a concentration level of 10^{-12} cm^{-3}, which in turn means chemical purity 99.9999999%. Achieving such chemical purity of input materials requires huge amounts of labor and energy. The use of a high vacuum of the order of 10^{-11} Torr means additional "dilution" in the dopant concentration in reactor, which allowed the use of input materials in the effusors—sources of elements in the MBE machine—with a chemical purity lower by one row: 99,999999%.

The MBE process was noticed in the late 1970s at Bell Telephone Laboratories by Arthur and LePore [6]. But, the main role of this method has become the production of quantum structures (QS) from the 1980s [7] and above all—heterostructures and quantum wells.

Another technology that also relates to high tech is the metal-organic chemical vapor deposition (MOCVD) in some ways competitive to MBE because it allows obtaining high-quality quantum structures also.

3.2 MOCVD technology

The MOCVD involves the use of gases—carriers of elements used in QS built from GaAs, AlGaAs, InGaAs, and others. We call these gases metal-organic, for example, three-methyl-gal ($Ga(CH_3)_3$), three-methyl-aluminum ($Al(CH_3)_3$), or three-hydrogen of arsenic (AsH_3). These substances are contained in bottles in a liquid state at about – 60°C, and are admitted as gases (still cool) to a reactor where the touching surface of the substrate at 500°C to immediately distributed to the constituent elements and relatively heavy metals as Al, As, Ga deposited on the substrate surface at this time how much lighter C and H are pumped out of the

DOI: http://dx.doi.org/10.5772/intechopen.93112

reactor. MOCVD technology is much cheaper to operate (no vacuum) and is used as industrial manufacturing technology for QS and devices on their base.

3.3 Two-dimensional electron gas (2DEG)

The improved interface quality of hetero-structures through the use of MBE technology has led to the discovery of the unique properties of two-dimensional electron gas. The point is that the quantum well, which is located at the interface in the conduction band, naturally fills with electrons. The electrons are located in a layer with a thickness less than 200 Å. This means that they are actually in a plane that adheres to the interface (parallel to the interface) with a negligible thickness compared to two other dimensions. That is, a two-dimensional electron gas is created at the interface, which we will denote as 2DEG.

One of the basic properties of 2DEG is that electrons occupy one of the energy states of the quantum well at the interface. We will call this state as the energy sub-band and the dispersion law—energy dependence from quasi-momentum $\mathcal{E}(k)$ for 2DEG can be written in the case where the interface plane is the plane (yz) and x—the direction of growth of the hetero-structure layers (as it is shown in **Figure 2**), in the following way:

$$E(k) = E_i + \hbar^2 \frac{k_z^2 + k_y^2}{2m^*} \tag{1}$$

where i is 1, 2, 3, ... number of the sub-band, $\hbar = h/2\pi$ the Planck constant, and $\mathcal{E}_i$ is the energy value of the sub-band i, m^* effective mass of electrons.

It is obvious that the dispersion law (1) is a consequence of the restriction of movement in the x direction, in other words, by the quantum confinement, which causes the energy quantization.

Graphically the expression (1) is presented in **Figure 4**. The parabolic sub-band corresponds to each value of i.

The density of states function shown in **Figure 5** corresponds to such a law of dispersion. In contrast to the bulk material where the function of density of states is proportional to $\sqrt{E}$, in the 2D case we have steps corresponding to each value of E_i: $D(E) = i\frac{m^*}{\pi\hbar^2}$, where m* is the effective mass of electrons.

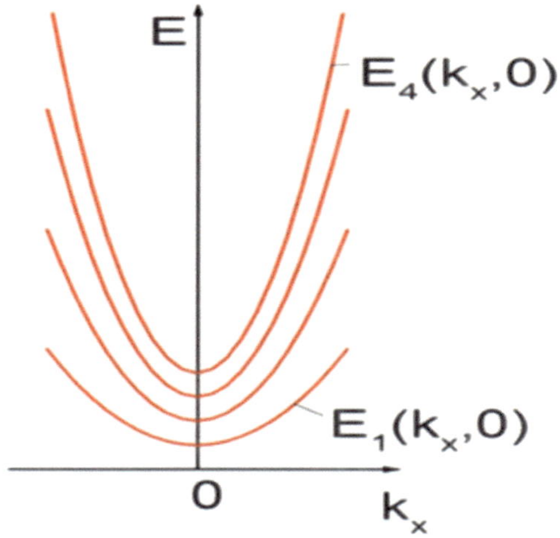

Figure 4.
The energy sub-bands corresponding to Eq. (1) where k_y = 0.

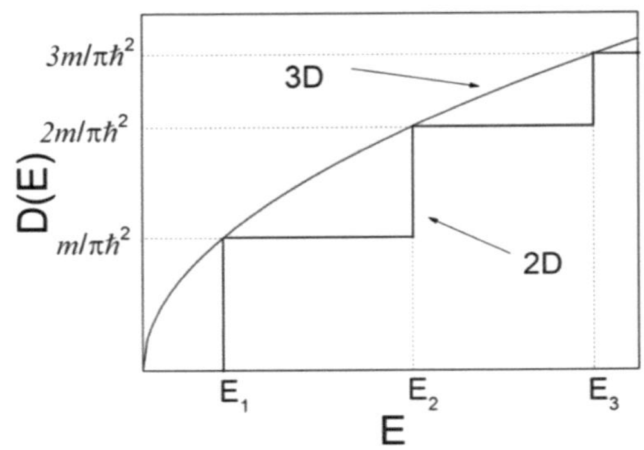

Figure 5.
The function of the state density for 2DEG in a quantum well.

This stepped nature of the function of the state density for 2DEG in a quantum well is manifested in a multitude of phenomena including the dependence of the current through the heterojunction on the gate voltage, on which the transistor operated on the GaAs/AlGaAs hetero-structure is based, the so-called high electron mobility transistor (HEMT) [8].

3.4 Quantum Hall effect

The most spectacular expression of 2DEG in QS is quantum Hall effect discovered by von Klitzing [9] in 1980. We can say that QHE is a manifestation of quantum mechanics on macroscopic scales [10].

Experimentally, QHE shows the remarkable transport data as it is shown in **Figure 6** for a real device in the quantum Hall regime which is the same as in classical Hall effect when magnetic field $\boldsymbol{B}$ is perpendicular to the plane of the sample xy and to the current I directed along the x-axis. Then, in the direction perpendicular to the movement of the charges (electrons), an additional transverse voltage is created, called the Hall voltage U_H. In classical Hall effect, the Hall resistance R_H is simply a linear function of magnetic field and resistivity also $\rho_{xy} \sim$ B. In QHE we see a series of the so-called Hall plateaus in which ρ_{xy} is a universal constant

$$\rho_{xy} = \frac{1}{\nu}\frac{h}{e^2} \tag{2}$$

(where e is the electron charge and ν = 1,2, ... an integer which means the number of the states occupied by electrons under the Fermi level and is called as *filling factor*) independent of all microscopic details (including the precise value of the magnetic field). Associated with each of these plateaus is a dramatic decrease in the dissipative resistivity $\rho_{xx} \rightarrow 0$ which drops as much as 13 orders of magnitude in the plateau regions.

QHE is a two-dimensional phenomenon because when the magnetic field B is perpendicular to the plane of the hetero-structure, the movement of the electrons in the plane of the quantum well is completely quantized. This quantization is universal and independent of all microscopic details such as the type of semiconductor material, the purity of the sample, the precise value of the magnetic field, and so forth. The growth of the magnetic field causes an increase of the distance between Landau levels, $\hbar\omega_c = eB/m^*$, and when the Fermi level is located between Landau levels then, for the electrons occupying the Fermi level, there are no states to

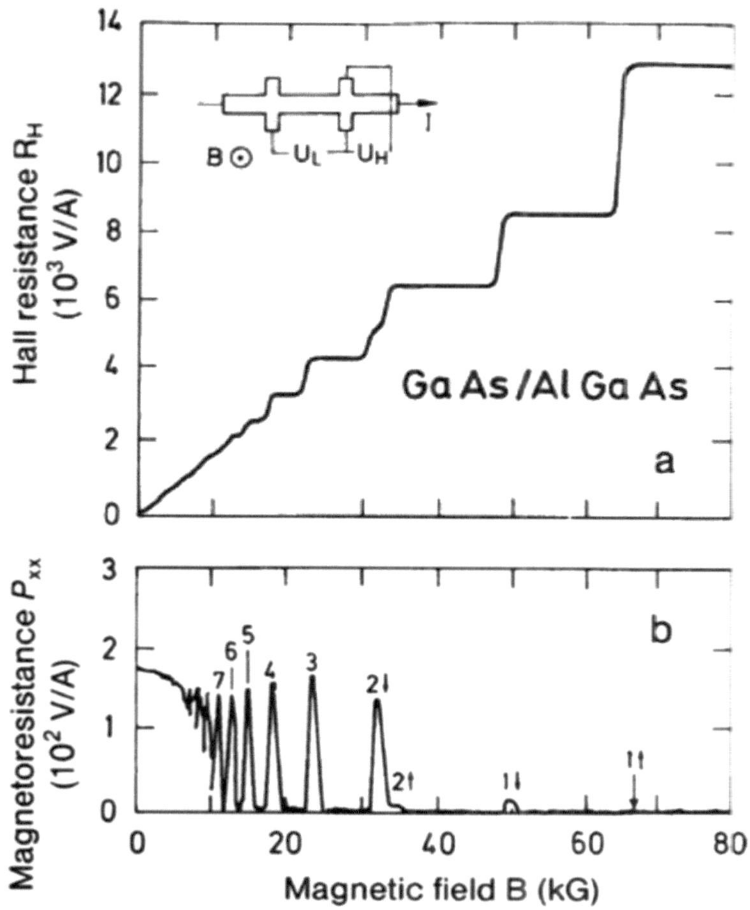

Figure 6.
(a) QHE (the Hall resistance R_H *as function of magnetic field* B*); (b) the Shubnikov-de Haas oscillations (magnetoresistance* $\rho_{xx}(B)$*) for hetero-structure GaAs/AlGaAs [11].*

dissipate, and it cannot move in the electric field—no electric current exists and ρ_{xx} is zero. In this situation the Hall voltage is constant until the Fermi level does not reach the next Landau level, and the next step of the Hall voltage takes place (in the case of bulk material, there are always scattering channels causing the presence of electric current in a strong magnetic field, therefore the dependence of the Hall voltage on the magnetic field is continuous and reflects the continuity of the density function of states (see **Figure 5**)).

As a result, the QHE is now used to maintain the standard of electrical resistance by metrology laboratories globally.

The magnitude h/e^2 = 25,812,80 Ω is so important as constant of the fine structure in the quantum electrodynamics.

3.5 Quasi-rectangular quantum well

To obtain a quantum well with a rectangular shape, it should be placed close enough to two heterojunctions as shown in **Figure 7**. How close? The experiment shows that at a 200 nm distance, two GaAs/AlGaAs heterojunctions exhibit the properties of a rectangular QW [11]. It can be seen from **Figure 7** that these two

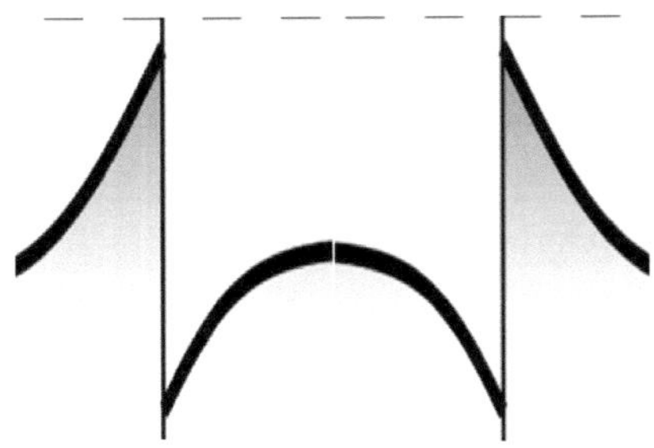

Figure 7.
QW formed from two heterojunctions.

heterojunctions must be a mirror image of each other: first is the growth of the GaAs layer—the QW—and next is of the AlGaAs layer, the *barrier* for QW. It is clear that such a well has a form still far from a rectangular well, but it is already known how to achieve the form of a rectangular potential: stretch the middle GaAs layer as much as possible. In this case we would have a very wide quantum well.

But there is another way of special engineering allowing to obtain a real rectangular QW considered in Section 3.7. Modeling of quasi-rectangular QW using heterojunctions InGaAs/InAlGaAs will be shown in the next paragraph.

3.6 Modeling of quasi-rectangular QW based on the InGaAs/InAlGaAs heterojunctions

Heterojunctions InGaAs/InAlGaAs are important advantage in comparison with GaAs/AlGaAs because low effective mass of electrons—adding the InAs to the QW material, i.e., to GaAs—allows effective mass to be significantly reduced from 0.65 m_0 for GaAs to even 0.4m_0 for InGaAs with 65% of the InAs. This means that the electron mobility is almost doubled, which is the main goal of the HEMT modeling. On the other hand, the composition of two solid solutions—$In_xGa_{1-x}As$ for QW and $In_yAl_{1-y}As$ for a barrier—can be selected so as to minimize mismatch of the lattice parameters. Such hetero-structures are isomorphic. In this way, the issue was the production of QW for HEMT based on isomorphic hetero-structures. That could be used in industry, so the production technology also had to be industrial. To implement this duty, the MOCVD technology was developed at the Institute of Electronic Materials Technology (ITME) in Warsaw for the production of isomorphic hetero-structures based on InGaAs/InAlAs heterojunctions. The structures are consisted from single $In_xGa_{1-x}As$ QW and from the two $In_yAl_{1-y}As$ layers—barriers. Four types (see **Table 1**) of different forms of structures with a single QW (SQW) were produced by MOCVD on semi-insulating GaAs at ITME by W. Strupiński group and tested at the Center for Microelectronics and Nanotechnology at the University of Rzeszów during the years 2005–2014 [12, 13]. After that, the program of producing double quantum wells (DQW) and multiple quantum wells (MQW) was developed in years 2015–2018 [14–16].

3.6.1 SQWs

In **Figure 8**, cross section of SQWs obtained by MOCVD on semi-insulated GaAs substrates is shown. If the δ-doping layer with Si is at the top above QW and below at QW then, the shape of the QW is symmetrical as in **Figure 9**, if and only at the top, a QW is asymmetric as in **Figure 10**.

Sample	Channel parameters			δ-doping donor concentration (10^{12} cm^{-2})
	Composition of In (%)	Thickness (nm)	QW profile	
1093	75	20	Sharp interface	2.5
1098	65	20	Changing composition in a channel	5.0
1607	65	23.5	Sharp interface	3.5
1088	53	20	Sharp interface	0.7

Table 1.
Parameters of the channels and descriptions of the interfaces for the SQWs.

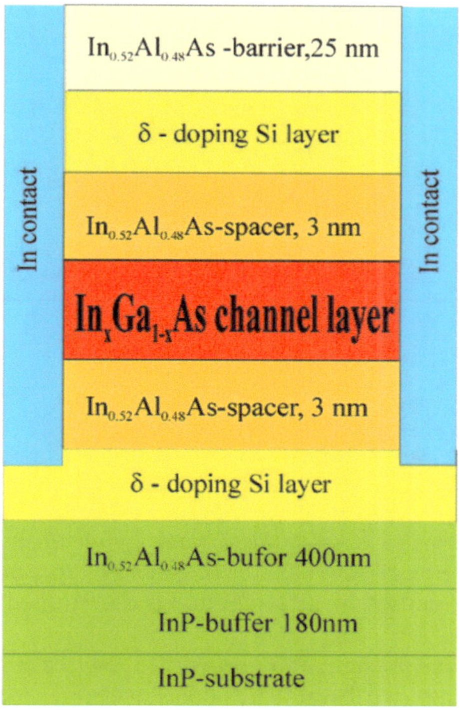

Figure 8.
The cross section of the SQW grown by the MOCVD.

The magneto-transport measurements, i.e., Hall's resistance curves R_H (B) or R_{xy} (B) and longitudinal magnetoresistance ρ_{xx} (B), were performed for all the presented SQWs. It is clearly seen in **Figure 9** the plateau on the curve R_{xy} (B) corresponding to the filling factors ν = 3, 6, 8, etc. Explanation of this values of filling factor is presented in **Figure 10** where the curve of R_{xy} (B) as well as R_{xx} (B) is interpreted.

In order to interpret the curves R_{xy} (B) and R_{xx} (B) for the QW 1088 which is practically a triangle QW (the electrons are located in the bottom left triangle), the theory developed by W. Zawadzki [17] was used

$$(a+b)a^{1/2}b^{1/2} + (b-a)^2 \ln\left|\frac{b^{1/2}-a^{1/2}}{(b-a)^{1/2}}\right| = \left[\frac{E_g^*}{2m_c^*}\right]^{1/2} \times 4eF\hbar\pi(i+3/4) \tag{3}$$

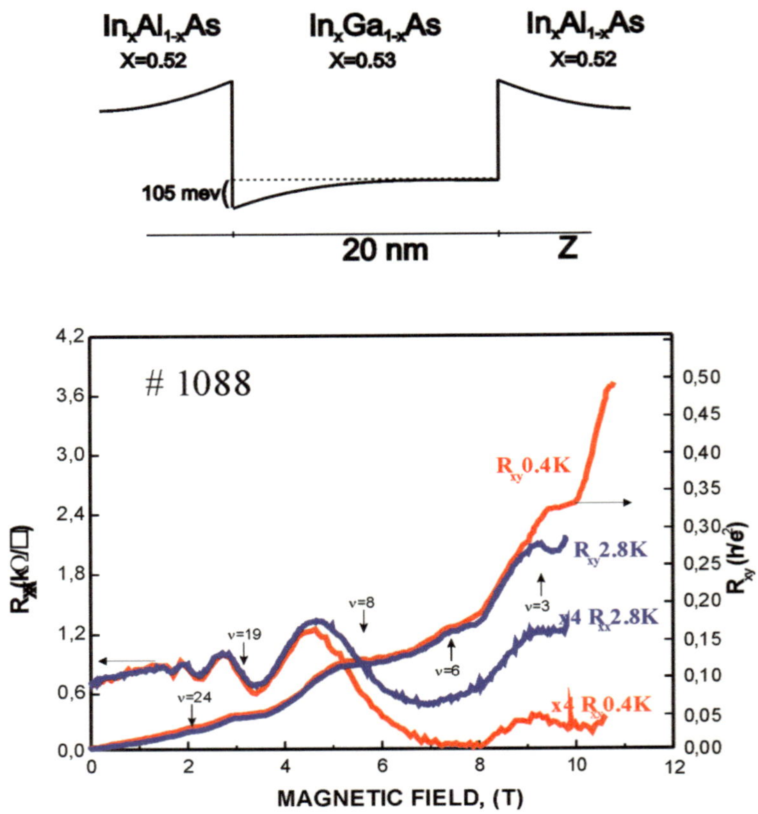

Figure 9.
*The Hall resistivity curve R_{xy} (B) and longitudinal magnetoresistivity curve R_{xx} (B) for the SQW 1088 (see **Table 1**) with asymmetric shape of QW [12, 13].*

where $a = E - E\perp$; $b = E_g + E + E\perp$; E is the energy of sub-band in QW, $E\perp$ is the Landau level energy sought, E_g is the energy gap, and F is the electrical field strength caused by interface and is determined by linear potential $U = eFz$.

Results of calculations presented in **Figure 10** show that the intersection of Landau levels of two energy sub-subbands takes place; hence the picture of QHE and SdH oscillations is more complicated but is perfectly explained by the theory for the triangle QW.

3.7 Special engineering of a rectangular QW

The special engineering of QW involves changing the composition of the solid solution in the well to compensate for the reduction in potential at the left and right corner of the bottom of the well. The schema of such compensation is shown in **Figure 11**: there is a change in the composition in the quantum well from the left side of the interface and the right side too. This mild change from x = 0.53 to 0.65 (on the left and vice versa from 0.65 to 0.53 on the right) accurately compensates for the value of the energy gap, as well as the decrease in the bottom of the well—the conduction band—so that it becomes almost flat.

This fact that we are dealing with an excellent rectangular quantum well confirms the experimental magneto-transport curves obtained for QW 1098.

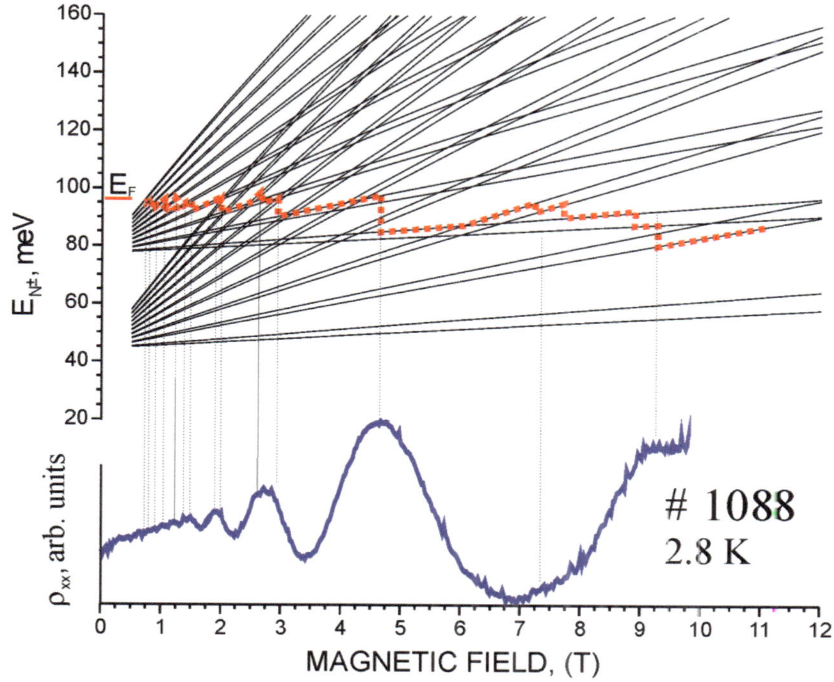

Figure 10.
Interpretation of the QHE curve and magnetoresistance curve for three-angle SQW 1088 [12, 13].

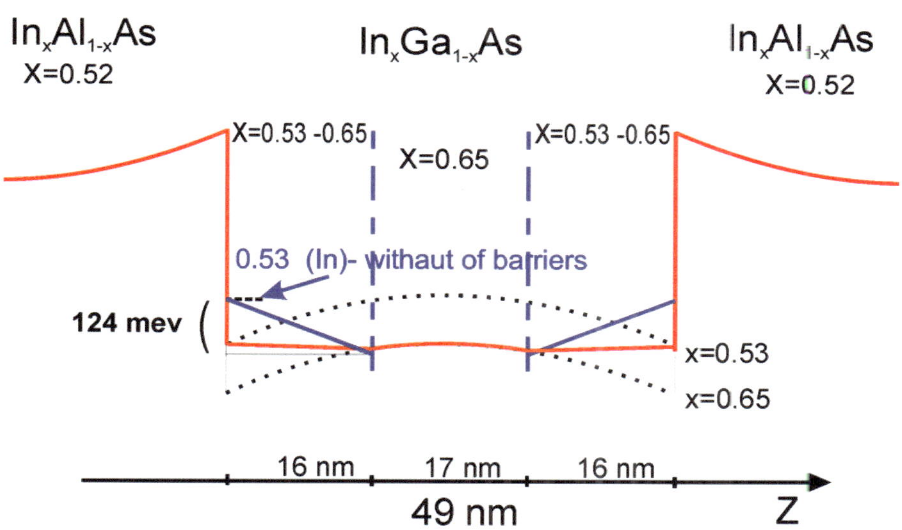

Figure 11.
Schema of the rectangular QW (1098) engineering [13].

The theoretical interpretation of experimental curves presented in **Figure 12** was performed by curves of the Landau level (LL) (presented above) calculated according the theory of Zawadzki [17]:

$$\frac{(E - E_{\perp})(E_g + E + E_{\perp})}{E_g} = \frac{\hbar^2 \pi^2 (i+1)^2}{2m_0^* a^2 k} \tag{4}$$

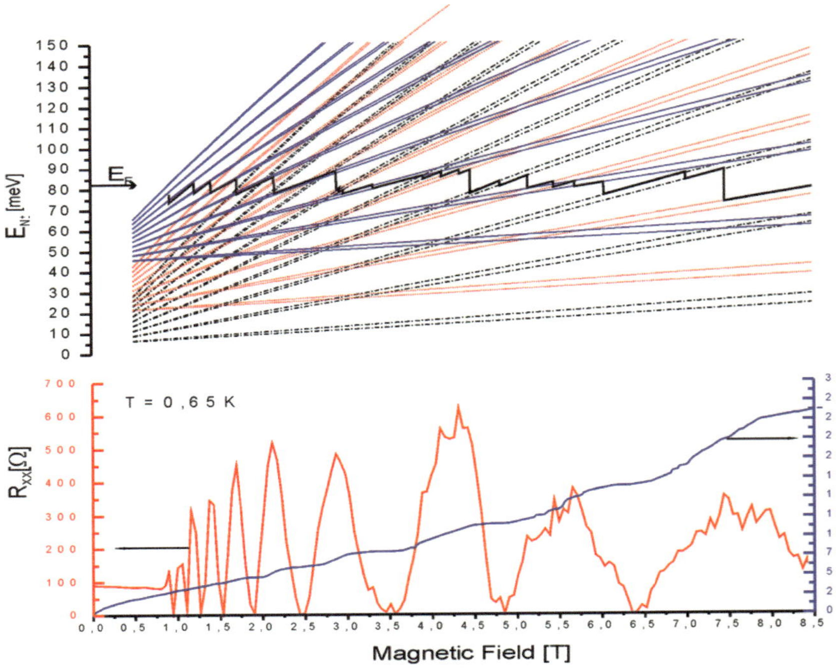

Figure 12.
The Hall resistivity curve R_{xy} (B) and longitudinal magnetoresistivity curve R_{xx} (B) for the SQW 1098 [13].

$$E_{\perp} = -\frac{E_g}{2} + \frac{E_g}{2}\sqrt{1 + \frac{4\mu_B B}{E_g}\left[f_1 \frac{m_0}{m_c^*}\left(n + \frac{1}{2}\right) \pm \frac{1}{2} g_0^* f_2\right]} \quad (5)$$

$$f_1 = \frac{(E_g + \Delta)\left(E_{\perp} + E_g \frac{2}{3}\Delta\right)}{\left(E_g + \frac{2}{3}\Delta\right)\left(E_{\perp} + E_g + \Delta\right)} f_2 = \frac{E_g + \frac{2}{3}\Delta}{E_{\perp} + E_g + \Delta} \quad (6)$$

where E is the energy of sub-band in QW, $E_{\perp}$ is the Landau level energy sought, and E_g is the energy gap, while Δ is the value of the spin-orbit splitting, i is the number of sub-band, n is the number of LL, μ_B is the Bohr magneton, and m_c^* is the effective mass of electrons on the bottom of the conduction band. As you can see in the right side of Eq. (4), the energies of states in a rectangular well with a correction for the finite potential through coefficient k are described.

It is seen that theoretical curve of the Fermi level in the course of the magnetic field reflected both the plateau of the R_{xr}(B) and the maxima of the R_{xx}(B) experimental curves: the QHE plateau positions correspond to the FL positions between LL that simultaneously correspond to the minima of the SdH oscillations.

In this way, it can be said that thanks to special engineering, it has been possible to make a *real rectangular potential of QW described by quantum-mechanical theory*.

3.8 Double quantum well

3.8.1 The SAS-splitting

Technology successes have allowed us to experimentally confirm one interesting quantum-mechanical phenomenon—it concerns the splitting into symmetrical and anti-symmetrical states thanks to the *Pauli exclusion principle*, in other words, *exchange interaction*.

Pauli's principle was known in relation to atoms, molecules, and crystal theory, while for the first time an artificial object was generated in which this principle was spectacularly confirmed—in an electron system consisting of two closely spaced QWs. In the inset of **Figure 13**, the potential profile of two QWs with narrow barrier between wells is shown. Due to narrow barrier, the tunneling among QWs is facilitated and electrons in these two QWs constitute the common electron system. This system is subject to Pauli's principle, as a result of which there are electron states in which the spin part of the wave function has the opposite sign—symmetrical and anti-symmetrical functions and correspondently symmetrical and anti-symmetrical states—separated by the energy gap, the so-called SAS gap.

3.8.2 Magneto-transport phenomena

For first time, this effect was considered in the work of G. S. Boebinger et al. [18] where the GaAs/AlGaAs DQWs produced by MBE technology were investigated. This fact was observed experimentally on the QHE curves: where quantum Hall states at odd integer ν (filling factor) were missing, the ν = odd quantum Hall states originate from the SAS gap [18].

Magneto-transport phenomena were studied also for the InGaAs/InAlAs DQWs. In addition to QHE and SdH oscillation, magneto-phonon resonance was also studied and interpreted using the LL energy theory for the DQW [19]:

$$E_{nj} = \left(n - \frac{1}{2}\right)\hbar\omega_{cj} + E_j + V_{nj}^F \tag{7}$$

$$V_{nj}^F = -\frac{1}{2\pi}\sum_{n'j'}\int_0^{+\infty} dq_{xy}q_{xy} \times \left\{\frac{1}{\exp\left\{\left[E_{n'j'} - \mu_c\right]/k_B T\right\} + 1}\right\} \times \left|A_{n'n}\left(q_{xy}\right)\right|^2 V_{jj'}^F\left(q_{xy}\right) \tag{8}$$

where j is number of the energy sub-band, n is number of the LL, $\left|A_{n\,n}\left(q_{xy}\right)\right|^2$ is matrix element for two Landau levels n and n', $V_{jj'}^F\left(q_{xy}\right)$ is factor of screening [19].

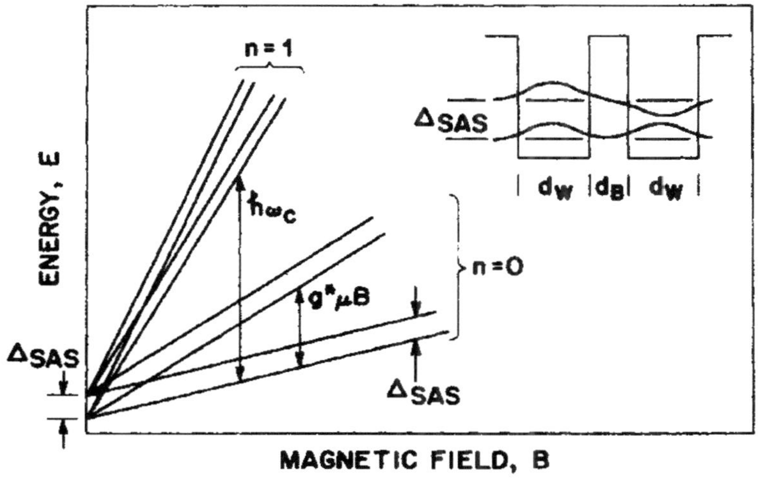

Figure 13.
Three kinds of splitting of energy states in DQW: cyclotron $\hbar\omega_c$, spin splitting $g^*\mu_B B$, and Δ_{SAS}—splitting on the symmetric and antisymmetric states.

The combination of this DQW theory with the Landau level theory presented above for SQW (Eq. (4)) gives us the following equation:

$$\frac{(E - E_{\perp})(E_g + E + E_{\perp})}{E_g} = \frac{\hbar^2 \pi^2 (i+1)^2}{2m_0^* a^2 k} \pm (\Delta_0 + 0.38 E_{\perp}) \tag{9}$$

Adding Eqs. (5) and (6) to this (9) allows the calculation of the LL energy in DQW. The value of Δ_{SAS} in Eq. (9) depends from magnetic field B as function of energy $E_{\perp}$:

$$\Delta_{SAS} = \Delta_0 + 0.38\ E_{\perp}. \tag{10}$$

where Δ_0 is the Δ_{SAS} value without magnetic field.

In **Figure 14**, the LL energies for DQW 2506 (see **Table 1**) and interpretation of the R_{xy} curve obtained for this DQW are presented. The splits caused by the SAS gap are clearly seen on the R_{xx} (B) curve. These are experimental data that indirectly indicate the SAS-splitting in DQW. But on the same DQW it was possible to make optical measurements from which the energy states were directly determined.

These optical measurements that concern the optical reflection in the infrared region were made using infrared microscope.

3.8.3 Direct determination of the energy states in the DQW

The experiment on the infrared reflection was performed at the Frascati National Laboratory in Italy. Synchrotron radiation served the brilliant infrared radiation source.

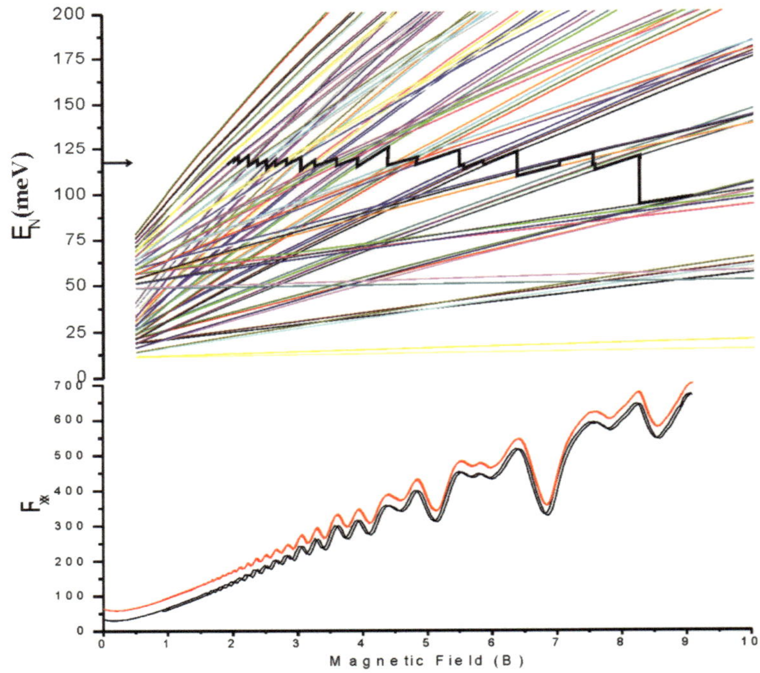

Figure 14.
Interpretation of Rxy (B) curve for DQW 2506 [15].

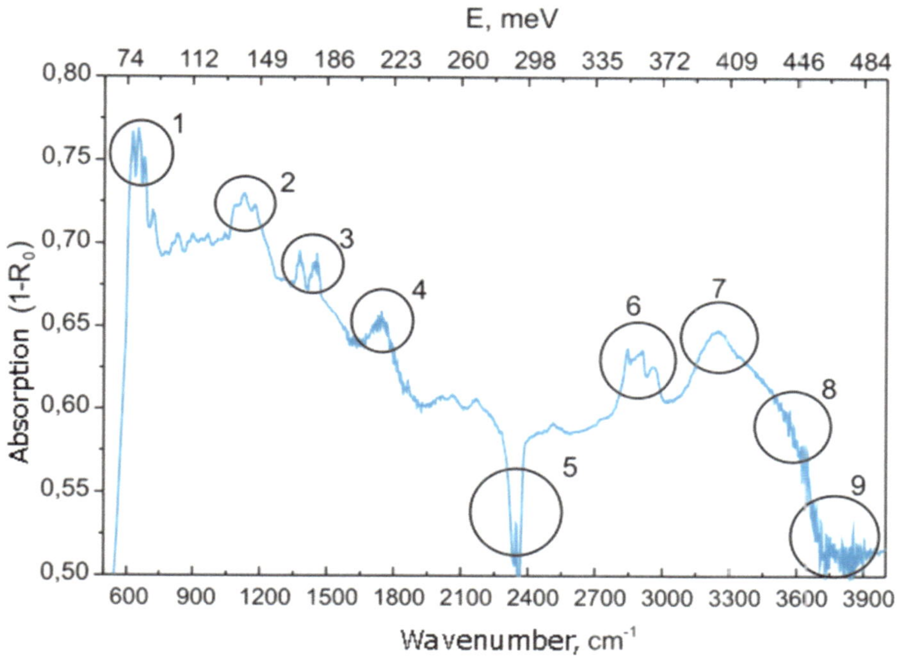

Figure 15.
Optical absorption curve obtained by averaging the optical reflection $\mathbf{R_0}$ *when scanning the sample surface 2506. The recorded absorption bands are renumbered (the double minimum 5 is due to the strong absorption of* CO_2 *in the atmosphere) [16].*

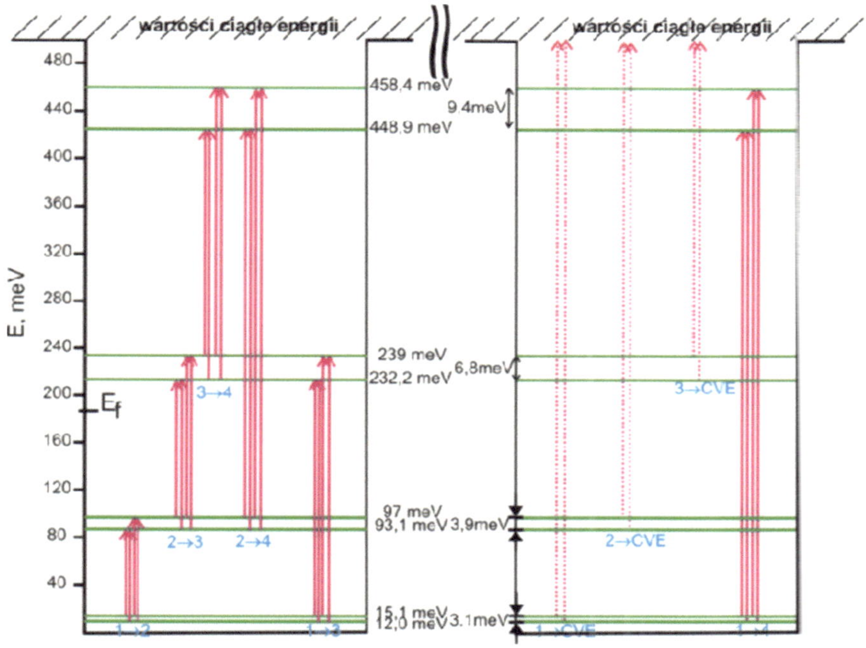

Figure 16.
The electron states in the DQW and optical transitions responsible for absorption bands shown in **Figure 15***:* *1 → 2 – band 1; 2 → 3 – band 2; 1 → 3 – band 3; 3 → 4 – band 4; 2 → 4 – band 6; 1 → 4 – band 7; 2 → CVE – band 8; 1 → CVE – band 9 [16].*

Supplying the infrared microscope with such a brilliant radiation source allowed for unique results [16]. For the first time, the energies of electron states were determined directly in DQW (see **Figures 15** and **16**) analogically as it was once done for the natural H_2 molecule [20]. From **Figure 16** it can be seen that the delta-SAS varies depending on the j number of the energy sub-band from 3.1 meV for $j = 1$ to 9.4 meV for $j = 4$.

3.9 Conclusion

The implementation of the quantum-mechanical problem of electron entrapment in a quantum well has been described. Various shapes of quantum wells—produced by advanced technologies as MBE and MOCVD, as well as based on different materials—are considered. Quantum wells based on GaAs/AlGaAs heterojunctions are especially important for the production of the semiconductor lasers, while the ones based on InGaAs/InAlAs heterojunctions are for the production of the HEMT transistors. Thanks to special engineering, it has been possible to make a *real rectangular potential of QW described by quantum-mechanical theory.*

Research into *double quantum wells* is a significant cognitive interest as an analogue of a two-atom hydrogen molecule in solid-state physics where essential role plays such quantum-mechanical phenomenon as *exchange interaction*. It can be predicted that their applications in electronics will also not make us wait long.

Author details

Eugen M. Sheregii
University of Rzeszow, Rzeszow, Poland

*Address all correspondence to: sheregii@ur.edu.pl

References

[1] Herman MA, Richter W, Sitter H. Liquid phase epitaxy. In: Epitaxy. Springer Series in Materials Science. Vol. 62. Berlin, Heidelberg: Springer; 2004

[2] Alferov ZI, Andreev VM, Korol'kov VI, Portnoy EL, Yakovenko AA. AlAs-GaAs Heterojunction Injection Lasers with a Low Room-temperature Threshold Soviet Physics Semiconductors. 1969;**3**:460

[3] Alferov ZI. Nobel lecture: The double hetero-structure concept and its applications in physics, electronics, and technology. Reviews of Modern Physics. 2001;**73**:767

[4] Zawadzki W, Pfeffer P. Average forces in bound and resonant quantum states. Physical Review B: Condensed Matter and Materials Physics. 2001;**64**: 235313

[5] Herman MA, Sitter H. Molecular beam epitaxy. In: Fundamentals and Current Status. Springer Series in Materials Science. Vol. 7. Heidelberg: Springer Verlag; 1996

[6] Arthur JR, LePore JJ. GaAs, GaP, and $GaAs_xP_{1-x}$ Epitaxial films grown by molecular beam deposition. Journal of Vacuum Science and Technology. 1969; **6**:545

[7] Sakaki H. Prospects of advanced quantum nano-structures and roles of molecular beam epitaxy. In: International Conference on Molecular Bean Epitaxy. 2002. p. 5. DOI: 10.1109/MBE

[8] Mimura T. The early history of the high electron mobility transistor (HEMT). IEEE Transactions on Microwave Theory and Techniques. 2002;**50**(3):780-782

[9] von Klitzing K, Dorda G, Pepper M. New method for high-accuracy determination of the fine-structure constant based on quantized hall resistance. Physical Review Letters. 1980;**45**:494

[10] Girvin SM. The Quantum Hall Effect: Novel Excitations and Broken Symmetries. New York: Springer-Verlag; 1999

[11] Weisbuch C, Vinter B. Quantum Semiconductor Structures. San Diego: Academic Press; 2007

[12] Tomaka G, Sheregii EM, Kąkol T, Strupiński W, Jasik A, Jakiela R. Charge carriers parameters in the conductive channels of HEMTs. Physica Status Solidi (A). 2003;**195**(127)

[13] Sheregii EM, Ploch D, Marchewka M, Tomaka G, Kolek A, Stadle A, et al. Parallel magneto-transport in multiple quantum well structures. Low Temperature Physics. 2004;**30**:1146

[14] Płoch D, Sheregii EM, Marchewka M, Woźny M, Tomaka G. Magnetophonon resonance in double quantum Wells. Physical Review B. 2009;**79**:195434

[15] Marchewka M, Sheregii EM, Tralle I, Ploch D, Tomaka G, Furdak M, et al. Magnetospectroscopy of double quantum wells. Physica E. 2008;**40**: 894-904

[16] Marchewka M, Sheregii EM, Tralle I, Marcelli A, Piccinini M, Cebulski J. Optically detected symmetric and ant-symmetric states in double quantum Wells at room temperature. Physical Review B. 2009; **80**:125316

[17] Zawadzki W. Theory of optical transitions in inversion layers of narrow-gap semiconductors. Journal of Physics C: Solid State Physics. 1983;**16**:229

[18] Boebinger GS, Jiang HW, Pfeiffer LN, West KW. Magnetic-field-driven destruction of quantum hall states in a double quantum well. Physical Review Letters. 1990;**64**:1793

[19] Huang D, Manasreh MO. Effects of the screened exchange interaction on the tunneling and Landau gaps in double quantum wells. Physical Review B. 1996;**54**:2044

[20] Woodgate GK. Elementary Atomic Structure. Oxford: Clarendon Press; 1980